THE ABDUCTION ENIGMA

THE ABDUCTION ENIGMA

KEVIN D. RANDLE

RUSS ESTES

WILLIAM P. CONE, PH.D.

A TOM DOHERTY ASSOCIATES BOOK

NEW YORK

To Deborah, Patricia, and Caroline, for their patience and understanding during the completion of this work.

THE ABDUCTION ENIGMA

This book is printed on acid-free paper.

A Forge Book
Published by Tom Doherty Associates, LLC
175 Fifth Avenue
New York, NY 10010

www.tor.com

Forge® is a registered trademark of Tom Doherty Associates, LLC.

Book Design by Lisa Pifher

Library of Congress Cataloging-in-Publication Data

Randle, Kevin D.
The abduction enigma / Kevin D. Randle, Russ Estes, William P. Cone.
p. cm.
Includes bibliographical references.
ISBN 0-312-86708-5 (hc)
ISBN 0-312-87270-4 (pbk)
1. Alien abduction. 2. Human-alien encounters. I. Estes, Russ.
II. Cone, William, Ph.D. III. Title.
BF2050.R34 1999
001.942—dc21 99-23888
CIP

First Hardcover Edition: June 1999
First Trade Paperback Edition: September 2000

Printed in the United States of America

0 9 8 7 6 5 4 3 2

CONTENTS

INTRODUCTION

The phenomenon of alien abduction is real. There are, literally, thousands of people who are suffering from the effects of alien abduction and who do not understand what has happened to them. Although they have sought treatment in a number of professional and private arenas, there is no doubt that they are telling what they believe to be the truth when they claim to have been abducted by alien creatures.

Research by both social scientists and abduction investigators suggest that between three million and six million Americans have been abducted. Support groups and abduction researchers can be found easily throughout the country. Abduction is so common, that many of the support groups are organized around specific aspects such as those who believe they were abducted by reptilian aliens, those who believe they were impregnated by aliens, and those who believe they are victims of an extremely long term, longitudinal study of their family histories.

What is happening today is a real-life horror story. We can present case after case in which people claim to have been taken from their homes, their cars, their pleasant surroundings, and then subjected to monstrous experiments by alien creatures on board interstellar spacecraft. We can offer eyewitness testimony from thousands of such reports from the United States, North America, and around the world. And we can present lie-detector tests and psychological evaluations that prove these people are telling the truth as they know it and are not the victims of some unreported or undiscovered psychopathology.

Although the concept of alien abduction might seem like a far-fetched and typically 1990s problem, the idea of contact with beings from other planets has been around as long as the UFO phenomenon itself. Almost from the moment that Kenneth Arnold, a Boise, Idaho, businessman first reported nine strange objects near Mt. Rainier, Washington, in June 1947, there have been tales of alien creatures piloting those strange objects. The first people to talk of contact with these alien beings were those invited on board the spaceships to engage in friendly conversation. Often they were given rides, sometimes to the moon or maybe to other planets in the solar system or even to locations far beyond our sun. Occasionally one of these "contactees" would speak of adventures on worlds light-years from Earth.

These people, George Adamski, George van Tassel, and Dan Fry, among many others, never had any real proof for their wild tales. They sometimes offered out-of-focus pictures that could have been taken anywhere and of anything. Sometimes they used technical jargon that smacked of 1950s science fiction. In one amazing case, a "moon" potato that resembled its earthly counterparts was offered as proof of the tale. Even without any real evidence, these contactees managed to find followers, often thousands of them. George van Tassel, who held yearly meetings at his isolated Giant Rock, California, airport in the 1950s, drew more than eighteen thousand people to one of his conventions.

UFO investigators, the leaders of the UFO research groups, the news media, and the Air Force, never took the stories of contact seriously. The tales of utopian societies that had survived the technological advances that threatened our world, the alien creatures who had messages of peace from throughout the galaxy or warnings about our impending doom, were nothing more imaginative than those that appeared in dozens of science fiction movies or stories already produced or published. Without any concrete or corroborative evidence, the reports of contact were rejected, and rightly so.

Beginning in the late 1950s, however, other stories of alien contact began to surface. These tales were different. Now the victims were dragged into the spacecraft and subjected to invasive examinations. Alien "doctors" collected samples of sperm and ova and performed minor surgeries before releasing the victims relatively unharmed. Alien doctors impregnated some of the women, allowing the babies to gestate for only short periods of time before they were removed by the aliens. These same extraterrestrial researchers were beginning experiments to create hybrids to walk among us undetected. Some researchers suggested that the government had entered into a treaty with the aliens allowing these abductions, and even the implantation by the aliens, of control devices into human hosts. The alien agenda was one of doom for the human race.

In a few of the more horrific cases, the abduction victims told of forced sexual intercourse on the ships. Male victims were forced to engage in activities with female aliens. Female victims were forced to perform with male aliens. Often the activities seemed to be little more than sex for the sake of sex with no scientific and research experimentation attached to it.

These new scenarios smacked of human scientists trapping and testing animals to gather data with little regard to the emotional trauma inflicted on the victims. It smacked of the way we have treated the "dumb" animals in the name of science. Now the tables have been turned and we are the victims of the same sorts of research conducted by "superior" intelligent beings.

All these new and horrifying theories and ideas were more acceptable to the UFO community because they were relutantly told. Often the means of recovery of the memories, that is hypnosis, was used to argue the validity of the memories. The emotion displayed by the victims of alien abduction was high. Polygraph examinations, when given, seemed to corroborate the validity of those horrible memories.

This growing group of victims of alien abduction is not one that anyone would join willingly, according to those doing the research. Why would anyone set himself or herself up as a target for ridicule by claiming alien abduction if it hadn't actually happened? How could they fake the emotions? Why do these stories, told by widely separated people who have never spoken to one another, seem to dovetail so nicely, according to these same researchers? According to them, there is no motive for inventing the tale, but there is solid evidence that abduction is fact.

Only a small part of the abduction phenomenon has been discussed in the UFO literature. All those books, magazine articles, and television shows published and produced in the last ten years have not looked below the immediate surface. They examine the pain of those claiming abduction, ask the standard questions about the events, but never delve into the real horror that lies hidden in the memories of the victims. Until the difficult questions have been asked and answered, and all the data are examined, we cannot draw intelligent conclusions about the reality of alien abduction.

We are not coming at this from the point of view of debunkers who believe that all UFO sightings can be explained in the mundane, or from the point of view that only those who wish to grab the spotlight are telling the tales of alien abduction, but from the point of view of researchers who, among the three of us, have nearly a century's worth of experience. We have been studying this aspect of the phenomenon almost from the moment that it was first reported by Coral Lorenzen, John Fuller, and a handful of other pioneering UFO investigators in the mid-1960s.

To understand this better, it might be beneficial to understand exactly who we are and what we believe about UFOs, alien visitation, and scientific thought. To that end, each of us has written a short biography that provides information to the reader about us. It explains how we came to the point in our research that we have.

WILLIAM P. CONE, PH.D.

One of my earliest childhood memories is of an extremely vivid dream. I was four years old, and living in the tiny town of Ontarioville, Illinois. This loosely knit patchwork of houses and fields was a farming community, and our small cinder block house on an acre of land was surrounded by endless miles of soybean and asparagus farms.

On that memorable morning, I suddenly woke up. It couldn't have been later than five or six o'clock, as the sun was just peeking over the horizon. While my family lay sleeping, I quietly crept out to the front yard.

What happened next will remain in my memory forever. As I entered the yard, something caught my eye. On the sidewalk near the wall of the house were three huge black ants. They must have been at least three feet tall. Although they were motionless, I was terrified, and immediately ran back into the house to awaken my mother. I shook her and cried out, but try as I might, I could not wake her. Eventually, I gave up and went back to bed. Later that day at the breakfast table, I recounted my experience, and was told, "It was just a dream."

Later that same year I had another intensely vivid dream. I was standing on the surface of what appeared to be an immense sheet of black glass. Two men were taking my mother down an elevator into a building. I was screaming in terror, but I could not reach her because two men were holding on to my arms. One of them said to me, "Don't be afraid. We aren't going to hurt her. She'll be back in a few minutes."

About this time in my life, I began to awaken with unremitting nosebleeds. The doctors told my mother not to worry, but the nosebleeds became worse. Eventually, they became so bad that I had to have my nose cauterized.

By the time I was five we had moved to Barrington, Illinois, a larger town several miles from Ontarioville. My new bedroom was in the back of our house, and had two walls completely covered by windows. A few weeks after we had settled into our new home, I awakened one morning to see an incredible sight. Outside, floating in the sky, were three rectangular objects colored like patchwork quilts. They hung motionless for a moment, and then were gone.

During this time I also had a recurring nightmare that a group of people were forcing me to take walks with an evil gray dog with a tube-like snout. I remember being terrified of the dog, and yet I was told that it was my job to take him to the park. As we would walk through the trees, the dog would render people temporarily unconscious by breathing on them through the tube. This allowed the owners of the dog to take the victims away.

When I was nine, our family moved to Golden, Colorado. My new room was in our basement. I was lying on my bed one afternoon when suddenly an orange ball of fire came shooting through my bedroom window. With a crackling sound, it raced down the wall, across the floor between the beds, and into my bathroom. I was dumbfounded by this bizarre happening, and still have no explanation for it.

Although I had never forgotten these experiences, I seldom thought about them. I chalked them up as childhood daydreams, and nothing else unusual happened to me for years. But early one spring morning in 1985, I was suddenly awakened by a rush of intense fear. As I became conscious, I could sense the presence of someone or something in my bedroom. As I began to open my eyes, I suddenly felt a weight on my chest. The mattress sank, and I could hear the bedsprings creak under the weight. It was hard to breathe. I began to panic. I struggled to get up, but to my astonishment, I was completely paralyzed. As I became more conscious, I heard a voice whisper, "He's waking up. It's time to leave." I then heard my front door open and close quietly. As the door closed, I fell asleep.

I did not think much about any of these things again until 1987. I was walking through a bookstore when a strange picture caught my eye. It was the cover of a new book called *Communion.* Mesmerized by those huge black eyes of the creature on the jacket, I bought the book immediately, hurried home, and read it cover to cover.

I found this book absolutely fascinating. On these pages was a description, and an explanation, for many of my experiences. Here was a man who had some remarkable experiences very similar to my own. After reading the book, I thought, "Is it possible? Could I have been abducted?"

I decided that day to begin researching the abduction phenomenon. I read everything on the subject that I could find. I wrote letters to Whitley Strieber, the author of *Communion*. I contacted all of the UFO organizations. In 1991 I attended my first Mutual UFO Network Convention. There I met Budd Hopkins, David Jacobs, and many "abductees."

During this time, I was also on the staff of five local psychiatric hospitals, where I worked as a staff psychologist. Three of the hospitals had recently opened new satanic ritual abuse (SRA) and multiple personality disorder (MPD) units. And as soon as the doors were opened, the patients started streaming in. At one time, I was seeing up to forty patients a week for group and individual therapy.

During my years of training in graduate school I was never taught anything about multiple personality disorder, satanic ritual abuse, repressed memories, or UFO abductions. There were no classes on dissociation, none on confabulation, and nothing on false memory. In retrospect, I also find it significant that there were no classes that invited a student to question his assumptions about the therapeutic process, or which invited alternative explanations for certain behaviors. Instead we were taught to revere the gods of psychotherapy, and to do what we were told.

Like every professional I knew, I got my education on multiple personality disorder, satanic ritual abuse, and repressed memory in the hospitals where I worked. There, while eating lunch in the doctor's dining room, we would eagerly discuss the day's bizarre encounters with the MPD and SRA cases we were working with. Clinicians would talk over cases and share new treatment techniques.

Again, as I take a more objective look back at this time, the techniques and ideas that were shared had never been empirically tested. No one ever questioned the assumptions being made. The information that was shared was based only on personal observation and experience. Through this interchange there came an agreement among us that what we were seeing was real. Not once during those years did I hear anyone disputing or questioning the assumptions and techniques.

Like my colleagues, I was fascinated by the gyrations and bizarre behaviors that occurred on the locked units of these hospitals. The contortions and sometimes violent behavior of the patients was shocking and truly frightening.

In order to work in this kind of environment, we all needed to feel that we understood the reasons for what we were seeing. We needed to feel that we knew how to handle it. And it was because of this need that we latched on to every new technique and idea without question.

Critics of the recovered memory movement claim that therapists mislead their patients deliberately for the purpose of making money. While this no doubt occurs in isolated cases, I do not believe that this is a prevalent motive. Most therapists that I know are devoted to helping the people they work with. No one devotes eight to twelve years of their life to training for the sole purpose of cheating people. Instead, I believe mistakes—when they are made—are due to two major factors: lack of critical thinking and the need to reduce anxiety.

Healers hate feeling helpless. I have heard countless stories about doctors getting angry at the patients that they could not help. Not knowing what to do makes a therapist extremely anxious. When someone offers a solution, a framework within which to work, naive and insecure therapists flock to it in droves. Such was the case with those of us who worked with MPDs. We were eager to listen to the experts. Armed with the new tools and techniques, we could walk into the consulting room feeling powerful and confident.

It is this feeling of being knowledgeable, of being able to take control, that draws the therapists to accept, without question, the techniques of recovered memory therapy. Once this belief system is in place, all work becomes self-verifying. Behaviors that fit the theory are seen as evidence that it is true. Behaviors that do not fit the therapy are seen as resistance, repression, and denial. Therapists are reluctant to give up the things that make them feel powerful. No one wants to hear that their deeply held beliefs are flawed. The need to feel that one knows the answer is what leads therapists into recovered memory work.

As I write these words, it is 1998, and I have spent ten years researching the abduction phenomenon. In my travels across the nation, I have talked to dozens of abduction researchers and hundreds of victims of alien abduction. For the most part these are sincere, honest people with a genuine interest in the phenomenon and a genuine belief they have experienced abduction.

While working with abduction researchers, I began to notice unmistakable similarities in the techniques being used with abductees and those used with satanic abuse victims. Hypnosis was the tool of the trade. Repressed memory was the answer.

This led into research that suggested a revelation about alien abduction. It was one that explained my own horrible childhood memories, and it was one that offered me some comfort about the human experience. And it made me want to share that knowledge with others in the hope that they could be spared some of the trauma that surrounds the abduction phenomenon as it is reported in today's environment.

RUSS ESTES

I was born in Chicago, Illinois, and raised in Southern California. My youth was spent in a typical middle-class tract home, in a typical middle class area of the San Fernando Valley. I went to public schools and was wrapped in the love of a not so typical middle-class family.

My father was a musician who worked many jobs to keep up our standard of living and my mother was an artist who also worked outside her chosen field. The remainder of my family was a younger sister, JoAnn, a dog named Mike, and the paste that held us all together was my grandmother, Sophie.

My grandmother was a widow at the age of twenty-six, her husband, an aspiring artist, died at the age of twenty-eight of a rare form of tuberculosis, leaving my grandmother alone with a two-year-old daughter.

It was a different time then. A woman of twenty-six with a child was considered old and not marrying material. So you see when my mother met the man who was to become my father, Grandma was a part of the package.

Grandma was the "not so typical" factor in our family, not that it was so odd to have what today would be called an extended family. In my generation I knew of many families who had a grandma or a grandpa who lived with them, but Grandma Sophie was someone who was very special in my life.

Grandma was the chief cook and bottle washer of our family, while my mother and father were at work Grandma was the woman of the house. She cooked, cleaned, and nurtured, and did it all in a loving, old-world way. When she came to the United States, at the age of eleven, she had already mastered three languages. She brought with her all of the old-world traditions, including one aspect that we tend to call New Age today.

Grandma practiced the old arts of spells, amulets, card reading, curses, and the evil eye. I learned very early in my life that when Grandma said, "Be careful," you should be very careful . . . and I was!

After my completion of high school I came to the first of many crossroads in my life. My heart was in the arts but my head was directed toward the sciences. I attended college with the goal of a life in medicine but my real goal was a life in show business, and the showbiz goal won out. After graduation I joined the U.S. Army and prepared myself for the unknown.

The year 1964 changed my life forever. I was stationed at Ft. Devens, Massachusetts. And it was then that I had my first brush with the world of the paranormal.

I was infused with the arrogance of youth. I was married, in love and lust, in the Army, and feared nothing, that is until I spent the weekend at the home of my in-laws in Connecticut.

My brother-in-law had taken a job in the state. Of course this meant moving his family. They found a rental home in a very scenic area outside of the city. It was a typical picture-postcard setting. A fully restored home that was built in the mid-1800s set on about two acres with a millpond and a large waterwheel.

The closest neighbors were at least a quarter of a mile away. Who could ask for more? The quiet, the scenic beauty, the perfect home—the family was more than pleased with the move. That is until they moved in.

Strange events started to take place as soon as the movers arrived. One of the first things they noticed was the calendars that were in each room. The calendars were turned to March of 1955. They found this fact to be very odd being that it was 1964 and three other families had rented the home in the past year. In the kitchen there was an overhead light fixture with a long pull string. A the end of the pull string was a small bottle that fell and broke during the move in. Inside the bottle was a parchment scribed with Arabic writing, it seemed very strange to my sister-in-law, so she took it to an interpreter to find out what it was. The parchment turned out to bear Arabic incantations to ward off evil sprites.

The fact that three families had lived in this house in the past year was not all that strange. The house was a rental on a month-to-month basis. Even the amulet could be considered normal. Many people have their house blessed or use good luck charms and religious symbols, but the calendars made no sense at all.

The house was occupied by three people, my brother-in-law, Pete,[1] his wife, Jan, and their thirteen-year-old son, Carl. Pete was employed in upper management at a local steel mill and was thrilled with his new position. Jan was a homemaker and loved the move to Connecticut. Everything about the move was positive for the entire family except for this house that seemed to have a life of its own.

---

[1]The names of my former in-laws have been changed.

From their first night in the house they started hearing strange noises. Groans and knocks that would wake them in the night. Being rational people they thought that the house was just settling. The question is, how much settling does a hundred-plus-year-old home do? As the months passed the noises became more pronounced and knickknacks started to fall off shelves.

By the fifth month in the house the noises were a constant companion but always at night. Now the residents were hearing the distinctive sound of an infant crying at the back door, pictures were coming off the wall nearly every day, and small objects would turn up missing.

The house was taking its toll. Pete was showing the strain; he was irritable, angry, short-tempered, and not acting like the man who had been married to Jan for the last fifteen years.

By their sixth month in the house Jan was convinced that something was trying to scare them out. Although Pete was functioning well at work he would be on edge at home. It was at this point that Jan asked if my wife and I would come up for the weekend. She had told my wife that she was doubting her sanity and she needed someone to experience the house.

We arrived early Saturday morning with sleeping bags in hand. The house was exactly as we expected. It was very pretty and well kept for a home that was over one hundred years old. The two bedrooms were upstairs, one was Carl's and the other was Pete and Jan's, so the living room floor was ours.

We spent a perfectly normal day sitting out by the millpond and having a barbecue lunch. Of course I had mentioned that everything seemed normal to me, but Jan said, "Wait till nightfall."

After dinner we sat at the breakfast nook in the kitchen and got deeply engrossed in a game of penny-ante poker. At about 9 P.M. Carl bid us a good night and went off to his room as we continued to play poker. It was about an hour later that the first noises started. Three distinct knocks came from the basement right below my feet. I got up from the table and moved across the kitchen. Wherever I stopped the three knocks followed. I would move a few feet and stop and three distinct knocks were at my feet.

This seemed harmless to me. I guessed that it might be mice in the cellar, but Jan just smiled and told me to wait and I would surely see more. She was right. No sooner had I sat back down when we started to hear an infant crying at the back door. I say an infant because anyone who has been around a newborn child can recognize that cry.

I went to the back door and opened it to see if there was any wind that could cause this sound. There was no wind at all. It was a perfectly still night. As I mentioned earlier, the closest house was a quarter of a mile away and the youngest child in the neighborhood was Carl and he was asleep in his room.

Jan was very excited that we had heard what she had heard. It was a verification that she wasn't crazy. Pete seemed just to ignore the noises as if they weren't even there.

We went back to our poker game and at about 2 A.M. we decided to call it a night. Pete and Jan went upstairs and my wife and I promptly fell asleep in our sleeping bags in the living room. A little after 3 A.M. we were wakened by a loud crashing sound in the basement. It was followed by the sound of something very heavy being dragged across the cement floor. The sounds were so loud that it woke everyone in the house.

Pete, Jan, and Carl all came down to the living room. Pete and I decided to have a look in the basement. He grabbed a flashlight and down we went. What we saw both amazed and confused us. A broken washing machine had a metal plate approximately two and a half by three feet leaning on the back of it. It seemed that the plate had fallen and had been dragged at least fifteen feet across the floor. There were no animals that we could find in the basement and I'm glad that there weren't any because I would not like to meet the rat that could drag that weight across a cement floor.

It took quiet a while to get back to sleep, and when we finally did, it was only for a couple of hours. Nothing else was said about the events of Saturday, so after breakfast my wife and I packed up and headed back to the Boston area.

My wife talked to Jan at least once a week after that. Jan related that Pete was getting more and more depressed. It was about two months after we visited the house when we got a call from Jan. She was in a panic, Pete was gone and so was the car. He had been very angry the night before and stormed out of the house. This was the first time that Pete had ever done anything like that and Jan was scared that he would harm himself, so she went to the local police. The police explained that when an adult left his home, even in anger, they could not be considered a missing person until seventy-two hours had elapsed. They also added that they would keep their eyes open for Pete and his car but officially he was not yet a missing person.

Over a week passed and when Pete had reappeared Jan became convinced that something horrible had happened to him. Oddly enough the noises in the house had stopped. It was on the morning of the third day of his return that Jan looked out the window and saw their car in the driveway. When she ran out to the car she found Pete sitting behind the wheel in a daze. The interior of the car was covered in a green slime similar to the pond scum that covered the waterwheel. Pete had doused himself with gasoline and was sitting with a lighter in his hand. It was obvious to Jan what Pete was about to do. She pulled him from the car and yelled for Carl to call the police.

Pete was hospitalized for a short time. By the time that he came home Jan had packed everything and scheduled the move back to Pennsylvania. Pete did not resist. After a couple of months back home everything returned to normal.

The interesting part of the story was that once the family moved back to Pennsylvania Pete's depression vanished. He was his even-tempered old self and he had absolutely no memory of any of the strange events that took place in Connecticut.

I will admit that after a weekend in that house I ran out and bought every book that I could find on ghosts and haunting. I was bitten by the bug. The quest for answers in areas of interest that had few, if any, answers was just too compelling to resist. My interest continued to grow, from ghosts and hauntings to other areas of paranormal anomalies.

I spent six years in the Army including a two-year stint in Vietnam. No matter where my military service took me, from Africa to Asia to Europe to Alaska, the one thing that occupied much of my free time was a search for local ghostly haunts and anomalous events.

In 1971 my interest in the paranormal grew even stronger. It started with a stomachache that led to a ruptured appendix. A trip to the emergency room and surgery led me to my very own near death experience. My logical mind told me that the combination of the anesthesia and the other drugs that had been administered to me could certainly cause hallucinations that would seem to give the impression of a near death experience. But no matter how much logic I applied to the situation, the one fact remained. To me it was very real. This event fueled my desire to find answers to the anomalies that we humans experience.

I have had a long and fruitful career in the arts. I started in the music industry and moved on to film and television. I have been involved in productions that run the gambit from comedy to cooking and from children's shows to travelogues. My favorite productions were the newsmagazine-format shows and my favorite month was October. Why October? Well, October is the month that every newsmagazine-format show does at least three segments on ghosts and haunting, and of course, it gave me the opportunity to research even deeper into the paranormal.

In 1984 I decided to become an independent producer and started my own production company. We produced two talk shows and a children's show on a weekly basis and also focused on commercials and industrial product.

But 1989 was a turning point in my life. It was then that I decided to stop taking on clients and devote my time to producing an in-depth documentary on paranormal anomalies. My original goal was to do a thirteen-part series on ghosts, goblins, and things that go bump in the night, but that goal quickly changed. As I started my preproduction research I found a great reference book that was published in 1980, titled *The Geo-bibliography of Anomalies.*

The author had done his homework, compiling anomalous events by state and city. The events cover from the mid-1800s to 1980 and this book must weigh ten pounds. In fear of being overambitious I decided to concentrate on California, so I got out my multicolored highlighting markers and set out to categorize the events. I started with green for all of the haunting and poltergeist phenomena. Brown was my choice for stigmata, rock falls, and fishfalls. Pink was used for ghost ships and monsters, both lake and sea.

Yellow was the color for UFO sightings. When I had finished you could fan the pages and the color that was overwhelming was yellow. In all of the years that I had looked into paranormal events I had never realized the shear numbers of UFO events. I must admit that it intrigued me to the point that my documentaries and research have been solely dedicated to UFOs and alien contact from that day on.

When my interest was first sparked I had to consider myself a true skeptic if not a nonbeliever. By the time that I had invested a few years looking into the unknown I had become a person who by his research and personal experiences knew that there is a realm that falls outside the arena of nuts-and-bolts science.

As unbiased as we all try to be it is impossible not to have personal opinions. A documentarian has the goal of documenting events pertaining to a specific topic in an unbiased manner. In the six documentaries that I have produced on the UFO phenomenon I have tried to be true to that goal. I present as many sides to a question as I can and supply the best information available at the time. If I have a bias, and indeed I do, it leans in the direction of the unknown. Thanks to Grandma Sophie I was taught to have an open mind at a very early age. I find it very difficult to slam the door shut on anything. To come right to the point, I have experienced the poltergeist phenomenon, I have documented numerous hauntings, and I have seen a UFO.

If asked the question, do I believe in alien contact? I would respond with a simple "Yes."

I certainly don't believe that every light in the sky is a UFO but I do feel that we have been visited. Most likely not on a daily basis but visited nonetheless.

I must admit that in the thirty-four years that I have researched all of the aspects of paranormal anomalies, the quest for answers in the UFO field is the most intriguing. I have more questions than answers, but the answers that I do have fuel my desire to keep searching with an open mind and a smile on my face.

KEVIN D. RANDLE

I usually blame my mother for my interest in UFOs. She had been a big science fiction fan and it isn't a great leap from reading science fiction tales of interstellar flight and civilizations to reading UFO books about the same thing. My interest in UFOs grew from an interest in science fiction and a desire to be able to travel in space.

Maybe I should point out here that as a kid I went to the Saturday matinees that featured science fiction films that dealt with trips to the moon. Now, I watch *Apollo 13* as history. Some of the scenes in those early science fiction films about a trip to the moon were re-created in the historical drama about a real trip to the moon. I'm sure that Ron Howard had that in mind when he included those scenes.

So my interest in UFOs began when I was a youngster. Beginning in the mid-1960s, UFOs hit the front pages of the newspapers and were reported on the national news programs. Interest in them reached a high and the Air Force, apparently because of public pressure, decided to form a civilian committee to study the problem scientifically. The result was the University of Colorado study headed by Dr. Edward U. Condon.

That study suggested that UFOs were nothing more than the misidentification of natural phenomena, or manufactured objects, such as aircraft, seen under unusual circumstances. Science had investigated UFOs and found the evidence lacking.

My enthusiasm didn't flag because, like so many others, I realized that about thirty percent of the sightings investigated by the Condon Committee were unexplained. They even found that one of the photographic cases, from McMinnville, Oregon, showed a large manufactured object and there was no evidence of a hoax. In other words, the Condon Committee found that one picture of a flying saucer apparently showed just that. They rejected this as proof of alien visitation, concluding despite their own findings that there was nothing to UFO sighting reports. The so-called scientific study had been anything but scientific.

I continued to study the phenomena and became a field investigator for the Aerial Phenomena Research Organization. I had also grown up and found myself in the Army as a warrant officer candidate and a fledgling aviator. While in training I found time to investigate UFOs, including a trip to the panhandle of Texas to interview contactee Carroll Wayne Watts. I was unimpressed with his sighting. He eventually failed a lie detector test blaming the Men In Black and government agents.

Eventually, I made my way through flight school and became a helicopter pilot with a tour in Vietnam. When my Army career ended, I attended the University of Iowa. My interest in UFOs continued, but I was beginning to examine the cases and reports with a more critical eye. I was becoming more skeptical in my point of view.

After college I was commissioned in the Air Force and served as a public affairs officer, a general's aide (for a short period), and finally as an intelligence officer. During that time, I was still interested in UFOs, but the cases were becoming too routine. How many sightings of lights in the night sky did we have to investigate before we realized that adding another such case to the pile was not going to advance our knowledge?

Not only that, I was learning that most people simply didn't know what was in the sky around them. I was called out by a man who believed he had a UFO hovering over his house, but was looking at the moon through low-hanging and quickly moving clouds that provided an illusion of movement. Another man said that a red UFO was landing in his backyard but had misidentified the lights on a radio tower that had been there for years.

Finally, while in the physics building at the University of Iowa, I saw a photograph of a meteor taken over the Grand Tetons in Wyoming. It was a spectacular picture and caused me to ask a question. If daylight meteors are extremely rare (and they are), and if UFO sightings are extremely rare, why are there no UFO sightings with pictures that mimick those of the meteor? Several people in widely scattered locations had taken photographs, and one or two even managed to film it. This seemed to suggest that UFOs were, in fact, misidentifications.

In 1980 Ronald Story put together *The Encyclopedia of UFOs*. I contributed a number of sections. One of them dealt with the rumors of crashed flying saucers and alien bodies. Based on the information that I had seen, and the evidence available at that time, I suggested that such stories were not true.

And, in a short profile, I provided a "position statement" suggesting that I found the evidence of extraterrestrial visitation to be lacking. There just wasn't sufficient proof to lead to the conclusion that aliens were visiting the Earth.

Ten years later I had changed my mind. First, I realized that the pictures taken over the Grand Tetons was of an object that was forty or fifty miles above the ground. It could be seen for hundreds of miles. A UFO usually flew much closer to the ground and couldn't be seen over a large geographical area. That doesn't, of course, prove that UFOs are extraterrestrial. It merely suggests that the comparison of the photographs of a meteor over the Grand Tetons and UFO sightings was not sufficient justification for rejecting the idea.

More important, however, was the beginning of my investigation into the UFO crash outside of Roswell, New Mexico. The project, conducted in association with the J. Allen Hynek Center for UFO Studies, was supposed to answer the questions about the evidence of the UFO crash.

I went into the investigation believing that we would spend a few days in Roswell and we would discover that the statements attributed to witnesses were not as strong as reported. I believed we would find few facts about the case as they had been reported. I fully expected that we would find a plausible answer and that we would leave Roswell in a matter of days having discovered that plausible solution.

Well, solve it we did, but not the way I expected. I am convinced that what crashed outside of Roswell was, in fact, a craft built on another planet and piloted by a crew from that planet. I believe this because of what a large number of people told me about their involvement in the case. I believe it because it is the only answer that makes any sense when all the facts are examined objectively. And I mean all the facts, not just those that support one of the many mundane explanations that have been offered over the years.

The point is, I believe now, that alien visitation has taken place. It is a conclusion I have drawn based on the interviews that I have conducted and the evidence I have seen. It is from my own study, some of which cannot be replicated simply because some of the witnesses are now dead. Most of the information is on either audio- or videotape, but some of it is merely handwritten notes. For some people that is not enough. They'd like to hear the tapes.

It was the Roswell investigation that led to my conclusion. It was sitting in the little restaurant in Carizozo, New Mexico, and hearing Bill Brazel Jr. repeat for my tape everything that had been reported about his involvement in the past. It was hearing him describe a metallic substance as thin as aluminum foil that when wadded into a ball would unfold itself, taking on its original shape with no sign of a crease or a fold.

It was hearing Major Edwin Easley, the provost marshal of the 509th Bomb Group tell me that the craft was of extraterrestrial origin. Unfortunately, Major Easley's comments were not recorded on tape, but I had planned to fix the situation. Major Easley died before I had a chance to sit down with him in person and have him repeat, for the recorder, his comment. I know what he told me, however, and though I can't prove it to the satisfaction of the debunkers, it doesn't matter to me.

So, I believe that we have been visited. I think the Air Force documentation, created as part of the Project Blue Book investigation of UFOs, proves the case. There are, literally, dozens of inexplicable sightings in the Blue Book files. And there are, again, literally dozens of solutions offered that make no rational sense. It is clear from the files that the Air Force was in the business of slapping solutions on sighting reports but not in the business of investigating those sightings.

This investigation, into alien abduction, was another that I was pulled into by others. I had, in the past, written articles about alien abduction. Budd Hopkins in his landmark book even mentions a case that I had investigated in the mid-1970s. Some of the first articles to report alien abduction were articles that I had written. In fact, some of the symptoms of alien abduction that are so common today were first reported by me in those early articles.

My investigation into the Roswell case, however, focused my attention on UFO crash reports. Alien abduction, while interesting, was not something that I pursued until Russ Estes and Bill Cone dragged me back into it. Then I was given the opportunity to review the dozens of videotaped interviews they had conducted. I had the opportunity, on my own, to speak to both the researchers and the victims of alien abduction. I had the chance, over the last several years, to review all of the work that had been done, and reach the conclusions outlined in this work.

And that's where I am today. I believe that UFOs represent, in rare cases, alien visitation. I believe that the majority of UFO sightings are of mundane objects and natural phenomena that have been misidentified. But the real point is that I believe, based on some very compelling evidence, and on some testimony from some very credible sources, that there has been extraterrestrial visitation.

So, as you can see, each us, Dr. William Cone, Russell Estes, and Kevin Randle believes that extraterrestrial visitation has taken place. Each of us has great respect for those telling tales of alien abduction. In sessions we've witnessed in person, the victims of abduction are suffering tremendously. The victims also attempt to relate their stories as accurately as possible. They are not consciously lying to us, to the researchers, or to themselves. They relate the truth as they see it.

The Abduction Enigma is the first attempt to understand the whole abduction phenomenon. *The Abduction Enigma* is a modern horror story in every sense of the word.

PART 1

ALIEN ABDUCTION

1.

INTERRUPTED JOURNEYS?

The concept of human interaction with alien beings is not one that is easily accepted. Although stories of contact have been told from the beginning of the UFO phenomena, they weren't believed by mainstream ufology. Contactees told tales of meetings with benevolent space brothers who provided messages of peace, who flew the lucky few to planets inside and outside of the solar system, and who promised worldwide prosperity. The contactees, however, failed to provide any evidence that they had traveled to other star systems, or had even met the aliens.

That perception began to change with the abduction phenomenon. The abductees were not contacted by kind aliens who worried only about human progress, but were accosted by creatures who cared nothing for human emotions or physical health. The aliens gathered their data in dispassionate research aimed only at collecting information. When finished, the humans were "thrown" back into the human environment with little thought to the trauma caused by the abduction.

Or they were taken by aliens who professed an interest in the fate of our planet. These benevolent aliens might be performing genetic experiments on their unwilling victims, but they were concerned with the health and well-being of the human race. They were also interested in human emotion and feelings.

Since the infamous Barney and Betty Hill abduction was first reported in the mid-1960s, hundreds, if not thousands or millions, have claimed similar experiences. In 1992, Las Vegas businessman Robert Bigelow commissioned the Roper Organization to study the abduction phenomenon. Although the results are open to interpretation, it seemed to indicate that as many as three or four million, possibly as many as six million, Americans had suffered experiences that suggested an alien abduction at some point in their lives.[1] The phenomenon was spread across the entire spectrum of American life.

---

[1]Budd Hopkins, David Jacobs, and R. Westrum, *Unusual Personal Experiences: An Analysis of the Data from Three National Surveys* (Las Vegas, NV: Bigelow Holding Co., 1991).

Ufological research in other areas slowed as more investigators began to study abduction. Resources that might have helped provide answers to the many questions about extraterrestrial visitation were sucked into abduction research. In fact, there were those who believed that nothing of importance could be learned from any other aspect of UFO research except that of alien abduction. Why waste time, effort and money studying the sightings of the ancient past when it was clear that abduction investigation would provide all the answers for which we searched.

The reports, studies, and investigations conducted by those who believed claimed many things about the aliens, their experimentation on board the flying saucers, their alleged manipulation of our DNA and genetics, and their partially understood motives. Thousands of abductees have been interviewed using a variety of memory enhancement techniques including hypnotic regression. A huge body of data has been collected and searched for patterns and clues about the alien creatures and their reasons for abducting so many human subjects.

THE FIRST OF THE ABDUCTIONS

Contrary to popular belief, the first abduction report was not made by Barney and Betty Hill in 1961. It wasn't even that of Antonio Villas-Boas in 1957, but was reported many years before either of those. What might have been the first recorded claim of attempted alien abduction came during the Great Airship sightings in the late nineteenth century, early on the evening of November 25, 1896. Colonel H. G. Shaw said he, with a companion, Camille Spooner, had left Lodi, California, "when the horse stopped suddenly and gave a snort of terror."

Shaw claimed that he saw three figures who stood nearly seven feet tall and were very thin. They looked human, didn't seem to be hostile, so Shaw tried to communicate with them. According to Shaw, they didn't understand him and responded with a "warbling" in a type of monotone chant.

Shaw continued his description, reported in the newspapers of the day, saying, "They were without any sort of clothing, but were covered with a natural growth as soft as silk to the touch and their skin was like velvet. Their faces and heads were without hair, the ears were very small, and the nose had the appearance of polished ivory, while the eyes were large and lustrous. The mouth, however, was small and it seemed . . . they were without teeth."

Shaw also said they had small, nailless hands, and long narrow feet. By touching one of the creatures, he discovered that they were nearly weightless. Shaw said he believed they weighed less than an ounce.

With Shaw close to them, they tried to lift him with the intention of carrying him away. When they couldn't budge either Shaw or his companion, they gave up and flashed lights at a nearby bridge where a large "airship" was hovering. They walked toward the craft using a swaying motion and only touching the ground every fifteen feet or so. Then, according to Shaw, "With a little spring they rose to the machine, opened a door in the side and disappeared . . ."[2]

This abduction, then, was a mere attempt to lift Shaw and carry him to the waiting ship. When it failed, they gave up, so it bears little real resemblance to modern abduction accounts. However, a tale from that same time, would contain some important elements of the abduction tales to follow. John A. Horen claimed that he had met a stranger who invited him to tour one of those turn of the century airships. They boarded the craft and toured southern California and then flew to Hawaii. Horen's wife laughed at the story, said that he was well known as a habitual practical joker, and, he had been asleep next to her on the night of the alleged journey.[3]

A week or so later, in early December 1896, two fishermen, Giuseppe Valinziano and Luigi Valdivia, said they had been held captive for a number of hours while the airship crew made repairs. The "captain" of the craft would only provide vague clues about the origin of the ship, but did say the invention would be announced to the world within weeks. When the repairs were completed, the men were allowed to leave.[4]

It would be nearly twenty years before there was another reported abduction that fit, at least marginally, into the category as it is understood today. In August 1915, in Gallipoli, Turkey, a British regiment was formed by their commanders to climb a hill near Suvla Bay. As they began to move, six to eight clouds were seen hovering near the hill. One of them slowly descended until it was resting on the ground.

Men of a New Zealand regiment watched as the British soldiers marched into the cloud. When the last of the soldiers had entered, the cloud began to lift. The British soldiers were never seen again.

At the war's end, British government officials demanded to know what had happened to the missing regiment. Turkey denied that they had captured an entire regiment, or had fought the regiment at the time and place indicated by the British army. Turkish officials claimed that they didn't know what happened to the men.[5]

---

[2]Jerome Clark, *The Emergence of a Phenomenon: UFOs from the Beginning through 1959* (Detroit, MI: Omnigraphics, Inc, 1992), 23.

[3]Jerry Clark, "Airships: Part I," *International UFO Reporter*, January/February 1991, 8.

[4]Jerry Clark and Lucius Farish, "The Great Unidentified Airship Scare," *Official UFO*, Nov. 1976, 56.

[5]Peter Brooksmith, *UFO: The Complete Sightings* (New York: Barnes & Noble, 1995), 29.

Airship sightings, mysterious cloud formations, and other such reports were infrequent during the next quarter century. There would be wildly scattered reports of strange things seen in the sky, but such reports, when made, received little attention. Not until June 1947 would tales of flying saucers again make headlines. Kenneth Arnold, a businessman from Boise, Idaho, told reporters that he had seen nine crescent-shaped objects moving through the air like "saucers skipping on the water."[6] Within days of that sighting, hundreds would be reporting that they too, had seen the flying saucers.

About a month after Arnold's story hit the press, the first report of an apparent abduction attempt would be made. On July 23, Jose Higgins was with a number of fellow workers near Bauru, Brazil, when they witnessed the landing of a large flying saucer. Higgins remained behind as his friends fled in terror. He was confronted by creatures seven feet tall who had rounded, bald heads, huge eyes, and long legs.

Although Higgins found the creatures strangely attractive, when they attempted to lure him into their saucer, Higgins refused. He fled the scene but while hiding in the bushes watched as the creatures spent an hour leaping and jumping around. They stopped long enough to draw what Higgins believed was a map of the solar system indicating they originated on Uranus. When they finished their play, they returned to the ship and took off.[7]

Another, similar case of almost abduction, but which hinted at some of the more sinister aspects of the growing phenomenon, came from the *Flying Saucer Review*. An unidentified woman said that she had been hurrying home to prepare dinner on May 20, 1950, at about four in the afternoon, when she suddenly found herself inside a brilliant, blinding light. Although she had been alone on the garden path and she was sure there had been no one near her, two huge, black hands appeared suddenly in front of her.

The woman described the hands as grabbing her from above, in much the way talons of a bird of prey would would grab a rabbit. Her head was jerked back hard, against a cold, iron chest. She felt the cold through her hair and on the back of her neck. The hands were also cold and she wasn't sure they were made of flesh. Big fingers seemed to choke off her screams, and pinched her nose so that she couldn't breathe.

She was now at the mercy of the being behind her, whatever it might be. She felt paralyzed when first touched by the bright light. As the hands touched her, she felt an electrical shock and she felt completely helpless and without reflexes.

---

[6]Kevin D. Randle, *The UFO Casebook* (New York: Warner, 1989), 4.

[7]Coral and Jim Lorenzen, *Encounters with UFO Occupants* (New York: Berkley, 1976), 148.

She was then pulled from the path and through bushes until she found herself in a small pasture. She tried to call for help, but she had no voice. Her cry was nothing but a tiny, shrill sound.

The attack stopped as abruptly as it began. The hand gradually slipped from her body. Eventually she was able to sit up, and finally she stood. When she heard noise to the left, in the bushes, she thought she would see her aggressors, but she saw nothing.

Terrified, she stumbled back to the path, trying to find some help. She felt nervous, exhausted and had a strange, metallic taste in her mouth. Her back hurt, as if she had been burned by something.

After several minutes, she reached a turn in the path and could see several houses. But before she could reach them, she heard a great noise, "like a violent wind storm." The trees, according to the woman, were bending as if under "sudden storm." There was a strong, bright white light, but she saw nothing behind the light or in the sky near it.

She reached the house described as that of the "lock-keeper" and opened the door. Those inside rushed to her and asked what had happened. They too, had seen the light. They told her they could see red marks on her face, looking like large red welts. The lock-keeper's wife used peroxide on the scratches on the woman's legs, and bathed her face in cold water. Feeling better after the treatment, the woman left.

There is one more interesting aspect to this case. According to the woman, the night before she had seen what she thought was some kind of shooting star. Rather than falling to the ground or burning itself out as was usual, it stopped, then climbed until it seemed to rejoin the other stars. It then grew larger, swung back and forth, and the bright light blinked on and off. It disappeared suddenly, curving upward at a great rate of speed.

An official investigation of the attempted abduction was made, one of the few times that happened. The investigators learned nothing and eventually dropped the case. However, it is reported that the case is still carried as an unsolved abduction attempt by police officials.[8]

This story provided some strong hints of what would come. It hints of something in the sky, a bright light that houses the unseen entity, and a hint of an examination of some kind. The horror felt by the victim foreshadows the terror of the abductions that would begin in the 1960s. The reactions of the victim, the paralysis, the loss of the ability to communicate are all predictive of elements of the abduction phenomenon.

To suggest, however, that this is a case of alien abduction is to make a

---

[8]Jacques Vallee, *Dimensions* (New York: Ballantine, 1988), 108.

number of wild assumptions linking the bright lights, the falling star, and events that would take place in the future. This might suggest an attempted abduction that went wrong but it is an assumption. The victim was dragged from the path, but was not taken into a craft.

But it should also be noted that it might be a case of a hysterical woman who imagined the attack on her. The physical evidence of the scratches could be attributed to the vegetation along the path. The real importance is not the attempted abduction, the physical evidence, but that the abduction attempt had been reported to authorities.

THE FIRST AMERICAN ABDUCTION ATTEMPT?

There was a somewhat similar circumstance in the mid-1950s in Nevada. A retired schoolteacher said that she had been paced by some sort of craft late one night. Her tale, of being followed for a long distance, and a trip that seemed to take longer than it should have, suggests to the UFO researcher today that this might be an abduction as we have now begun to understand the phenomenon. The Air Force was alerted, but they wrote the case off as coming from an unreliable observer. This was not the first time the teacher had reported seeing a flying saucer. Under Air Force criterion, this made the woman unreliable. The Air Force officers believed that seeing a flying saucer was such a rare event that no one would be fortunate enough to see one more than once. If a witness reported multiple sightings, then clearly that witness was unable to identify the mundane such as airplanes, stars, and other natural phenomena. That being the case, the witness was unreliable and further investigation was not required.

But the case, like many of its predecessors, foreshadowed the phenomenon that was to come. There was the hint of missing time and a fear that something else had happened. But, in 1955, no one was looking for tales of alien abduction. The clues were ignored.[9]

VILLAS-BOAS—1957

The first reported case of alien abduction, as such, was made on February 22, 1958, by a Brazilian farmer named Antonio Villas-Boas. (It must be noted here, however, that others have now claimed to have been abducted before Villas-Boas, but none reported the abduction until years after this report was made.)

[9] *Project Blue Book Files* (Washington, D.C.: National Archives, 1976), Microfilm, Roll 23, T-1206.

He was interviewed by Dr. Olavo Fontes, the Aerial Phenomena Research Organization (APRO) representative in Brazil, and a newspaper columnist, Joao Martins. This case would become the model for abductions reported for the next twenty years. The point that must be made here is that Villas-Boas couldn't have read about other claimed abductions and then designed his to follow those leads. Nor could Fontes guide him subtly to the tale of abduction, because there had been no other abductions reported in the ufological literature. In 1958, this case was unique. That doesn't mean that it is grounded in reality, only that it was the first "officially" reported alien abduction.

According to Villas-Boas, "It all began on the night of October 5, 1957." Villas-Boas was in his room after a party. Because of the heat, he decided to open the window. Outside, over a corral, he could see a bright, fluorescent light. It seemed to sweep upward, into the night sky, but there was no object behind the light that he could see. He called to his brother Joao, who refused to get out of bed to look.

Villas-Boas then lay down, but soon got up again. The light was still then, and as he watched, it began to move toward the window. Villas-Boas slammed the shutters closed. The light bled through the slats on the shutters, and Villas-Boas told Fontes that he and his brother "watched the light appear through the crevices of our shutters . . . and shine through the tiles of the roof, lighting up the darkness of our room."

A few days later, on October 14, Villas-Boas was in one of the farm fields late at night with another brother. They saw a bright light, so bright that it hurt their eyes to look at it. This time he tried to approach the light, though it was hovering in the sky. As he neared, it moved off, evading him. He chased it from field to field until he was too tired to keep at it.

The next night, October 15, Villas-Boas was working in the fields late. At one in the morning (October 16), he noticed another extremely bright red star overhead. As he watched it, he realized it was moving, growing larger as it approached him. He hesitated, wondering what to do, and in those few seconds, the light changed into an egg-shaped craft descending toward his freshly plowed field. It came to hover over him, its light so bright that Villas-Boas could no longer see his own headlights.

Villas-Boas thought of trying to drive away, but the tractor had no real speed and the object could easily overtake him. Running through the plowed field would have been difficult and if he stepped in a hole, he could have broken his leg. Escape seemed to be impossible.

A bright light came from the front of a craft that resembled an elongated egg. There were purple lights near the large red one, and there was a small red light on a flattened cupola that spun rapidly. As the object slipped toward the ground, three telescoping legs slid from under it.

Villas-Boas realized that the legs, like those of a camera tripod, were for landing. He also realized that it was now time to escape. He turned his tractor and stepped on the accelerator. Before he had driven far, the engine sputtered and died and the headlights faded. He tried to start the engine, but the ignition didn't work. He then opened the door on the side away from the alien craft, and tried to run.

He had taken only a few steps when something touched his arm. He spun and faced a short humanoid being. Struggling to escape, he put a hand on the creature's chest and pushed. The alien being stumbled back and fell. But three other creatures grabbed him and lifted him. He twisted, kicked, jerked with his arms. He shouted for help, but the beings held him tightly, moving slowly toward the craft.

A door opened and a narrow ladder extended to the ground. The aliens tried to lift him into the ship, but he grabbed at the narrow railing, holding on to it. One of the creatures peeled his fingers from the flexible metal, and he was forced upward, through the door.

He found himself in a room with a metal rod, which he assummed held up the roof, running from floor to ceiling. The room itself was square shaped with silvery, polished metal walls. To one side was an oddly shaped table surrounded by backless chairs.

For several minutes, Villas-Boas and the aliens stood in the room. The creatures talked among themselves in a series of low, growling sounds while they held on to Villas-Boas. Eventually they began to strip his clothes from him carefully so they didn't tear anything. When he was naked, one of the aliens began to "wash" him with an oily-looking liquid that made him shiver as it dried.

Villas-Boas provided a fairly complete description of the alien beings. While it is detailed in regards to what they wore, it provided few clues as to what they actually looked like. He indicated that they were all small, no more than five feet tall. He said, during his long interview with Dr. Fontes, "I must declare that up to that moment I hadn't the slightest idea as to how those weird men looked nor what their features were like. All five of them wore a very tight-fitting siren-suit, made of soft, thick, unevenly striped gray material. This garment reached right up their necks where it was joined to a kind of helmet made of a gray material . . . that looked stiffer and was strengthened back and front by thin metal plates, one of which was three-cornered, at nose level. Their helmets hid everything except their eyes, which were protected by two round glasses, like the lenses in ordinary glasses. Through them, the men looked at me, and their eyes seemed to be much smaller than ours, though I believe that may have been the effect of the lenses. All of them had light-colored eyes that looked blue to me, but this I cannot vouch for."

He went on to describe the large helmets that had three tubes extended down from the helmet and blended into the suit. He saw no evidence of any type of tank, and when asked about it, had no explanation for the system of tubes or the lack of tanks.

He said that when he was forced deeper into the ship, into another, smaller room, blood samples were taken from under his chin by two figures holding two rubber-looking pipes. After the samples were taken, he was left alone for an hour or more, sitting on a large couch. Then, feeling tired, he sat down, and noticed a strange odor in the room. From the walls, about head high, he noticed a gray smoke pouring into the room. Its thick, oily odor made him physically ill. He fought that feeling for several minutes but finally vomited.

Feeling better, Villas-Boas sat back on the couch, waiting. Eventually there was a noise at the door and he turned to see a woman entering. Like Villas-Boas, she was completely naked. She moved slowly, walked toward him, and embraced him, rubbing herself against his body.

The woman was short, reaching only up to his chin. She had light, almost white, hair that looked as if it had been bleached heavily. In their 1967 book *Flying Saucer Occupants,* Coral and Jim Lorenzen wrote that the woman Villas-Boas described had blood red hair under her arms. Lorenzen told Randle in a 1972 interview that she had changed that part of Villas-Boas's testimony because she didn't want to write that it had been red pubic hair.

The woman had blue eyes that were slanted to give her an Arabian look. Her face was wide with high cheekbones but her chin was very pointed, giving her whole face an angular look. She was slim, with high, very pointed breasts. Her stomach was flat and her thighs were large. Her hands were small, but looked like normal human hands.

When the door closed, the woman began to caress him, showing him exactly what she wanted. Given the circumstances, Villas-Boas was surprised that he could respond at all, and then felt sexually excited. Later he would suggest that his state of arousal was induced by the liquid that had been rubbed on him.

She kept rubbing him, caressing his body, and in moments they were together on the couch. Villas-Boas responded to her, and before he realized what was happening, they were joined. According to him, she responded like any other woman.

When they finished, they stayed on the couch, petting, and in minutes, both were ready again. He tried to kiss her, but she refused, preferring to nibble his chin. When the second act of lovemaking was over, she began to avoid him. As she stood up, the door opened and one of the alien men stepped in, calling to the woman. Before she left, she smiled at Villas-Boas, pointed to her stomach and then to the sky, southward, according to Villas-Boas.

One of the men came back and handed Villas-Boas his clothes. While dressing, he noticed that his cigarette lighter was missing. He thought that it might have been lost during the struggle in the field. Or, it might have disappeared on the ship, taken as a souvenir by the alien beings.

The creature directed him out of the small room and into another where the crew members were sitting, talking, or rather growling, among themselves. Left out of the discussion, and with the aliens ignoring him, he tried to fix the details of the ship's interior in his mind. On a table near the beings was a square box with a glass lid and a clocklike face. He decided that if he should survive his ordeal he would need some proof of his strange experience and tried to steal it. Almost before he could move, one of the creatures jumped toward him and pushed him away.

At last one of the aliens motioned for him to follow. None of the others looked up as he took a quick tour of the ship. The door was open again, with the ladder extending to the ground, but they didn't descend. Instead, they stepped onto a platform that went around the ship. Slowly they walked along it as the alien pointed out various features. Since he didn't speak, Villas-Boas didn't know the purpose of any of the things he was shown. He glanced up at the cupola, which now emitted a greenish light and was making a noise like a vacuum cleaner as it slowly spun.

When the tour ended, he was taken back to the ladder, and the alien motioned toward the ground. When Villas-Boas was at the bottom, he stopped and looked back, but the alien hadn't moved. Instead, he pointed to himself, then to the ground, and finally toward the sky. He signaled Villas-Boas to step back as he disappeared inside.

The ladder telescoped into the craft and the door vanished. When it was closed, there was no sign of a seam or a crack, at least none that Villas-Boas could see. The lights brightened as those on the cupola began to spin faster and faster until the ship lifted quietly into the night sky.

As the ship disappeared, Villas-Boas walked back to his tractor. He noticed that it was 5:30 in the morning. He had been on the ship for more than four hours.

Villas-Boas didn't tell anyone about the encounter, except his mother, until February 1958. After reading several articles about flying saucers in *O Cruzeiro,* a Brazilian publication, he wrote to the author, Joao Martins. He also carved a model of the UFO and sent it to Martins. Not long after that, Martins arranged for Villas-Boas to be brought to Rio, where he and Dr. Fontes interviewed Villas-Boas and produced a sworn statement.

Both men were impressed with Villas-Boas. He was a sincere, intelligent young man, who would eventually become an attorney. Martins, however, thought the story was too strange for *O Cruzeiro.* The only evidence of the tale were two scars on each side of Villas-Boas's chin. Fontes, who gave the young farmer a thorough physical examination, reported he found two "small hyperchromic spots, one on each side of the chin . . . scars resulting from some superficial lesion with associated bleeding under the skin. . . ."

The first mention of the Villas-Boas case in an American publication came in 1962 review of Coral Lorenzen's book *The Great Flying Saucer Hoax.* The book was a compendium of UFO sightings and evidence, and mentioned nothing about Villas-Boas. Max B. Miller, writing in the magazine *Fate,* suggested that Dr. Olavo Fontes might not be reliable because he had circulated a report of an "alleged rape of a Brazilian farmer by a somewhat uninhibited female from space."

Coral Lorenzen, understandably irritated at the reference, responded, "Dr. Fontes has earned a reputation for thoroughness, objectivity, and originality of thought. . . . The so-called 'rape' case . . . was never published in the *APRO Bulletin* [a UFO newsletter edited by Lorenzen], nor was it mentioned in my book for the simple reason that we do not feel that it was sufficiently authenticated."

Of course, in 1962, the Villas-Boas report was one of a kind. There were the contactees, but Villas-Boas didn't really fit into the category. And there had been occupant reports, and Lorenzen was one of the first to publish reports of alien beings. But in 1962, this case was too bizarre for anyone to accept as authentic.

The case is interesting because of its groundbreaking nature. It set the tone for the abduction reports that would follow in the next several years. The victim would be isolated, it would be late at night, and there would be a period where the victim was involved in some sort of experimentation on the alien craft. When the experimentation was finished, there would be a tour of the craft before the victim was released. Once all that was finished, the craft would disappear into the sky.

Although Lorenzen was angry when the case was mentioned in the review of her book in 1962, five years later, she and her husband would publish in *Flying Saucer Occupants* the entire transcript of the Villas-Boas interview with Dr. Fontes. By this time, with the Barney and Betty Hill case well publicized, apparently the Villas-Boas case was no longer thought to be insufficiently authenticated.[10]

It is also important to note that in 1967, the same year that Lorenzen finally published the Villas-Boas account, a science fiction novel, *The Terror Above Us*, was published by George Wolk under the pen name Malcolm Kent. It was an

---

[10]Jerome Clark, *The Emergence of a Phenomenon* (Detroit: Omnigraphics, 1992), 392–4; Coral and Jim Lorenzen, *Flying Saucer Occupants* (New York: Signet, 1967), 42–72; Kevin D. Randle, *The October Scenario* (Iowa City, IA: Middle Coast Publishing, 1988), 107–12.

account of sexual experiences on UFOs, somewhat similar to the Villas-Boas tale, and the novel, like the Villas-Boas abduction, was supposed to be true. The novel was, in fact, fiction.

BARNEY AND BETTY HILL

The landmark case in the UFO abduction phenomenon came almost six years later, though the reports on it wouldn't reach mainstream ufology until 1965. Because a reputable psychiatrist had conducted a number of hypnotic regression sessions with the two witnesses, and because a large part of the testimony from both Barney and Betty Hill was recovered under hypnosis, the UFO community accepted the report as authentic almost from the beginning. The case would be widely reported, would be the subject of one book, reviewed in dozens of others, would be the topic of many magazine articles, and even the force behind one network movie. It would become the landmark case of the abduction phenomenon.

The story began on September 19, 1961, as the Hills were returning from vacation, and driving along a deserted section of Highway 3 in the White Mountains of New Hampshire. They had stopped for coffee about ten that night and believed they would arrive home before three the following morning without having to push themselves.

As they drove through the mountains, Betty Hill noticed a bright star near the moon. She was sure that it hadn't been there earlier, and she was sure that it was getting steadily brighter. Finally, she pointed it out to Barney and he told her he thought it was nothing more than an artificial satellite.

The bright "star" intrigued both of them. Several times during the next hour, they stopped. Once or twice they got out a pair of binoculars to try to see the star better so they could identify it. Betty was now convinced they were looking at something out of the ordinary, but Barney kept insisting that it was nothing more unusual than an airplane or a satellite, or maybe just a very bright star they had failed to notice earlier.

During one of the stops, Barney used the binoculars to study the star. Now he saw red, amber, green, and blue lights rotating around it. To Barney it appeared to be an aircraft fuselage with no wings. He could hear no sound from the engines. When he returned to the car, he realized that he was frightened. He didn't want Betty to know that and tried to tell her again the object was nothing more exciting than an airplane flying slowly along the terrain.

The object then swooped down and began to pace the car. Betty watched through the binoculars and now she could see a double row of windows. She demanded that Barney stop the car but he refused for a few minutes. Finally he stopped in the middle of the road and when Betty handed him the binoculars, he got out, car engine running. Now he got a better look at the object, seeing for the first time that it was a large disc and not an aircraft as he had tried to convince himself and his wife. Again he told Betty, "It must be a plane or something."

Although he was now badly frightened, he stepped away from the car and began to walk across the road, toward the object. He kept walking until he was about fifty feet from the craft, which was hovering just above the trees that bordered the field. Through the binoculars he could clearly see the double row of windows, and behind them were six beings. One, who Barney thought of as "the leader," wore a black leather jacket.

As he watched, five of the six turned their backs and seemed to manipulate controls. The saucer began a slow descent. Fins holding red lights spread along the craft, and something, possibly a landing gear, lowered from the belly of the object.

Barney now focused on the one remaining face and was overwhelmed with the feeling that he was going to be captured. He jerked the binoculars from his eyes, spun, and ran back toward Betty and the car. Shouting that they were going to be captured, he threw the binoculars on the backseat and slide beneath the wheel. He slammed the car into gear and roared off as fast as possible. He ordered Betty to watch for the thing, but it had apparently disappeared.

As Barney began to calm down, he slowed the car and they heard a series of strange electronic beeps. Both seemed to feel drowsy. The beeps came again, they believed, almost immediately and they saw a road sign that told them Concord, their hometown, was seventeen miles away. They continued home, arriving about five in the morning.

Once they arrived, they unpacked before going to bed. Betty took a bath and, according to Coral Lorenzen, "for no reason whatsoever, bundled up the dress and shoes she had been wearing and shoved them into the deep recesses of her closet."

Six days after the event, Betty Hill sent a letter to Major Donald E. Keyhoe, a writer and the director of the National Investigations Committee on Aerial Phenomena (NICAP), describing what had happened to her and her husband. She also suggested that she was thinking of finding a reputable psychiatrist to perform hypnosis on Barney because he was having trouble remembering parts of the story.

In December 1963, after Barney began to have stomach trouble, and after he had consulted with two different doctors, he was sent on to Dr. Benjamin Simon, a well-known and highly qualified neurosurgeon. Simon believed that Barney's illness might be psychosomatic. He eventually used hypnosis on both the Hills in his attempts to treat Barney.

Under hypnosis, Barney Hill told of what happened that night in 1961 after they heard the first set of beeping sounds. For some reason, Barney turned down a dirt road and drove up to a roadblock, where the car's engine quit. Several men appeared around the car who guided the Hills through a wooded area to the craft, which was sitting on the ground.

Betty Hill later described the beings as having a Mongoloid appearance, with broad, flat faces, large slanting eyes, and very large noses. Barney added to the description, saying that the "leader" had very large, almond-shaped eyes that seemed to wrap around to the side of his head. The mouth was a slit with a vertical line on each side. The skin, according to Betty, had a bluish-gray cast to it.

According to most accounts, Barney is said to have kept his eyes closed during most of the time they were on the craft. Some researchers have suggested that this is why his tale was never as rich in detail as that told by his wife. He did mention an examination by the alien beings and that he was put on a table that was too short from him.

Betty described the examination in much greater detail, saying that unusual instruments touched her body in various places, that samples of skin and fingernails were taken, and that hair was pulled from her head. A long needle was pushed into her navel. She screamed at the "examiner" and the "leader" passed a hand over her eyes, stopping the pain and calming her.

Betty communicated with the "leader," though there is no real indication that they spoke out loud. She thought that they used some form of telepathy for communication. She also had the impression that the "leader" was keeping the rest of the crew away from her. She was told that she didn't have to worry about Barney, that he was being well cared for and he would be all right. During this discussion, she asked where they came from, and the "leader" showed her a star map but rather than describe it to her, he asked Betty where the sun was. When Betty failed to identify it, he told her that the map would do her no good.

After the discussion, Betty was escorted from the ship and joined Barney in the car. There was the second series of beeps, and they "awakened" traveling down the road, nearing Concord.

The hypnotherapy by Dr. Simon took several months. When it was over, Simon said that he thought the Hills were recounting a fantasy. He believed that Betty had originated the tale in her dreams that followed the event, and then shared it with Barney by telling him, over a period of weeks, of her dreams and as he listened to Betty describe the events to various UFO investigators. Simon believed this because Barney's account was less detailed than Betty's.

There is one additional and important fact. The letter that Betty sent to Don Keyhoe detailed much of the story. Later, investigators, reported that Barney sat in as Betty discussed the case with them. While it may be said truthfully that Barney and Betty didn't discuss the case between themselves, Barney was there as Betty was talking about the sighting. He certainly had plenty of opportunity to learn the details from her.

The star map that Betty saw on the craft became an important point of corroboration. Marjorie Fish, an Ohio schoolteacher, spent years trying to find a pattern in the stars that matched the one drawn by Betty under hypnosis. She created a three-dimension model, or rather, a series of models of Earth's section of the galaxy and then examined them from all angles, searching for the pattern. Her first attempts failed, but in 1972 after six years of intensive work, she finally discovered a pattern that matched the map Betty Hill had drawn. To her surprise, it was the only one that she could find that matched Betty's map.

Fish discovered that the main stars in the map were Zeti I and Zeti II Reticuli, star systems about 37.5 light-years from Earth. The map showed what might be interpreted as lines of communication among the stars. There were many lines between the two closest stars, suggesting close communication between them. Heavy lines were drawn among the closest stars and lighter lines among those farther away. A double line connected the sun into the mix, suggesting that the sun was a relatively unimportant star.

Walter Webb, at one time APRO's consultant in astronomy, wrote an analysis of Marjorie Fish's work. He was impressed by the fact the lines on the map, as developed by Fish, connected stars that were exclusively the type defined by astronomers as those suited for the development of planets that could contain life. A random pattern, Webb believed, would not generate that sort of subtle yet corroborative evidence. He wrote, "The pattern happens to contain a phenomenally high percentage of all the known stars suitable for life in the solar neighborhood."

What was interesting, according to Webb, was that Fish had believed that dozens of patterns would emerge as she began her work. Instead, after six years of concentrated effort, she found only a single pattern that met all the criterion she had established. If the map was accurate, if Betty Hill remembered the map correctly, if the two-dimensional representation could adequately convey the three dimensions of the map, then a good clue had been found about the location of one group of alien visitors. Of course, those were big ifs.[11]

---

[11]Jerome Clark, *High Strangeness: UFOs from 1960 through 1979* (Detroit: Omnigraphics, 1996), 235–53; John G. Fuller, *The Interrupted Journey* (New York: Dial, 1966); Lorenzen, *Flying Saucer Occupants*, 73–86; Mark Rodeghier, "Hypnosis, and the Hill Abduction Case," *International UFO Reporter* (March/April 1994), 4–6, 23–24.

THE SCHIRMER CASE—1967

After the Hill case, other reports of abduction were made, though they were still rare. One of the most credible came from Ashland, Nebraska, where the victim was an on-duty police officer. Surely a policeman would not risk his career, or his reputation, by telling tales of alien abduction if they were not true. It was also another tale where there were surface memories of some type of event—that is, the landing of a strange and seemingly alien craft—in which the witness knew there was something else, something more important, buried deep in his or her mind.

On December 3, 1967, Patrolman Herbert Schirmer was driving on the outskirts of Ashland, feeling that something wasn't right. He saw, for example, an agitated bull charging the gate of its corral and Schirmer stopped to make sure the gate would hold. A couple of hours later, he was near the intersection of Highway 63 when he saw something ahead of him. He thought it was a truck, but when he flipped on his high-beam headlights, the truck flashed up into the night sky.

When he returned to the police station early in the morning, he noted the "flying saucer" in his police report and thought nothing more about it. At home, he had a headache, a red welt on his neck, and a buzzing in his ears that kept him from sleeping.

That was the end of it until his case came to the attention of the University of Colorado's Air Force–sponsored UFO investigation, the Condon Committee. Someone on the committee noticed that twenty minutes on Schirmer's police log seemed to be unaccounted for. It was suggested that hypnotic regression be used to see if the missing time could be explained.

Under hypnotic regression, Schirmer told of seeing a flying saucer on the ground. As it landed, he had seen legs telescoping from the bottom of it. Although he wanted to get out of there, he felt he was prevented from starting his police cruiser. There was something in his mind that made it impossible for him to drive off.

As he sat there, a hatch or door opened and alien beings began to get out. Schirmer tried to draw his weapon, but again, he was prevented from acting. The beings surrounded the car, and one of them pointed something at him. There was a bright flash and he passed out momentarily. The next thing he remembered was rolling down the cruiser's window.

The aliens, according to him, were four and a half to five feet all. All wore close-fitting uniforms made of silver-gray material with boots and gloves. The uniforms all had a hood, like that of a pilot's helmet with a small antenna over the left ear. The visible skin of the face was gray-white, the nose flat, and the mouth merely a slit. The eyes were slanted with strange pupils that irised opened and closed like a camera lens.

Schirmer said that the beings communicated with him by voice and telepathy. The English spoken by the beings was broken and sounded strange. Schirmer was told that they, the aliens, had been studying many human languages.

While on the ship, Schirmer was provided with some technical information. Their ships, according to what he was told, were vulnerable to radar because of ionization. They had hidden bases on Earth, including one under water near Florida and one in the polar region. According to Schirmer, they came from another galaxy not far from our galaxy. It was assumed here that Schirmer meant solar system instead of galaxy. The "leader" showed Schirmer a "view screen" and he saw three "war ships" flying through space against a background of stars that included the Big Dipper.

The aliens also told Schirmer about stealing electricity from power lines, but at such a low level that detection by power companies was prevented. He was told that the aliens had been coming to Earth for a long time, observing human history. He was told the same things that many other abductees would eventually claim to have been told, and his information mirrored that provided by the other contactees.

Before Schirmer was released, the "leader" looked into Schirmer's eyes and said, "I wish you would not tell you have been aboard this ship." Schirmer was given instructions about what he was to say about the incident. "You are to tell that the ship landed below in the intersection of the highways, that you approached and it shot up into the air. . . . We will return to see you two more times."

Investigation by the scientists of the Condon Committee provided some interesting conclusions. Schirmer was given a polygraph examination by an official agency. There were no indications of deception. Schirmer sincerely believed that he had seen a flying saucer. Psychological tests were also administered, with Schirmer's permission. Finally, Dr. Leo Sprinkle, Professor of Psychology at the University of Wyoming, used hypnosis on Schirmer and discovered the "hidden memories." Sprinkle concluded that Schirmer believed in the reality of the events he described.

The Condon Committee scientists concluded, "Evaluation of psychological assessment tests, the lack of any evidence, and interviews with the patrolman, left project staff with no confidence that the trooper's reported UFO experience was physically real."

In defense of the Condon Committee scientists, it must be said that their conclusions were the only ones that could be reached scientifically. No physical evidence was reported (other than that noted by one writer, which has not been substantiated) and there was no corroboration for the report. Clearly Schirmer believed he was telling the truth, but in scientific inquiry additional evidence is needed.

Schirmer's case is interesting because he was alone on a deserted portion of highway late at night when abducted. Many of his abductors' physical traits matched those described by other abductees. A pattern was being established that would suggest these abductions were more than hallucinations. Researchers began to take these reports seriously.[12]

A CANADIAN ABDUCTION—1967

David Seewalt, a teenager living in Alberta, Canada, also reported an abduction in 1967. He was alone in a vacant lot about 6:30 P.M. on November 17 when his attention was drawn to a silvery-grayish object about thirty or forty feet overhead. Seewalt tried to run, but a beam shot from the ship, stopped him, and then lifted him up into it.

While on board, he was subjected to a medical-type procedure that included a physical examination and a shot in the arm. When he described an area as like that of an operating room, one psychologist suggested that the event was a "flashback" that mixed a real hospital stay with too many science fiction movies on television.

The aliens who conducted the examination were described as "monsters" by Seewalt. The creatures had scaly skin, holes for noses and ears, and slits for mouths. They had two arms and two legs, but the hands had only four fingers and the feet had only four toes. The skin, like that of a crocodile's, was rough and brown. These beings wore no clothes. It should be noted that Seewalt's description of the aliens would be mirrored by others in the years to come.

They spoke among themselves in a language that Seewalt didn't understand. To imitate the language, Seewalt made a high-pitched, continual buzz like that of a large bee. Seewalt mentioned that it seemed one of the creatures was giving the others instructions.

When they finished the examination, they put Seewalt's clothes back on him and then "beamed me down." He landed in the empty lot where he had begun the adventure. He raced home, ran to his room, and told his sister that he had been chased by a flying saucer. He remembered nothing about the examination until five months later. Then, after a nightmare, he knew what had happened. The nightmare brought the story into his conscious mind.[13]

---

[12]Ralph Blum, with Judy Blum, *Beyond Earth: Man's Contact with UFOs* (New York: Bantam, 1974), 107– 21; Roy Craig and John Ahern, "Case 42," in *Scientific Study of Unidentified Flying Objects*, ed. Daniel S. Gillmor (New York: Bantam, 1969), 389–91.

[13]B. Ann Slate, "Contactee Supplies New Clues to the UFO Mystery," *UFO Report*, Apr. 1976, 26–30.

Like those who had reported their abductions before him, Seewalt was alone when taken. He was subjected to a single examination, samples were taken and he was sent home. Unlike some, he didn't remember being on the craft until later. Like those including the Hills and Schirmer, he remembered seeing a flying saucer but didn't know something else happened to him or that he had been on board it. Like Betty Hill, he remembered that aspect of the event after a dream. Hypnosis clarified and solidified his memory.

UFO researchers tended to accept these tales because of the use of hypnosis. Researchers also noticed the emotion displayed by the victims under hypnosis. To many, this underscored the validity of the memories and the research being conducted.

THE YEAR OF THE ABDUCTION, FALL—1973

Abduction research was about to change. The similarities exhibited by the cases would begin a slow process of evolution. We had noticed certain patterns about the cases reported before the occupant sighting wave in 1973. But with new researchers entering the arena, with more people reporting they had been abducted, and with more information being discovered, new patterns began to emerge.

The first of the 1973 abductions to be reported came from Pascagoula, Mississippi when two men, alone at a fishing hole, were taken by robotlike creatures into their craft. Hypnosis would again be used to learn all the details of an examination on board a flying saucer. Calvin Parker, nineteen, and Charles Hickson, forty-two, were fishing from an old pier on the west bank of the Pascagoula River when a bright blue light attracted their attention. At first it was high overhead, but then it dropped toward the ground. Hickson later told investigators that it stopped only a few feet above the surface of the bayou. There was a buzzing noise coming from it but there was no wind or blast like that from a jet engine.

As he watched, one end of the object, an egg-shaped craft, opened and three creatures floated from it. Hickson believed them to be about five feet tall, covered in wrinkled gray skin, long arms that ended in lobster-like claws, and legs that seemed to be fused together. The creatures, which seemed to be buzzing slightly, headed for the men. The three beings separated so that one could pick up the now unconscious Parker and the other two lifted Hickson, floating him toward the dimly glowing ship.

A door seemed to appear in the side of the craft, and they were floated through it. According to Hickson, he and Parker spent about twenty minutes on the ship. The interior was bare except for a device that Hickson thought looked like a big eye. Eventually he was floated back off the ship and deposited on the riverbank with Parker.

Hickson didn't see the UFO take off. He reported there was a buzzing sound and the UFO vanished. For a few minutes, Hickson didn't know what to do. The event was so fantastic that he didn't think anyone would believe him. Besides, Parker, now conscious, was still terrified.

After several drinks in a local bar, Hickson decided that someone in authority should be told. He tried the newspaper office but it was closed. Next he tried Kessler Air Force Base and received the now standard reply that the Air Force no longer investigated UFO sightings. It was suggested that possibly the local sheriff or university would be interested.

Hickson opted for the sheriff, telling him that he had no desire for publicity, despite the fact he had visited a newspaper office first. Within hours, however, the story was national news and Hickson had more publicity than he could have dreamed possible.

The next morning, Jim and Coral Lorenzen of the civilian Aerial Phenomena Research Organization, learned of the abduction and began making plans. All three of their consultants in psychology were busy, but James A. Harder, their Director of Research, could travel to Pascagoula. Harder, although a civil engineer, was versed in the use of hypnosis.

Dr. J. Allen Hynek had also heard the news reports and called APRO headquarters to find out if they planned to investigate. Learning that Harder was being dispatched, Hynek made his own plans to fly to Pascagoula.

On Saturday morning, the whole cast was assembled in the offices of the J. Walker Shipbuilders Company. Along with Hickson, Parker, Hynek, and Harder were Hickson's attorney, Joe Colingo, Dr. Julius Bosco, Deputy Sheriff Barney Mathis, and police detective Thomas Huntly.

Over two days the details of the abduction were examined. Hickson confirmed much of his story, adding that the alien spaceship was sixteen to eighteen feet long and had a "trap door" in the back. He described the creatures for the assembled men.

Harder was impressed with the tale and told APRO Headquarters that it was his opinion that it would be nearly impossible for the men to simulate their feelings of terror (an erroneous conclusion) while under hypnosis. Newsmen asked both Hynek and Harder for their opinions. Hynek suggested the men had a frightening experience but that didn't translate into an extraterrestrial event.

Reporters weren't as reserved in their opinions as Hynek and Harder had been. Headlines the next day screamed, "SCIENTISTS BELIEVE UFO STORY." That wasn't exactly right, but before the claim could be corrected, reporters were off chasing another UFO event.

Late in the month, after Harder and Hynek had completed much of their investigation, a polygraph examination was arranged in New Orleans for Hickson. According to several sources, Hickson passed the test, which took nearly two hours to complete. Newspapers claimed that the "lie test" proved the case.

Philip Klass, who launched his own investigation some days later, believed there was something wrong with the test. He wondered why the operator, a man who had yet to finish his training, conducted the test, when other, more qualified polygraph operators could be found closer to Pascagoula. To Klass, this suggested the test had been skewed in Hickson's favor by those with a desire to believe his story.

But Klass, Harder, and all the rest overlooked a couple of important points. The lie detector test would only suggest to a reasonable certainty that Hickson was telling the truth as he believed it. It would not prove that Hickson and Parker were abducted and it certainly would not suggest the abduction had been at the hands of alien beings.

Two years later, in October 1975, Hickson attended a UFO conference held at Fort Smith, Arkansas. He had been invited to tell his story, on the condition that he submit to another lie detector test arranged by the conference organizers. Hickson agreed, but when the time came, Hickson said his attorney had advised him not to take the test. Although many attorneys routinely advise their clients against taking such tests because they are so open to human interpretation, the controversy was stirred again.

There were other, minor controversies in the story. Hickson seemed to have no real idea when he had seen the UFO, telling researchers and investigators at different times it was about seven in the evening, or maybe about eight, or even as late as nine. Hickson told author Kevin Randle that he never wore a watch because it would not keep the proper time. No matter how many he tried, they either ran fast, slow, or stopped altogether. That does provide him with a good excuse for not having a good feel for the time.[14]

The second of the 1973 incidents to gain national attention took place on October 17, but was not reported until two years later. Pat Roach wrote to *Saga's UFO Report*, telling editor Martin Singer that she believed she and her family had been abducted. Although it would become standard in abduction accounts later, this seems to be the first time that anyone had reported that the aliens came into a house to remove the occupants. Her report set a new standard that would soon be followed by hundreds of others.

Under hypnotic regression also conducted by Dr. James A. Harder, Roach

---

[14]Jerome Clark, *Emergence*, 389–95; Philip J. Klass, *UFO Abductions: A Dangerous Game* (Buffalo, NY: Prometheus, 1989), 18–19; Coral and Jim Lorenzen, *Abducted!* (New York: Berkley Medallion, 1977), 132–134; Randle, *Scenario*, 11–16;

said that she was awakened in the night and saw the beings inside the house. She was taken out to a ship landed in the vacant field next to her house. Two of her daughters and one of her sons were also taken. The youngest of the daughters said that she saw neighbors on the ship, standing in line, waiting for their turn "on the machine."

After twenty or thirty minutes, they were all taken back into the house. As the aliens left, Roach said that the house erupted with the kids crying and the cat screaming. She didn't know what had happened, but called the police, afraid there had been a prowler. The time, according to police files, was about 12:40 A.M.

For more than two years, Roach would say nothing about the home invasion, though she discussed it with her family. Finally, after reading about Dionisio Llanca in *Saga's UFO Report*, she wrote to the magazine, telling of her abduction.[15]

DIONISIO LIANCA—1973

Llanca, a truck driver living in Argentina, reported that at the end of October 1973, while on an overnight trucking assignment, he was stopped due to a flat tire. As he attempted to change it, the ground around him lit up. He believed, at first, it was an approaching car, but the light changed to a bright blue. When he tried to stand, he found that he couldn't. It was as if he were paralyzed.

Hovering above the roadway, not far from his truck, was a domed disc-shaped object. Standing under it were three beings, two male and one female. They all had long blond hair and elongated eyes. They were wearing one-piece, silver flying suits with long gloves, high boots, but no belts, helmets, or weapons.

The beings talked among themselves for a few moments with the sounds of a badly tuned radio, full of chirps, buzzes, and bursts of static. Llanca felt panic spreading through him and he struggled to stand, but couldn't. One of the men reached out, grabbed Llanca by the sweater and lifted him to his feet. Although he tried to scream, no sound came from his mouth.

While one of the beings held him, another put a small black box next to his hand. At the touch of the box, peace began to replace the panic. When the box was removed, he noticed a few drops of blood against his finger. At that

---

[15]Kevin D. Randle, "The Family That Was Kidnapped by Ufonauts," *UFO Report*, June 1976, 21–3, 50–2, 54.

point, Llanca lost consciousness. As far as he knew, consciously, nothing else happened.

Llanca, found by motorists the next morning stumbling along the road, was taken to the hospital. For three days, doctors worked to restore his memory of the missing hours. Under hypnotic regression, Llanca began to tell a tale of a flying saucer and alien abduction.

Llanca told the doctors and researchers that he had been taken on board the craft. He was able to communicate with the aliens and learned they had been communicating with people since 1960. It wasn't clear if that meant they had been abducting people since then, or if that was when they entered into some type of communication with humans.

He also said that the aliens were trying to learn if we, humans, could live on their world. Under probing by the doctors, he gave up additional information, eventually telling them that he had been given a message but couldn't reveal it. He told them, "I have a message from the beings on the craft but I can't tell you what it is. No matter what you or any other Earth scientists do, there will remain a memory lapse while I was on the ship. I was there for forty to forty-five minutes."[16]

THE HIGNON ABDUCTION—1974

As October 1973 drew to a close, the number of UFO sighting reports and claims of abduction dropped off dramatically. There were continued sightings, but national attention was drawn into other arenas. The next abduction case to receive any national publicity was that of Carl Hignon, an oil-field driller, who, on October 25, 1974, encountered the aliens.

Hignon, to supplement his income, hunted the Wyoming fields. On one such expedition, Hignon saw five elk standing together. He raised his rifle, and fired, but instead of the kick from the high-powered weapon he had expected, there was nothing. The bullet, according to Hignon, seemed to float from the barrel, falling to the ground fifty feet from where he was. There was no bang as he fired. In fact, there was no sound from anything around him.

Hearing a twig snap, Hignon turned and saw a humanoid standing near him. It was over six feet tall and weighed, according to Hignon's estimate, about 180 pounds. The skin tone was like that of an Asian, and the face seemed to blend into the neck. Like so many other commentators, Hignon said that the

---

[16]Kevin D. Randle, "The UFO Kidnapping That Challenged Science," *UFO Report*, Spring 1975, 15–17, 54.

creature was dressed in a one-piece suit that reminded him of a scuba-diving suit. The metal belt worn around the waist of the being had a six-pointed star in the middle.

The first thing the creature asked was "How you doin'?"

Hignon said, "Pretty good."

He was then asked if he was hungry but before he could answer, a small package was floated toward him. Inside were four pills. He was told they would last for four days. Unable to control himself, he took one of the pills and put the other three into a pocket.

In the distance, Hignon spotted what he believed to be the creature's ship. It looked like a box with no landing gear or hatches, or anything else. When the creature realized that Hignon had seen the ship, it asked if Hignon would like to take a ride.

Before he could answer, he found himself in the ship. They took off and soon Hignon found himself on a planet "163,000 light miles" from the Earth. From inside the ship, he was able to see buildings on the alien planet.

One of them he described as looking like Seattle's Space Needle. It was surrounded by rotating, intense spotlights. Hignon threw up his hands, shouting that the lights were hurting his eyes. They were burning him.

One of the aliens responded, "Your sun burns us."

The next thing that Hignon remembered was walking down the road in Wyoming, confused. He still had his rifle, but he didn't know who he was or where he was. In the distance he saw a truck parked in a stand of trees. He decided to try to use it for shelter, not realizing that it was his own vehicle.

Dazed and cold, he sat in the truck, and heard a voice on the two-way radio. He found the microphone hooked to the dashboard, unhooked it, and called for help. By staying in constant communication, he was able to assist his rescue searchers.

Once Hignon was found, his family was alerted, and they joined him at the hospital, where he was still disoriented. He kept asking for his pills, the ones given him by the alien. It wasn't until late the next day that he began to regain his memory and equilibrium. He was released from the hospital and told to get some rest.

Dr. Leo Sprinkle, of the University of Wyoming in Laramie, interviewed Hignon and conducted a number of hypnotic regression sessions with him. Under hypnosis, Hignon provided more details, telling Sprinkle that the name of the alien leader was Ausso One. Sprinkle also learned more about the trip to Ausso One's home world.

There was a single piece of corroborating evidence. The bullet fired by Hignon at the beginning of his adventure was recovered from the field. Hignon had somehow located it and put it into his canteen pouch, where he later found it. He couldn't easily identify it and took it to the Carbon County Sheriff's Office. There it was identified as having come from a 7mm Magnum rifle. The deputy who examined it said it looked as if it had been turned inside out. He couldn't explain the condition. The bullet was given to investigator Walter Walker for examination. Later it was turned over to Sprinkle. In 1994, Sprinkle told Randle that the bullet had somehow disappeared.

What is interesting about the Hignon case is that it parallels more closely the tales of contact told by authors Dan Fry, George Adamski, and George van Tassel than it does those of alien abduction. Hignon was not forced on board the craft, but was given a view of another world. Although he was examined, there didn't seem to be the same sorts of invasive procedures that are described by abductees. Hignon had the opportunity to "chat" with the aliens, while most abductees suggest that any discussion is the result of alien attempts to gain information about human life.

Given that, however, it seems that many of those who would have rejected out of hand a contactee's tale have accepted Hignon's. They have said that those who have interviewed him find him to be a sincere, seemingly honest man who had a strange adventure. Of course, they point to the physical evidence, and the fact that others saw strange lights on the night Hignon was abducted. These observations seem to suggest that the Hignon tale, unlike those of the contactees, is authentic.[17]

The abductions, as reported to researchers, continued at an increasingly rapid pace. In August 1975, Sandy Larson, who lived in North Dakota, reported an abduction. Larson, like Roach, had witnesses to the event. She was with her daughter and a friend when the abduction took place from a highway early in the morning.[18]

THE WALTON ABDUCTION—1975

But the case that claimed the most headlines, and the biggest interest in 1975, was the abduction of Travis Walton. Independent witnesses reported Walton was hit by a beam of light from a UFO hovering a few feet above the ground near Heber, Arizona. They had driven close to the object and Walton had leaped

---

[17]Timothy Green Beckley, "Kidnapped by Aliens! The True Story of Carl Hignon's Incredible Contact," *UFO Report, Fall 1975, 40–43, 71–9;* Clark, *High Strangeness, 231–4.*

[18]Jerome Clark, "The Bizarre Sandy Larson Contact: UFO Abduction in North Dakota," *UFO Report*, Aug. 1976, 21–3, 46–53.

from the cab of the truck, walking toward the craft. When the beam struck him, seeming to lift him from the ground, Walton's friends panicked and fled the area. They believed that he had been killed by the blue beam.

A police search later that night, and during the following day, failed to find a trace of Walton or his body. He reappeared five days later, dirty, tired, and slightly confused. He called his sister for help and spoke to Grant Neff, his brother-in-law.

Although he'd been gone for five days, Travis seemed to think that only a couple of hours had passed. At his sister's home, he talked to friends and family about the ordeal. He remembered little, other than seeing the UFO and being struck by the beam of light.

His first memory, after being hit by the beam, was to awaken in what he thought was a hospital room. All the details of his surroundings didn't register right away. Slowly he became aware of three creatures, about five feet tall, standing around him.

These creatures, according to Walton, had high, domed heads, large eyes, and tiny noses, mouths, and ears. They were dressed in what has become the standardly reported uniform of a one-piece coverall. Walton noticed that none of them seemed to have fingernails.

Eventually the creatures left the room, and when they did, Walton climbed down from the table. He walked to the door, which he described as "normal" height and rectangular, with rounded corners.

He exited the room, walked down a corridor until he came to a room on the right. Looking in, he saw that it was round and that he could see the stars through the ceiling. This is, of course, one of the things that both Dionisio Llanca and Pat Roach said about their abductions as well. They thought they could see the stars through the walls of the ship.

In the room was a chair and Walton entered. He sat down and found a lever on the left arm of the chair. When he moved it, the stars seemed to move. On the other arm were buttons, but Walton didn't experiment with them.

He was discovered by a man dressed in a blue coverall. To Walton, he appeared to be a normal-looking human being. Again, the observation is reminiscent to the Roach abduction. It should be noted that the Llanca abduction was reported in spring 1975, before Walton's abduction, but the Roach case was not in the public arena until after November 1975.

Together, they walked through the ship and finally walked out onto what seemed to be a hangar deck. He looked at the craft he'd just exited and thought that it looked like the one he had seen near Heber. The hangar deck held another three or four of the craft.

They crossed the hangar deck and entered another small room, where there were two men and a woman dressed like Walton's "guide," though they weren't wearing helmets. The guide crossed the room and exited, leaving Walton with the three beings.

It was at this point that Walton was apparently examined by the aliens. They had gestured for him to climb up on the table, which he did. They put an oxygen-mask–like device over his face and he lost consciousness again.

That was all that he remembered until he surfaced later, trying to obtain some help from his family. When he reappeared, his brother Duane called Bill Spaulding of the Ground Saucer Watch, trying to find a doctor to examine Travis. An appointment was arranged, but the doctor wasn't an M.D., and apparently, according to Coral Lorenzen, wasn't even a Ph.D. At any rate, no physical was performed.

Jim Lorenzen, International Director of APRO, arranged for a polygraph examination. Walton, apparently, flunked this test, though the Lorenzens and others suggested that the test itself was flawed. According to Coral Lorenzen, three psychiatrists who had examined Walton on the day he took the test said that the test was meaningless because of Walton's agitated state of mind. Because of that, Lorenzen arranged for a second test. Walton apparently passed it.

Those who had been with Walton when the abduction took place had also been given polygraph tests. Again, there is a dispute about the results, though the Lorenzens reported in *Abducted!* that all of them passed, with a single exception. The consensus seemed to be that none of the participants was lying about his individual role in the case, though the tests had been designed to learn if they had murdered Walton. The single exception was the sixth man, who had joined the crew only a short time before the abduction took place.

Leo Sprinkle, in February 1976, was able to travel to Arizona to interview Walton. Sprinkle used hypnotic regression on Walton to gather additional details of the experience. In consultation with Lorenzen and Dr. Harold Cahn, a physiologist and APRO Consultant, Sprinkle helped polygrapher George J. Pfeifer design the test that Walton would be given.

Walton, who hadn't been in the room when the test questions were drafted, reviewed them with the men. He then asked that for a couple of wording changes and suggested a few additional questions. In his report, Peiffer wrote that Walton had dictated the questions. Coral Lorenzen said that it was more of a suggestion.

Another problem with the Walton case arose when it was leaked that Walton had a minor criminal record. Lorenzen said that Walton himself had supplied the information about it because he believed that it was necessary to understand the case. He also admitted to using drugs a couple of years earlier, but all that was in his past and had no relevance to the abduction case.

It didn't help anyone's credibility when Duane was caught in certain lies about Walton's background. When Philip Klass, the UFO debunker, asked

Duane if his brother had ever been in trouble with the law, Duane denied it. Klass used this to suggest the whole abduction was a hoax.

Coral Lorenzen, in response, wrote, "Klass lingers on this lie told by an older brother who is obviously trying to protect his brother from the slings and arrows of individuals, who, like Klass, want to discredit Travis for a deed which was done when he was a young, impressionable man. Klass also leaves out the fact that the two young men who accomplished the check forgery were also aided and abetted by an older man who had a criminal record and who suggested the forgery."

Duane also denied there was an earlier polygraph, as had Jim Lorenzen. Coral wrote that it was Duane's attempt, as it had been Jim's, to protect the confidentiality of that first polygraph.

Coral Lorenzen believed that the lies had been justified, given the probing nature of some of the questions being asked. The lies and misrepresentations were the result of the pressure from the outside. The real point is that these lies had relatively little to do with the case. They had been told to protect Walton.

Both Lorenzens believed that Walton had been abducted by aliens. The story he told under hypnotic regression reinforced this attitude. The careful probing by Sprinkle, as well as the second polygraph, provided additional proof. Lorenzen wrote, "And so the mystery remains. Where was Travis Walton for five days?"[19]

THE CASEY COUNTY ABDUCTION—1976

While the Walton investigation was being conducted, three women from Casey County, Kentucky, were abducted late at night from a lonely stretch of road. According to Coral Lorenzen, the women saw an object in the sky, believed that an aircraft accident was imminent, and rushed forward to help. The interior of the car was bathed in light. A moment later the engine stalled. A split second later, they were eight miles away with no idea how they had gotten to that point.

When they arrived at home, all three began to experience a burning sensation on the skin. A red rash developed, which disappeared a couple of days later. And they noticed that it was nearly 1:30 in the morning. They should

---

[19]Jerry Clark, "The Travis Walton Kidnapping Controversy," *UFO Report*, Dec. 1976, 8, 68–9; Lorenzen and Lorenzen, *Abducted!*, 80–113; Curtis Peebles, *Watch the Skies* (New York: Berkley, 1995), 273–81; Kevin D. Randle, *The Randle Report* (New York: M. Evans & Co., 1997), 11–30; Travis Walton, *Fire in the Sky* (New York: Marlowe Co., 1996); Travis Walton, *The Walton Experience* (New York: Berkley Books, 1978).

have been home by midnight, at the latest. Nearly an hour and a half was unaccounted for.

Once again APRO learned of the UFO report and the missing time, and sent Leo Sprinkle to conduct an investigation. He was assisted by APRO field investigator Bill Terry. But before they could begin, a "jurisdictional" dispute developed. Both the Center for UFO Studies and the Mutual UFO Network had investigators on the scene. No one wanted to cooperate or share results with anyone else. Eventually it was worked out so that the results would be shared between APRO and CUFOS.

However, the dispute took up too much of Sprinkle's limited time, so he was unable to complete the investigative work he was sent to do. The *National Inquirer* entered the case, offering to underwrite the costs of the investigation, including polygraph examinations. Although the three women had been reluctant to become involved with the *Enquirer,* they all eventually agreed.

One of the things that was finally accomplished was a polygraph of the three women. The pertinent questions asked of all three suggested only that they had an experience with a UFO, that the car seemed to "go" out of control, and that they were not engaged in a hoax. It didn't suggest that an actual abduction had taken place. The polygrapher believed, based on the test results, that all were telling the truth as best they could.

Under hypnotic regression performed by Sprinkle, each of the women described an abduction experience. They were taken from the car and apparently examined on the flying saucer. One of the woman, Mona Stafford, said that she had been alone on a white table and that a large "eye" seemed to observe her. She believed her right arm was pinned, and it felt as if her left leg had been doubled under her, causing her some pain. Her fingers on her left hand were manipulated.

She also reported a number of humanoids, "four or five" sitting around her wearing "surgical" masks and clothes. These beings were short. At one point during the examination, she felt that she had an "out-of-body" experience, or she had been taken from one room and allowed to see another woman lying in another.

Elaine Thomas also spoke about seeing the small humanoids. She said they had dark eyes and gray skin. She also remembered something was placed around her neck and each time she tried to speak, or think, she felt as if she were being choked. She also believed that she had been separated from her two friends.

Coral Lorenzen, as she investigated the case, talked extensively with Louise Smith. She wrote in *Abducted!* that, "when one listens to the actual tape recording [of the regressions], it is instantly obvious that the women are remembering a very frightening and painful ordeal."

There would be other abductions of others to follow. Many of them reported to have been on a deserted part of the road, just as the Hills had said so many years before. The abductees, it seemed, were targets of opportunity. They were away from civilization, traveling late at night, along a deserted portion of road. They would be paced by the UFO for a period of time, before the abduction would take place. Then, they would arrive at their destinations aware that they were later than they should have been. It was a pattern that suggested itself over and over again.[20]

THE HOPKINS ERA BEGINS—1980

Abductions were about to be radically altered again. Prior to 1980, the events seemed to be one-time happenings. That is, the victim, or victims, would be grabbed, examined and returned, but not sought out for additional abductions. Pat Roach, in 1973, was the first to report that the aliens were in the house, but again, that was a single event. After the publication of *Missing Time*, by Budd Hopkins, the abductees began to report multiple experiences going back into their childhoods. To them, the abduction phenomenon began long before Villas-Boas in 1957. The victims, interviewed after 1980, began to suggest they had first been abducted in the 1950s or the 1940s, and in a few cases, the 1930s.

Suddenly, the selection of the "victim" of abduction wasn't a random choice or an apparent target of opportunity, but someone who had been selected early in his or her life. A few of the abductees began to hint that older family members had been abducted as well. Cases began to show up now suggesting abduction in the 1930s, or 1920s, or even at the turn of the century. In one case, a woman reported that her great-grandfather might have been abducted in 1888. It was almost as if the aliens were conducting a longitudinal study of specific families in their attempts to understand the human race and our emotional makeup.

A good example of this new information was reported by Hopkins in *Phenomenon,* edited by John Spencer and Hilary Evans. Hopkins said that typical of the abductions he was investigating was the case of the woman he called "Angie."

Angie, according to Hopkins, was a woman in the military service when she wrote. She held an important job with a high-security clearance. She wrote that she had enjoyed walking in the woods near her home, but after seeing a hovering UFO, she became frightened whenever she was in the woods alone. Both her father and her brother had seen the UFO.

Hopkins wrote, "The various gaps and confusions in the sequence of events described in the original sighting, the odd phrasings in Angie's account, were all immediately suggestive to me."

---

[20]Lorenzen and Lorenzen, *Abducted!* 114–31.

Angie eventually visited Hopkins at his studio in New York, where she was regressed by Dr. Don Klein. Later, Hopkins did his own hypnotic regressions, with Angie's cooperation. She remembered walking into the woods, where she met a small, gray-skinned humanoid. She was escorted to the the landed craft, taken inside, and then subjected to a medical procedure.

Prior to 1979 or 1980, the medical procedures reported by the abductees had been passive and noninvasive. The abductees, originally, had talked of being scanned by some kind of machine, not unlike that of an X-ray or an MRI. A big "eye," for example, was reported by some such as Charles Hickson. Reports of invasive procedures were rare, and usually involved nothing more than a needle pushed into the body, as described by Betty Hill and eventually by Pat Roach.

After 1980, the abductees began to talk of much more invasive procedures. Angie, for example, told Hopkins that a circular incision was made on her thigh. Hopkins speculated that some sort of "coring" device had been used to gather cellular material, though there seems to be no medical reason that it was taken from her thigh. Fingernails and hair would have supplied the cellular material without the trauma of cutting into her leg.

Angie also reported, eventually, that she had been abducted at least three other times. She believed, according to Hopkins, that she had been abducted from her earliest childhood, and she believed that her mother, and possibly her sister, had also been abducted. It was the pattern that would emerge again and again.

Hopkins reported that as he researched the UFO abduction phenomenon, he made a number of discoveries. First was that the aliens were sometimes able to block almost every memory of an abduction in the mind of the victim. He postulated that there could be a significant number of people in the United States who had no conscious memory either of seeing a UFO or being abducted by the crew of a spacecraft.

He reported that Steven Kilburn only had a fear of a short stretch of highway. Under hypnotic regression, Kilburn detailed his abduction from that stretch of highway. Hopkins wrote that Kilburn's tale showed that anyone could be abducted at any time and have no memory of it.

Hopkins discovered that many, and he believes most, abductees were first taken early in life. They are then abducted periodically throughout their lives. The alien beings obviously have developed a method of tracking the people they want to study. In fact, Hopkins, who also pioneered research into implants, believed that the implants might be tracking devices not unlike the tags human biologists put on animals in the wild.[21]

---

[21]Jerome Clark, *UFO's in the 1980s* (Detroit: Apogee, 1990), 134–136; Budd Hopkins, *Intruders* (New York: Random House, 1987); Budd Hopkins, *Missing Time* (New York: Richard Marek Publishers, 1981).

Since the publication of Hopkins's books, there has been an explosion in alien abduction. Dozens of books, some written by mental health professionals, have detailed their specific research into alien abduction.

Notables, such as John Mack, a Harvard psychiatrist and Pulitzer Prize–

ABDUCTION RECORDED ON VIDEOTAPE

One of the new broadcast networks, UPN, aired a program that they had advertised as "a one hour special centered around an alleged videotaped account of a family's purported encounter with what may be extra-terrestrial life forms . . ." UPN reported. "The recently acquired videotape is the sole testament to the fate of the McPherson family, missing since last Thanksgiving."

UPN noted, "During the special, several people who claim to have had similar experiences relate their ordeals, plus experts on aliens discuss the authenticity of the videotape."

According to UPN, "A series of strange occurrences, caught by the camera, culminated into what appears to be a frightening encounter with strange creatures." Here, it seemed, was the type of evidence that the skeptical, the scientific, and the journalistic communities demanded. The press release suggested that this was a real event that had taken place in Lake County, Minnesota, and more important, there was a videotaped record of it.

But almost from the moment the show ended, there were those who suggested it was a hoax, or to be more polite, an entertainment program that had no basis in fact. It was a representation of what an alien abduction could be. Some of those who watched it reported that it wasn't an amateur home video, made by the family's sixteen-year-old son, but a well-lighted and well-produced video production. It smacked of a docudramatic without the trappings of the truth.

Many of those on the Internet, or in the UFO community, were angry about the program. They believed it to be an exploitation of what they consider to be a very real problem. By making the show, UPN had moved alien abduction from the realm of reality into the realm of science fiction. It was a real disservice to those who have patiently conducted alien abduction research for the last two decades.

Reporters, apparently intrigued by the story, began calling Lake County sheriff Andy Haugan. He told them that no one named McPherson had been abducted. According to the telephone directories, and a 1995 database, no one named McPherson, or MacPherson, had lived in the Lake County area

winning author, not only endorsed the idea of abduction, but wrote a book about his own investigations into claims of alien abduction. His research convinced him that alien abduction was real.

Now hundreds have been identified as having been abducted. Researchers

when the abduction took place. There have been no missing persons reports concerning anyone named McPherson.

Those who appeared as the experts, such as Stan Friedman, began to complain about the show as well. Friedman said that he had only answered generic questions about UFO abductions. Under attack by some ufologists, Friedman asked if anything he had said on the program was in error. Although his words reflected the state of abduction research, they were applied to the Lake County case.

In fact, it is clear from the program that none of the experts, which included California-based abduction researcher Yvonne Smith, had seen the videotape before answering questions about abductions. All were answering questions about alien abductions in general rather than this case in particular. These were cut in to provide an authenticity that the tape would have otherwise lacked.

John Velez, of the New York–based Intruders Foundation, was asked to appear as one of those who had been abducted. Velez questioned the producers and decided that he didn't want to participate. He felt that it was not going to be an accurate representation. In fact, he warned many of us to be aware of this program.

Others suggested that any exposure of the problem of alien abduction helped alert the public to the problem. And others still insisted that this was simply an entertainment program, just like *Star Trek* and *The X-Files.*

Of course, neither *Star Trek* nor *The X-Files* ever suggested that they were reality based. That is the problem with the UPN show. Their own press release and their promotions for the program suggested that it was real, though they carefully used words like "alleged" and "purported."

But all the arguments about this case could have ended before they started. The reporters didn't have to bother the Lake County sheriff, the databases didn't have to be searched, and the telephone company records were unnecessary. If everyone had read the closing credits they would have easily realized the truth. The names of the actors, including those who played the aliens, were clearly visible. UPN knew from the start that they were producing a program of entertainment.

around the country provide a forum so that those who believe they have been abducted can meet with others with similar experiences to share their experiences and their trauma. There are group sessions and survivors networks and communication among those who believe they are victims of alien creatures. Nearly every city with even a moderate-sized population boosts one or more abduction "survivors" group.

And there are forums, national and local, for the abductees. Although many say they want no publicity, many eventually tell their stories publicly. When that happens, they have the opportunity to meet with others who have had similar experiences.

The Roper Organization conducted a poll in the early 1990s to determine the extent of the abduction phenomenon. The results of that poll, according to many of the abduction researchers, was that more than three million Americans might have been abducted since the beginning of the modern UFO era in June 1947. At least that seemed to be what the results, as interpreted by Hopkins and two others, suggested.

The abduction researchers are continuing their work, but they have few answers. They have theories about the reasons for alien abduction and why it is continuing. After Betty Hill produced her star map, and Marjorie Fish produced her star map, they believe they have located the home system of some of the aliens. Though the researchers claim to have uncovered some facts, the answers they have are few and far between. All they know for certain is that alien abduction is real because too many people are involved in it. Not everyone could be a liar, deluded, or crazy. Besides, there is just too much corroboration among the various sighting reports.

There are some areas of corroboration. Remember, Hopkins noted that Angie saw the UFO hovering in the air and alerted both her father and her brother. They too, saw the craft, though it seems that neither of them were abducted.

To Hopkins, as well as others, this is an important bit of corroboration. Not only is there a tale of abduction, but others, not involved in the abduction itself have reported seeing the craft. This provides a connection between the UFO phenomenon and the alien abduction, according to some researchers. In fact Mack, in what he calls his "five basic dimensions," said that skeptics must account for the association of UFOs witnessed independently while the abductions take place.

The Walton case comes to mind immediately. Walton, and his friends saw the UFO hovering near their truck. Walton got out to approach and seemed to have been injured in the process. His friends fled to alert local authorities. The point, however, is that each of them passed a polygraph examination about the UFO sighting, which provided some independent corroboration for the UFO event.

Budd Hopkins has told various investigators that he, as well as other abduction researchers, routinely withhold specific information from the public as a way of cross-checking the validity of an abduction story. These "secret" clues, when mentioned by someone claiming abduction, provide some validation for the new tale. How could anyone know these specific things since they hadn't appeared in the abduction literature if that witness had not been abducted?

Coral Lorenzen told Randle in the mid-1970s that APRO and its researchers were doing the same thing. All the researchers used by APRO were aware of these specific clues, but Coral said that she had kept the information out of her books on the topic and refused to let the information seep into the *APRO Bulletin,* the organizational newsletter. In the interest of science, it was important to withhold some information.

Each of the researchers is careful to point out that people can, and do, lie under hypnosis and there are those who have faked their abduction experience. But the tales told by those inventing the stories do not display proper emotion or the proper fear, according to the abduction researchers. That is the one thing that seems to cement these tales together. The trauma of the experience is obviously real to the researchers attempting to learn more.

Coral Lorenzen, for example, wrote about the Casey County, Kentucky, abduction, "These excerpts from the transcripts of the hypnosis sessions are stunning enough when in print, but when one listens to the actual tape recordings, it is instantly obvious that the women are remembering a very frightening and painful ordeal."

During the investigation of the Roach abduction, Harder was concerned during the first session that there was not the appropriate emotion displayed by Roach. In the later sessions, her fear of the ordeal, and the trauma of it as well as her concern for her children became one of the most convincing aspects of it, at least for Harder.

Of course, all these points are interesting in corroborating alien abduction, but the real proof comes from the multiple abductions. Shared delusion or hallucination is extremely rare. How does one witness communicate a hallucination to a second, or third, and in some cases fourth and fifth witness? The rare cases of multiple abduction would seem to confirm the reality of the situation.

Barney and Betty Hill told, under professionally directed hypnotic regression, stories that were astonishingly similar. Even the differences were astonishing because they seemed to reflect the points of view of two separate individuals each telling the same tale from his or her own perspective. It is the sort of subtle detail that underscores the validity of the experience, at least in the minds of the researchers.

And although abductees have a terrible record at obtaining the "cigarette lighter," that is, a piece of physical evidence for scientific research and investigation, they have obtained valuable information. Betty Hill, for example, saw the star map on the alien ship. It wasn't until the star catalogs were updated, long after her abduction, that Marjorie Fish was able to pinpoint the proper stars in the three-dimensional model she created. While not exactly the physical evidence demanded by the skeptics and the scientific community, it is an important piece of corroboration. How could Betty Hill "invent" such a detail when our scientific knowledge was inadequate to provide the clues until after her abduction.

The abduction phenomenon, then, is not made up of only single-witness testimony. Yes, the majority of the abductions are single witness, but there are enough multiple-witness abduction accounts to suggest something real. Yes, there are some instances of corroboration in the form of advanced knowledge, which also suggests this is something real. And there is the testimony of others, not involved in the abduction, about the UFOs in the area.

The real stumbling block is a lack of physical evidence, though there have been landing trace cases related to abduction reports. The Schirmer case of 1967 provided that sort of corroboration. Warren Smith, a writer from Iowa, claimed that he had found the landing site complete with landing-gear markings and swirled grass and debris. This provided, for Schirmer, indirect physical evidence that a real event had taken place.

Also remember that Schirmer, a police officer, was given a polygraph examination. There were no areas of deception. In fact, in a great majority of the cases where the polygraphs are used, no deception is found. If nothing else, it shows that the witness believes the tale being told.

While the indirect evidence is interesting, it is not always persuasive. The implants now being reported in some cases would represent that direct physical evidence. While not the same as grabbing a cigarette lighter, the implants are, according to the researches, alien artifacts. There are a number of cases where the artifact, or implant, has been detected under the skin. There are a smaller number where X rays have been taken showing the implant, and an even smaller number still in which an implant was recovered. Again, this evidence is not conclusive, but it shows that progress is being made. Physical evidence is being recovered by the researchers.

As we look at the abduction phenomenon, it appears that the physical evidence is being collected. It seems that the researchers are objective in their methods, are careful in their investigations, and conservative in their conclusions. What Budd Hopkins finds in New York is being repeated by Leo Sprinkle in Wyoming, John Carpenter in Missouri, and Richard Boylan in California.

The researchers tell us that something is happening out there. Something is going on. We've gone beyond the question of whether or not it is real. We must now begin to learn exactly what it is. Further study will give us those answers.

2.

THE BEGINNING OF ABDUCTION RESEARCH

The abduction phenomenon was introduced, in this country, to a large extent by Coral E. Lorenzen, the founder of the Aerial Phenomena Research Organization. Her pioneering work contributed to our understanding of UFOs and the surrounding phenomenon. At one point, she was nearly a lone voice advocating an acceptance of occupant reports, the abduction phenomenon that included the Barney and Betty Hill case, and other aspects of UFOs that had been overlooked and ignored by nearly everyone else in mainstream ufology.

Her dedication and belief in UFOs didn't come from a profound scientific background or understanding, but from her own observations. Born in 1925, she was still a child of only nine when she saw what she would later understand was a UFO. The hemisphere-shaped object crossed the sky in an undulating movement. Three years later, she mentioned the sighting to her family doctor, who provided her with one of the books written by Charles Fort, an author who had collected tales such as that told by Lorenzen.

On June 10, 1947, four years after she married Jim Lorenzen, who would be her partner in APRO, she saw another UFO. This was a small sphere that rose from the ground near Douglas, Arizona, and disappeared into the night sky. This reawakened her interest in UFOs. She began to correspond with others who had reported seeing a UFO, or who shared her renewed curiosity concerning the topic,

In January 1952, Lorenzen founded APRO and began to publish the *APRO Bulletin.* In 1964 Jim Lorenzen became the director and Coral was the secretary-treasurer. They would keep the organization and the *Bulletin* alive for the next three decades.

APRO was important because of Coral's investigations into the tales of alien encounters. Others in the UFO field rejected such stories as impossible and often confused them with the claims of the contactees. Coral, however, realized that an alien spacecraft would require a crew, and she not only collected such reports but published the details in the *Bulletin.* Although others were to follow her lead, hers was the first of the "mainstream" UFO organizations to investigate the so-called occupant reports.

More important, she was one of the first to investigate and report on tales of alien abduction. She often spoke to Betty Hill, interviewing her about her experiences. One of the most exotic of the abduction stories, told by Antonio Villas-Boas, was first published in detail in this country by Lorenzen. By the time of the Travis Walton abduction in 1975, Lorenzen was convinced that alien abduction was a reality. Her work led other UFO researchers, including Leo Sprinkle and James Harder, to the investigation of abductions.

Jim Lorenzen died in 1986 and Coral died just two years later. With their passing, their organization fell into inactivity and finally ceased to exist. Their legacy, which was the tirelessly collected files on the UFO phenomena, photographic and physical evidence cases, occupant reports, and alien abductions investigations, were passed to their children. Attempts by UFO organizations and others to secure those records have failed. A few chosen researchers have been allowed to review portions of the records since the deaths of Coral and Jim, but the files remain in storage and are slowly disintegrating.[1]

The legitimacy of the abduction phenomenon got a big boost in 1966 with the publication of *The Interrupted Journey*. It was condensed in *Life* magazine at about the same time. It is, of course, the Hill story told at length and in great detail. For the first time a book by a respected writer, John Fuller, would be given treatment that suggested there was something real and perhaps frightening about UFO abductions. No longer were reports of being taken on board the alien craft left in the hands of the contactees. Now the tales were being told by people who would be considered normal by their friends and family. The concept of alien abduction was working its way to the forefront of ufology.

As noted, during those early years of the abduction phenomenon, there were few who embraced the concept. Dr. Leo Sprinkle, a psychology professor at the University of Wyoming until he retired in 1989, was one of APRO's scientific consultants, and one of the first to seriously investigate alien abductions.

He was born on August 31, 1930, attended the University of Colorado, and eventually received his Ph.D. from the University of Missouri in 1961. He taught at the University of North Dakota for three years, and served, during his final year as the director of the school's counseling center. In 1964, he moved to the University of Wyoming in Laramie.

Sprinkle's interest in UFOs, like that of Coral Lorenzen, came from his own sightings, one in 1951 and the other in 1956. Jerry Clark, in his massive en-

---

[1]Jerome Clark, *UFO's in the 1980s* (Detroit: Apogee, 1990), 153–5; Coral and Jim Lorenzen, *Encounters with UFO Occupants* (New York: Berkley, 1976), v–vi, 1–6, 416; Coral and Jim Lorenzen, *Flying Saucer Occupants* (New York: Signet, 1967), 16, 213; Frank B. Salisbury, "Introduction," in *Encounters with UFO Occupants,* by Coral and Jim Lorenzen, (New York: Berkley, 1976), ix–xvi.

cyclopedia about UFOs, noted that Sprinkle became an influential figure in "the contactee movement of the 1980s [as opposed to just the abduction phenomenon]." He, again according to Clark, "entered into correspondence with contactees who had written to him about their psychic communications with beings who Sprinkle good-naturedly call 'UFOlk.' "

During the Air Force–sponsored Condon Committee investigations of UFOs, it was Sprinkle who discovered the missing time segment of the Schirmer case. Later he would be the hypnotist used in Hignon and Larson abductions.

Sprinkle also began a series of UFO and abduction-related conventions, the first held at the University of Wyoming. It was poorly attended, but Sprinkle continued with his Rocky Mountain Conferences on UFO Investigations despite the poor turnout. He did notice that a different crowd appeared each year rather than the same, familiar faces coming back again and again.

Sprinkle has told researchers that he believes UFOs to be physical craft piloted by biological creatures, but that they can manipulate time and space. Because of this ability, we would regard them as angels or demons. But his attitudes also reflect a New Age orientation that includes an interest in reincarnation.[2]

In the beginning of the abduction phenomenon, in the 1960s, there were few who were interested in those sorts of reports. Dr. James A. Harder, a civil engineer from California, was one of them. In addition to his training as a civil engineer, he learned the process of hypnosis. The Lorenzens would use him on a number of occasions for hypnotic regression of abductees. Harder would be the first to use hypnosis to investigate the tale of Charles Hickson and Calvin Parker.

In fact, Harder would eventually become APRO's director of research, often called in to use hypnosis. Harder has a strong belief in the physical reality of the UFO phenomenon and believes abductions are a physical manifestation caused by the extraterrestrial beings. He would periodically revisit the abduction accounts, searching for new information. He was looking for validation of the Hill case.[3]

It was during the early 1970s that the abduction phenomena began to bloom. UFO-related publications such as Saga's *UFO Report* and *Argosy's UFO*

---

[2]Clark, *UFO's in the 1980s*, 198–9; Daniel Kagan and Ian Summers, *Mute Evidence* (New York: Bantam, 1983); R. Leo Sprinkle, "Hypnotic and Psychic Implications in the Investigation of UFO Reports, in Coral and Jim Lorenzen, *Encounters with UFO Occupants* (New York: Berkley, 1976), 256–329.

[3]Lorenzen, *Encounters*; Kevin D. Randle, "The Family That Was Kidnapped by Ufonauts," *UFO Report*, June 1976, 21–3, 50–2, 54.

recounted many abduction cases. One of the first to write about these reports was Kevin Randle. But at that time, tales of alien abduction were looked on, for the most part, as an extension of the contactee phenomenon, when they were reported at all. In fact, *UFO* magazine, published by Vicki Cooper, would, in the late 1980s and early 1990s, contain a column labeled "Contactee," chronicling stories of alien abduction.

So while Coral Lorenzen, with a minor assist from Fuller, Sprinkle, and Harder, can be credited with introducing the concept of alien abduction, it was Budd Hopkins who turned it into an international phenomenon. Hopkins, of course, didn't start out to be the leading authority on alien abduction. In fact, and according to his own work, he didn't believe in UFOs or alien visitation and was uninterested in them.

Hopkins was born on June 15, 1931, in Wheeling, West Virginia. He was graduated from Oberlin College in 1953 and moved to New York. Hopkins, it should be noted, is considered by some to be a world-class artist. His works are part of the permanent displays in some of the biggest and best art museums in the world. He has lectured at various universities, colleges, and art museums. His vocation, at least in the beginning, was that of artist.

His interest in UFOs began in 1964 when he, along with two others, sighted a disc-shaped craft in broad daylight. It remained in sight for two to three minutes, giving him and his companions a good look at it. Because of the sighting, he joined NICAP (at the time the leading UFO organization) and started to read about UFOs. In 1975 he investigated a multiple-witness sighting in New York and reported his findings in *The Village Voice.*

Because of that article, he began to receive letters about other, similar sightings. Working with a psychiatrist, a psychologist, and ufologist Ted Bloecher, he began to investigate these cases, which led to his first two books, *Missing Time* and *Intruders.* These two works were responsible for alerting the world to the phenomenon of alien abduction. He was responsible for coining many of the terms used by today's researchers, and he was responsible for bringing the other leaders into the field. Those who worked with him, or learned from him, include David Jacobs, Ph.D. and John Mack, M.D.

Hopkins said that he had discovered some disturbing patterns in the alien abductions. He believes that the phenomenon is widespread and could involve hundreds of thousands, if not millions, of victims. He believes that many victims come from families of other victims going back decades, and that individual abductions begin in early childhood and continue throughout the lifetime of the victim.

Unlike some of the other abduction researchers, Hopkins believes that the aliens lack emotion, and that they are interested only in their scientific research, which revolves around human reproduction and genetics. He has found what he believes to be proof that there have been widespread cases of artificial insemination of human women by the alien scientists. The pregnancies are not allowed to develop to term, but the women are abducted at some point so that the embryos can be removed and grown in the UFO laboratories. There are reports that the infants and children are shown to their human parents to allow for the natural bonding that takes place between child and parent. These meetings are, of course, arranged on board the alien craft, are the result of still another abduction, and, of course, are traumatic for the human victims because of the feelings of loss.

Hopkins reported the ultimate abduction, involving not only alien beings, but governmental secrecy in his book, *Witnessed*. It is the story of a woman he identifies as Linda Cortile, who was taken from her twelfth-floor apartment in 1989. According to Hopkins, what makes this tale exceptional is that the abduction was witnessed by police officers, secret service agents, and a high-ranking member of the United Nations. Hopkins also interviewed others who claimed to have seen the UFO, and who said they saw Cortile floating toward a hovering UFO, escorted by the alien creatures.

Hopkins has also reported on a new and disturbing trend. He suggests that some abductees have claimed they were forced to participate in abducting other human victims. In other words, the aliens have abducted one person, given him or her a uniform, and then required them to perform some task during the abduction of another victim.[4]

Hopkins inspired others to begin investigating tales of alien abduction. One of the first was David Jacobs, a history professor at Temple University, author of *The UFO Controversy in America*, which was considered by many to be one of the classic UFO books. Jacobs was born in 1942 in Los Angeles, California, and he received his Ph.D. from the University of Wisconsin at Madison in 1973. He is credited with giving the first paper on abductions to a scientific organization.

Jacobs has published two books on abduction, *Secret Life* in 1992 and the wildly controversial *The Threat* in 1998. In the latter, Jacobs reveals the alien mission as he has deduced it from his interviews and hypnotic regression sessions with abductees. Jacobs claims that the aliens are creating a race of hybrids who have the mission of taking over the Earth.[5]

---

[4]Budd Hopkins, *Intruders* (New York: Random House, 1987); *Missing Time* (New York: Richard Marek Publishers, 1981).

[5]David Jacobs, *Secret Life* (New York: Simon and Schuster, 1992); *The Threat* (New York: Simon and Schuster, 1998); *The UFO Controversy in America* (New York: Signet, 1975).

Hopkins also inspired John Mack. Of those doing abduction research, Mack has the most impressive, and relevant, credentials. He is a Harvard-trained psychiatrist and a Pulitzer Prize–winning author. He is a professor of psychiatry at Harvard Medical School. His Pulitzer Prize came from his biography of T. E. Lawrence (Lawrence of Arabia).

Mack, by his own admission, had no interest in UFOs and abductions, criticizing those who reported on or studied the phenomenon as being out of tune with the world. But in January 1990, he agreed to meet Hopkins and was impressed by the evidence that Hopkins had amassed. He was also impressed with Hopkins's sincerity and intelligence. When asked if he wanted to meet some experiencers (the new term for abductees) he agreed. That started Mack on the journey that would produce *Abduction,* his book about "Human Encounters with Aliens."

Mack was criticized by his Harvard colleagues, not for studying alien abduction, but for his methodology. In a controversial attack at a CSICOP (Committee for the Scientific Investigation of Claims of the Paranormal) convention, Mack was confronted by one of his "experiencers." She claimed that she had invented her tales of abduction, which included seeing both Kennedy and Khrushchev on board a UFO in the early 1960s. She claimed that Mack had believed her story, but nothing about it appeared in his book.

To the credit of the membership of CSICOP, many were appalled by the "ambush" of Mack. They believed it unfair to bring the woman to the meeting to confront Mack without telling him that she would be there. And since he had published nothing about her tale, and had not made reference to it in his work, there seemed to be little point to the attack.

Mack's work is considered by many to be the most important contribution to the abduction phenomenon. His impressive credentials, and his scientific training, provided a note of authenticity and respectability for the tales of abduction. Mack was, as a psychiatrist, one of the few people with the training to recognize psychological aberrations, if such existed in the abductees.[6]

But he is not the only credentialed investigator of alien abductions. Several psychologists have been actively treating the trauma of abduction. These include Dr. Richard Boylan, Ph.D. and Dr. Edith Fiore, Ph.D.. Medical doctors including Raymond Moody and Richard Neal have researched the abduction phenomenon. Dr. Neal was especially interested in claims that alien abductors had impregnated humans and later removed the fetuses for their own experimentation. Dr. Neal's study of that aspect of the abduction phenomena is one of the best examples of a truly scientific survey.

---

[6]John Mack, *Abduction* (New York: Charles Scribner's Sons, 1994).

Others with postgraduate degrees who have entered the abduction arena include William P. Cone, Ph.D.; Dr. Aphrodite Clamar, Ph.D.; Barbara Levy, Ph.D.; Jeffrey Mishlove, Ph.D.; Mary Ellen Trahan, Ph.D.; and Dr. Roberta Fennig. Like John Mack, Fennig is one of the few psychiatrists who is investigating and treating those who have been abducted.

The point here is that there is no shortage of those with impressive credentials who not only believe that alien beings are abducting humans, but who are attempting to deal with the psychological trauma of those events. Many of these people, though by no means all of them, have written little about their experiences and their clients.

What is interesting, however, is that many of them have become abductees themselves. Richard Boylan, for example, told one of his patients that he was going to make a trip to the Southwest, and during that trip, he thought he might be abducted. His prediction came true, so he is not only a researcher of alien abduction, but a victim of it.

We had thought that his claim made him rare. Boylan was someone who studied the abduction phenomenon as a scientist but who also had the experiences of the victim. It should provide a unique insight into alien abduction. But then we learned that many of those who conduct this sort of research also have been abducted.

Leo Sprinkle, for example, in the introduction to Edith Fiore's *Encounters,* wrote, "However, I claim to be a UFO contactee [that is, an abductee] as well as a UFO observer."

Raymond Fowler, who gained widespread fame in the UFO community for his tireless and competent investigations of UFOs, including a number of abduction accounts, came to believe that he was an abductee. He believes that the many sightings by his family, beginning in 1917, when his mother saw a huge, hovering object over Bar Harbor, Maine, are the result of an ongoing alien interest in his family.

Fowler has written a number of books about alien abduction. He was responsible for uncovering the series of abductions by Betty Andreasson who claimed not only abduction, but in a tale more reminiscent of the contactees than the abductees, received messages from the aliens about the human condition and was taken to the alien's home world. He wrote two books about Andreasson and her adventures.

He also wrote two books, *The Watchers: The Secret Design Behind UFO Abduction* and *The Watchers II: Exploring UFOs and the Near Death Experience.* He also wrote *The Allagash Affair,* in which he detailed the simultaneous abduction of four friends in Maine.

The most popular of those writing about alien abduction, or claiming to have been abducted, is Whitley Strieber, who was a bestselling author before he began to write about abductions. His earlier books included *The Hunger* and *The Wolfen,* both of which were made into movies.

Strieber was born in San Antonio, Texas, on June 13, 1945. He attended the University of Texas, and even attended law school, but quit to travel in Europe. When he returned, he settled in New York to begin his writing career. In 1985, his life took a turn when a series of encounters began in upstate New York. Under hypnotic regression he would realize that he had been abducted. He wrote *Communion* and then *Transformation,* which told of his encounters. He also wrote *Majestic,* a novel that dealt, in part, with the Roswell, New Mexico, UFO crash of 1947.

Strieber learned of the work of Budd Hopkins, and discovered that they lived near one another in Manhattan. Through hypnosis, conducted by Donald Klien, Streiber learned that he had been interacting with aliens since he had been a child. To tell his tale, Strieber wrote *Communion.*

Although Hopkins and Strieber had been working together, they eventually parted company. Part of the reason was their differing interpretation of the abduction phenomenon. Strieber believes that the aliens are essentially benign, or even benevolent, despite the terror and trauma he reported about his experiences. Hopkins, on the other other hand, believes the aliens are cold, calculating scientists who have no interest in the trauma suffered at their hands. At best the experiences, according to Hopkins, could be viewed as negative.

Strieber's book led fellow San Antonian Ed Conroy to write *Report on Communion.* Conroy investigated Strieber's story, his relation with the UFO community, and other aspects of the case, concluding, to his mind, that Strieber was telling the truth. According to Conroy, human visitors to Strieber's cabin

FARRAKHAN AND UFO ABDUCTION

There are very few reports of abduction by alien creatures from the African-American community. There are a few of them, however, including stories circulating around Nation of Islam leader Louis Farrakhan. The tale told by Farrakhan and his followers more closely reflects that of the contactees than that of the abductees.

It was on September 17, 1985, that Farrakhan visited the half-mile by half-mile mother ship carrying fifteen hundred smaller baby planes with bombs designed for the destruction of the world. While on this craft, apparently built on Earth, Farrakhan heard the disembodied voice of then

also claimed to have witnessed some sort of paranormal experience, though not necessarily an alien presence.

These tales have inspired many others to investigate alien abduction. English professor Alvin Lawson, with the help of his friend, Dr. W. C. McCall, a physician versed in hypnosis, conducted a number of interesting experiments about alien abduction.

Lawson became interested in UFOs in the 1970s and established a hotline for the reporting of UFO sightings. Then, in the fall of 1975, Lawson received a telephone call about an abduction case. His investigation of it lead him to the conclusion that this particular case was a hoax, but it gave him an idea for his experiments.

Lawson decided to hypnotize students who had no abduction experiences to learn if those experiences, with all their various and complex traits, could be induced in his subjects. Although he learned to his satisfaction that such abduction scenarios could be implanted in those with no knowledge of UFOs or abductions, he eventually developed a theory that UFO abductions were, in reality, memories of the birth trauma. He found in his review of these tales what to him was compelling evidence that birth trauma explained alien abductions without the need to "invent" interstellar flight or a race of superintelligent alien beings.

The big disappointment for Lawson was his failure to convince abduction researchers that he had stumbled onto a viable theory. Worse still, to his mind, was the lack of scientific respect his theory had received. He had developed, according to him, the only falsifiable theory about alien abduction, and the scientific community, as well as the UFO community, were universally unimpressed.

President Ronald Reagan telling of his plans to bomb Libyan leader Qaddafi.

Those claiming to be members of the Farrakhan faction have announced on various radio talk shows, including that of Rush Limbaugh, that these craft contain the means for returning control of the world to those who have long deserved it. They report that all attempts to shoot down the mother ship have failed just as all attempts to crush the Nation of Islam have failed.

Farrakhan's UFO connection, and his visitation to the mother ship, is one of the few times that a black, national leader or otherwise (with the notable exception of Barney Hill), has suggested any type of contact with the UFOs.

But Lawson's theories did make their way into the scientific literature. His studies were cited in a number of referred psychological journals. However, the interest was not in his birth trauma theory but in his attempts to induce false memory of abductions in the minds of his students who had reported no such abductions.[7]

Others also began to investigate alien abduction. Some of them, such as the Mutual UFO Network's (MUFON) Director of Abduction Research, John Carpenter, have training in the mental health care fields. Others, such as the Close Encounters Research Organization's (CERO) Yvonne Smith, have trained in a respected school of hypnotherapy. Others still have no specialized training in hypnosis, psychology, or science, but have opened clinics for the investigation of abductions just the same.

In fact this research and investigation of abduction has moved into a public arena with nearly every community of any size having those who investigate abductions and those who treat clients suffering from the trauma of alien abduction. Some of these groups have become very specialized. In the Midwest there is a support group for those who claim to have been abducted and raped by reptilian creatures.

There are those who have attempted to find the physical evidence demanded by so many skeptics and scientists. There are those who claim that aliens have "implanted" witnesses with some kind of tiny device. Recovery of these objects has become another aspect of the abduction phenomenon. Hopkins reported that Linda Cortile had X rays showing an implant in her nasal cavity.

Derrel Sims, an abduction researcher living in Houston, Texas, and Dr. Roger Leir, a podiatrist from California, have teamed in an attempt to recover a large number of implants. These, again identified on X rays, were located in the victims' hands and feet. They have successfully recovered several and have taken them to various laboratories for analysis. The analyses are continuing.

There are now, literally, hundreds of people engaged in abduction research, trying to learn what is happening. They are building a huge base of data that, when examined properly and scientifically, could provide the clues to explain what is happening. They have entered into this work, not for the glory that has surrounded some UFO researchers, nor for the money that has bypassed so many other researchers. They do it because they have a desire to discover the truth and they believe that they have something important to contribute to the investigation.

---

[7]A. H. Lawson, "Perinatal Imagery in UFO Abduction Reports," *Journal of Psychohistory* 12, no. 2, Fall 1984.

The amount of data that has been recovered is staggering. There are dozens of books, magazine articles, television shows and documentaries, and even movies about alien abduction. These have been written or produced by those of professional qualifications, by those who have conducted the research and those who have had the experiences. All have provided important clues about alien abduction. A close reading or viewing of them allows all of us a glimpse into what is happening in the world around us.

3.

TWO STORIES OF ALIEN ABDUCTIONS

We have seen how the history of the abduction phenomenon has developed. We have seen how others have reported on alien abductions, and what they believe to be the universal truths about it. We have completed, to that point, what in science is called a review of the literature. It is something that is done as an investigator begins to study a problem that has fascinated him or her. But to this point, it seems that we have not performed any real, outside investigation. Studying the problem in the library is one thing. Studying it in the field, with people who have lived the experiences, is something that is quite different. "Book" learning is often different from "field" learning, or as some would call it, "street" learning.

Following are two tales of alien abduction that we have personally investigated. They are not, by far, the only tales with which we have had a personal involvement. We began studying the topic in the 1970s when abductions occurred outside the home, often from deserted highways. We have watched the transition, as the alien abductors moved from "targets of opportunity," that is, people who were available because of their location, to abductions that seem to be planned far in advance with a specific individual as the subject. In fact, the first tale of alien abductors inside the house to be told in a national forum was one that we had investigated and one that we reported. At the time, we didn't realize the significance of what we were learning.

But the point is, we have been studying the problem for years. We have heard many tales of alien abduction from those who experienced it. We did our review of the literature, but we also did our fieldwork. What follows is typical.

THE STORY OF SHERRY

Sherry[1] had just moved into her new home in a rural desert area of Southern California. An attractive, single mother in her late thirties with an eight-year-

---

[1]The names of some of the witnesses have been changed. These will be noted where appropriate.

old child, Sherry had searched for a quiet area with low crime to bring up her daughter.

Sherry had always felt that she was in tune with the paranormal. She had played with Ouija boards and throughout her life she had what she would call "many strange experiences." These were nothing that one would consider earth-shattering, but strange nonetheless. Things around her home might be moved. She would catch a glimpse of something out of the corner of her eye. But that was the extent of her paranormal connection until she bought her new house.

Soon after Sherry had moved to her new home out of the city, she discovered something wondrous overhead. The sky was filled with stars. The night was bright with stars. For someone who had seen a night sky only washed out by big-city lights, the sight of bright stars or the dim splash of the Milky Way was breathtaking.

Sherry's bedroom is on the second floor of her home. There are two picture windows facing the mountains and a window seat perfect for stargazing. Every night Sherry would sit for hours at the window watching the sky. What she saw both astonished and frightened her. The seeming billions of stars simply amazed her. But the strange globes of bright golden light dancing over the distant mountains frightened her. "How odd," she thought. She had seen nothing like it in the city. The golden globes were as new to her as the stars filling the sky. As the months went by, she found herself spending more time at the window. She saw the dancing lights often and they seemed to be coming ever closer.

When Sherry moved from the city, she was a well-established escrow officer in the lucrative Southern California real estate business. But the more she watched the dancing lights, the less interested she was in her work. It got to the point that she didn't want to leave her new house for fear that she would miss something important. The lights kept dancing, only now they were much closer. At times she would see them between her house and some neighbor's house about three hundred feet away. The houses are separated by some cattle pasture, flat grazing land covered with nothing more than grass, giving her an unobstructed view.

Sleep became less important to her than the lights. Work became less important than the lights. The only thing more important than the lights was Sherry's daughter.

Kimberly was eight years old when they moved into their new home. She was a well-adjusted child with a big smile for one and all. Kimmy, as she was called, was a true innocent who did not seem to have been affected by the evils of the city or the move to the country, and that's just the way Sherry wanted it to stay.

Kimmy attended a private religious school. It was the only alternative to the public schools in the area. Sherry felt that Kimmy would get more attention and a better education there. Kimmy was happy and healthy. Sherry believed that all was well with the world. All was well except for those dancing lights coming closer and closer every night.

Sleep was becoming ever more difficult to catch. Not only were the lights still there but now there were strange noises in the house as well. Sometimes out of fear, or maybe just seeking comfort, Kimmy would share her mother's bed. On many occasions, not only was Kimmy sleeping with her mother, but there were also two dogs and four cats. All of the life in that house was huddled in the upper bedroom. Everyone in a single room, seeming to take comfort in the presence of the others.

Sherry had become used to the sounds of her house and the animals that shared it. The new sounds that Sherry was hearing in her new home at night were not the sounds of her pets. At least not of earthly pets.

Nearly every night she would hear noises, bumping, thumping, and groaning. Her dogs would respond in fear to those sounds. It got to the point that Sherry would leave the intercom on downstairs, and she kept a tape recorder on her night stand ready to record any of these strange new noises.

As Sherry watched night after night the lights continued dancing. Inside her house, the sounds became more menacing. Sherry started to see movements out of the corner of her eyes. Strange nonhuman faces would look into her windows at night. As Sherry's paranoia grew she searched the grounds around the house. The searches might have allayed her fears except she found strange-looking footprints. Her curiosity became an obsession and then her obsession became a fear.

On a warm night in the summer of 1992, Sherry went through the nightly routine that had become the norm for her. She sat at the window watching the golden globes dancing from the mountains to the pastures near her house. Exhausted by the routine, she finally went to bed, eventually dropping off. She was restless that night, drifting in and out of sleep. In what she thought was a dream, she felt a warm sexual tingle between her legs. The tingle grew to waves of orgasmic passion as she felt something hard penetrating her in a way that could not be mistaken for anything other than raw sex. In her sleep, she reached for whatever it was between her legs, pushing it deeper into her. She had an orgasm so explosive that she remembered little else.

When Sherry got up the next morning, the event of the nighttime visitor was fresh in her mind. She felt a burning between her legs and across her thighs. When she looked down, she saw something that shocked her. Both of her thighs were red with weltlike scratches. It was clear to her that the dreamed sexual activity of that night was more than just her imagination.

It was also clear to her that whatever it was that had whipped her into a sexual frenzy did it without waking Kimmy or her pets with her in the bed. The strange events had just become even stranger.

Over the next week Sherry shared her unusual story with her closest friends. Their responses ran the gambit from shock to laughter. One friend just shrugged it off by saying it was just Sherry's kinky imagination. But another friend told her that she had heard of people who had the same things happen to them. She went on to say that she had been told that it could be an alien contact.

"Alien contact. Now that's a concept!" Sherry mused. That was something that had never crossed her mind. Oh sure, she had seen some of those "crazy people" on the TV talk shows, but she wasn't like that. And anyway it wasn't the same! Those people were different. They were abducted in the night and taken away by little gray beings for medical experiments. Sherry's experience wasn't like that at all. Then again, she thought, there are those globes of light and the face at the window, and it didn't even wake Kimmy.

Sherry was even more confused now. The more she thought about the event, the more it seemed to be something from outer space. But why sex and why her? And why had they waited so long? She'd been watching the dancing globes for weeks.

Sherry started looking for answers. Through other friends she heard of a hypnotherapist who led a group of people who claimed they had been abducted by aliens. Sherry felt that she had to check this out, so she visited one of the group sessions. Such sessions were held at the home/office of Lorna, a hypnotherapist and founder of an organization that specialized in alien contact and abduction research.

In the state of California, very little is required for a person to hang out a shingle as a hypnotherapist. There are no laws or board of examiners that govern the profession. This is no licensing procedure or certification for those practicing hypnotherapy. In fact, after taking a weekend course in hypnosis, anyone can go directly into business the following Monday. This, however, was not the case with Lorna.

Lorna had attended one of the few long-term schools of hypnotherapy. It was a strict and stringent yearlong course of study with an internship. The school has a good reputation for turning out quality hypnotherapists and Lorna was not the exception to that rule.

Lorna is a caring, well-structured, well-intentioned therapist who has the proper attitude and empathy toward her clients. She also has a strong belief that the alien abduction phenomenon is real. Her research, as well as that of others such as Budd Hopkins, John Mack, David Jacobs, and John Carpenter, has convinced her of the validity of alien abduction.

Lorna's home is found in an upper-middle-class area of the San Fernando Valley, about one hundred miles from Sherry's house. The distance didn't stop Sherry from joining and becoming an active member of Lorna's abduction survivors group.

From Sherry's first meeting with Lorna and the group she had felt a kinship with all of the people involved. Sherry noticed from the very beginning that these were, what seemed to be normal people, who were having very unusual experiences. The group was not made up of strange people. It was composed of normal men and women who had had the abnormal experience of alien abduction.

The group sessions were more social than therapeutic. All the members shared a fellowship. The telling and retelling of stories about contact with alien creatures was the primary reason for being there, but it wasn't the only reason. There were refreshments and snacks and at times potluck feasts.

Sherry noticed that as different as each client's story was, there was an odd similarity among all of them. Sherry's strange, late-night visitation was not an exception. The creatures described varied but some had similar characteristics. Some of the events were blatantly sexual but some were sterile medical procedures with reproductive overtones that seemed to border on the sexual. Some of the people were abducted once a year, some were abducted once a month, and some claimed to be abducted every night.

With all of their differences and all of the similarities she heard in their stories, Sherry had found a group of people not unlike herself. They had a common bond and were desperately looking for meaningful answers. Sherry felt that the terror she had experienced no longer separated her from the human race. She had found a group of people who understood her, who wouldn't reject her tale as science fiction, and who could identify with her. It was almost as if she had found a family.

The group meetings were very cordial and open to all interested parties. There was no cost other than donations for coffee and the occasional snack. Lorna's hypnosis sessions, however, are billed at fifty dollars an hour and are private. Lorna exerted no pressure on any of the group members or guests to submit to the hypnosis. Sherry was with the group for many months before she was comfortable enough to be hypnotized. Once she started the regressive sessions she felt a need to continue.

Russ Estes met Sherry at a group gathering that Lorna invited him to attend. At the time Estes was in production of a documentary on the UFO phenomenon, including the alien abduction. The group was very friendly and open. Estes taped a chat session with a few of the members who consented to be on camera. Some of the group stayed to the side, off the camera. Sherry was one of the members who was camera shy. She didn't want to be seen as an oddity and Estes respected that. He interviewed a few of the members individually while the others sat by and listened intently. Lorna had mentioned that Sherry's experience was particularly interesting because she had so much conscious memory of what took place. The memories were reinforced by the external lights and sounds she reported. Estes asked Sherry if he could interview her on tape if he would conceal her face and alter her voice electronically. She agreed to that.

At the time of their first interview, Sherry had not started her regressive hypnosis sessions, so all of her memories were of what she consciously remembered. Sherry told Estes about her new house and the lights she saw in the distance. She talked of the sounds at night and the strange faces peering in through the window. She talked of the intense sex with the unidentified visitor. Estes asked her how she knew that she had had sex with the entity and if it was possible that it was just a very vivid dream.

Her answer was crisp. "I know the smell of sex."

At this point in her life Sherry seemed torn between what she knew had happened and the inconceivability of it. On one hand it was very real and on the other it couldn't possibly be real. She was at a point where she would say things like "I must be crazy" or "This can't be real." But then her beliefs would be reinforced by the group with statements like "It's happening to all of us." And there were the materials generated by the UFO community that quoted statistics for alien abduction. One study claimed the number of alien abductions to be as high as one in four people in the United States, though many suggested it was only three to four million.

Sherry and Estes kept in touch and talked many times over the course of the following year. After she started her regressive hypnosis program Sherry became less skeptical of the alien abduction phenomenon and more convinced that she was indeed a subject of alien abduction.

Estes was frequently amazed at how easy it was to get "abductees" to sit in front of the camera. He was also puzzled because here was a group of "normal" people sharing accounts of events in their lives that could cause them to be labeled as "kooks," or worse, suggest that they were psychologically abnormal. There were the constant statements by researchers that abductees did not want publicity, yet Estes noticed they seemed to have a need to tell their stories in a public forum.

"Why?" he asked many of them. "Why go on TV or radio and share such a private part of your life?"

Their answers were always similar. "If the sharing of my abduction will help one another abductee, it is worth it."

Many of the abductees (or experiencers as they are now called by some researchers) have gone public. On the television tabloid talk shows, the same abductees appear over and over again. The scenario is always the same. There are three or four abductees, one UFO specialist who has photos, or video- or audiotape, and last but not least, a skeptic. There is a patronizing host who keeps the pot stirred when it seems that the program is beginning to drag.

What usually happens is a yelling and screaming match with the abductees ganging up on the skeptic. And the audience members, sometimes siding with the abductees, and sometimes with the skeptic, join in at every opportunity.

Sherry was one of those who shied away from the spotlight and the celebrity, or at least that is what Estes had thought. Every so often she would mention some exciting new information that she had recovered through hypnosis, but basically her story remained unchanged.

In January 1995 Estes was invited to speak at Disney World on the topic of the UFO phenomenon from a documentarian's point of view. The Disney people were kicking off a new attraction in Tomorrowland called "Alien Encounter." They wanted a group of UFO researchers and abductees to lend a note of reality to the attractions.

Estes had talked to Sherry after he was invited and mentioned it to her. To his surprise Sherry told him that she had also been invited. Estes asked her why she would go public with her story after avoiding the press and the cameras for so long. Sherry giggled and said that she couldn't resist the free trip to Florida and the weeklong, all-expenses-paid vacation on Disney. Estes told her that he understood and, yes, it would be a nice vacation. He too was looking forward to it.

Estes was at Disney World on week one, Sherry was there on week two so they did not run into each other. About four weeks after his return from Orlando he noticed a promo on a local television station. It was a program hosted by Robert Urich shot at Disney World and called "Alien Encounters." Sherry was not only appearing but telling a story about her daughter, Kimmy.

Sherry didn't mention the lights or the smell of sex. She didn't tell the story Estes had heard many times over the past two years. Instead, she told a story of seeing her daughter being abducted by little gray aliens. She told of her daughter being taken to a spaceship and put onto a table. She told of medical experiments on her daughter, as she, Sherry, stood by paralyzed and helpless. And then she added an important fact. The memory was recovered while Sherry was under regressive hypnosis. It had not been in her conscious memories.

Sherry had never mentioned any classic gray aliens to Estes, nor had she ever mentioned any involvement with Kimmy. To top it off, there she was on national TV, using her real name and adding to the horror of alien abduction.

Estes called her as soon as the show ended. She was pleasant but a little distant. She said that she had remembered the abduction of Kimmy during her last hypnotic session and said: "I haven't told you *everything!*"

In the time since the Disney World event, Sherry and Estes still talk often. Sherry has not been back to the group for many months, and there is a very good reason for that. Sherry fell into a deep clinical depression. She cried day and night. The depression became so bad that she went to her doctor, who promptly put her on Prozac.

Several weeks after beginning the treatment, Sherry, while talking to Estes, said, "I haven't been to the group in over six months."

"Do you miss it," asked Estes, thinking about the social setting.

"No, not really. The funny thing is I haven't had any contact with the aliens in a long time."

Estes asked, "How long?"

"Oh, about six months or so."

What could have happened in Sherry's life to stop all of the activity? After thinking about the last six months Sherry told Estes something very interesting. "You know, since I've been taking the Prozac all of the strange stuff has stopped." No more lights, no more noises in the night, and, she added, "No more sex."[2]

THE STORY OF JOEL

Joel[3] is a handsome man in his early forties. He is of normal height, normal weight, and has a broad smile. As a matter of fact, Joel seems normal in every way. That is, every way but one.

Joel has had a lifetime relationship with an alien race.

Born and raised in Tennessee, Joel presents himself as one of the true Southern gentlemen. Joel says he was raised in a loving home with a brother, two sisters, and parents who have been happily married for nearly fifty years.

There is good reason to mention Joel's loving home. It goes against nearly all of the stories of dysfunctional home lives that Estes has encountered during his research with abductees. When Estes asked what his parents think about his alien abductions, Joel said, with a smile, "They have their own ideas."

"What does that mean?"

"My parents feel that I believe what has happened to me is real and they are very supportive, but they have their own ideas."

Joel's conscious memory of alien abduction goes back to the age of four. He remembers reptilian beings that played games with him and took him to wondrous places. Most of the time they would travel to cavelike structures or underground facilities. Joel remembers that no matter where the abductors took him the rooms were very large and clean as if they were in a modern laboratory or hospital. They would play games, electronic, high-tech games, and they would belay Joel's questions and fears by telling him that he would understand it all later.

---

[2]Sherry, interviews by Russ Estes, videotape, Crystal Sky Productions.

[3]The name has been changed.

The visits continued throughout his childhood. At about the age of ten, Joel met a new creature. A very tall white being covered with a thin veil. According to Joel, it was seven to eight feet high and had very large, human-like eyes. Joel believed, at first, this must be the Easter bunny. He told Estes that the creature didn't have a beard and wasn't wearing red and white, so he couldn't be Santa. Joel hadn't lost any teeth, so the being couldn't be the tooth fairy. Therefore, to Joel's ten-year-old mind, the creature must be the Easter bunny! Easter bunny or not, this is the one creature that would remain with Joel on all of his abductions from that first meeting until the present. Joel said that this creature had a calming effect on him.

His memories of the childhood abductions are vague but for the most part pleasant. A photograph of Joel at the age of four shows what seems to be a scoop mark on his leg. Joel assumed that it was a product of his first abduction.

His first memory of a medical procedure was at the age of six. Joel remembers being placed on a table, naked, in the fetal position, with his knees to his chest. He remembers an anal probe that was very traumatic. He says that it was the most embarrassing thing in his life, lying there, naked, with a probe up his anus, with the beings coming to the doorway to examine him.

Aside from the embarrassment of that early medical encounter, his childhood memories are filled with games and the smell of flowers. After Joel met the tall white being at the age of ten, he remembered a trip in what seemed to be a spacecraft. These were childhood memories, mostly fragmented images, but through hypnotic regression Joel was taken back to the flight. He recalled flying over his hometown, able to recognize many landmarks.

By the time that Joel had reached puberty a new group of creatures had appeared. They were very short, three and a half to four and a half feet tall and very robotic in their movements. They were very dark, nearly black with the look of what is now called the classic gray alien. They had large heads with big black eyes. They moved in unison and in silence, almost as if their actions had been choreographed.

Joel reports various noises like humming and buzzing during his abductions. The creatures communicate with Joel, but the communication is telepathic. No sounds are uttered through the mouths of the aliens but Joel hears every word they "say."

When Joel reached puberty the contact with the creatures changed drastically. The reptilians were gone as were the pleasant memories. The games he had enjoyed were also gone. Only the the medical procedures remained, now the focus of the abduction. The creatures were no longer friendly.

On most occasions, Joel is taken by force. He has been dragged, pulled, and carried into the craft by the small black creatures. At other times he is levitated to the ship. All of this is done against his will. The term "kicking and screaming" would readily apply. Now a calming effect is provided by the tall white creature who stands next to Joel while he is on the table. While the small black creatures probe Joel with all sorts of instruments, the white creature pets him as one might a kitten and tells him, telepathically, that it will be all right.

Over the years, the procedures have become less scientific and more erotic. The creatures have removed sperm from Joel by placing a funnel-like object over his penis or inserting what seems to be a catheter into his penis. All of this is very traumatic. Joel is disgusted and overwhelmed by embarrassment. Disgust, embarrassment, and fear, are elements that have always been present in all of Joel's adult contacts with the alien creatures.

But more important is another element that also has always been present. This is the sexuality, either in the form of the probing medical procedures or with actual sexual contact with the alien beings. It is an element that seems to be present in the majority of all alien abductions but one that is rarely reported by the researchers.

The burning question, when the abduction phenomenon is examined, is always, "Why do they want or need human ovum and sperm?" The most common and accepted answer given by the abductees, and by the researchers involved, is that the aliens must be missing some vital gene. They are trying to revitalize their race by creating "hybrid" or alien/human mixed children.

Nearly all of the abductees who claim to have had sperm or ovum removed have been introduced to their hybrid offspring by the aliens. Joel tells of many abductions over the last twelve years where he was introduced to his hybrid children. He describes the meetings as grand events where he is brought into a very large room similar to an auditorium. Joel is placed in the center of the room with many alien observers surrounding him. One by one the children are brought into the room and presented to him. The children are frail and anemic looking, with fine white hair and overly large, very human eyes. They are smaller than human children, with a distinct look of the alien crossbreeding, and yet they retain many of their human features.

Joel has seen his "children" from infancy to what seems to be the age of twelve. At each meeting the aliens allow interaction between Joel and the children. As a matter of fact, the aliens seem to want it. Each meeting is similar to the one before. The children are very shy and keep their distance. The emotional impact of these meetings is very hard for Joel, and after the event is over Joel finds himself in a deep state of depression that lasts for many days.

In most cases of abduction, the gender of the alien is in question. When an abductee is asked about gender or visible genitalia, the answer is almost always, "They have none, but I think they were male."

Or, in other cases, "One was female and the others were male."

When asked how they could tell, most say that they could sense it. Or that the clothing worn by the aliens suggested their gender.

In most of the male abduction cases reported, the collection of sperm is very similar. It is presented as a medical procedure where a cone-shaped device is placed over the penis or a catheter device is placed into the urethra and the sperm is removed. This is done without erotic stimulation or the cooperation of the abductee. Most male abductees are very surprised and embarrassed that it could be done to them, and they have drawn the analogy to actual physical rape. The third and less often reported method of hybridization is actual sex with a female alien.

No matter where he is living, on a regular basis, Joel is visited by the group of aliens that have become familiar to him. The tall, white-veiled alien is there with two or three dark "worker" types and one female. The actual abduction begins as it always has. Joel is taken to a clean room, either on board a craft or at some kind of underground facility. He resists them as best he can, but always to no avail. They remove his clothing and place him onto a table in a seated position. Then the female enters the room and mounts him, her back to his chest. The aliens make him watch over her shoulder as he penetrates the female.

Joel remembers the sexual experiences with disgust and revulsion. He told Estes that all he can do is scream, "I'm sorry, I'm sorry," over and over again. He doesn't know if he yells it out loud or if it is just in his mind. He feels great sorrow and can't understand why they continue do this to him.

When Joel told Estes about the sexual contact, Estes had a number of obvious questions. With all of the unisex descriptions of the aliens, Estes asked how Joel could tell that his partners are female?

Joel said, "It's clear she's female even though she's alien. She has a softer look and she has breasts."

Estes also wondered how Joel could maintain an erection without erotic stimulation and in the fearful state that he was in. Joel said that he didn't know, "It just happened."

When Estes asked what he penetrated, if it was female genitalia, Joel said "It was just an opening, an opening, not a vagina."

Although Joel has been abducted numerous times in his life, and nearly all of the abductions contained the element of medical procedures with the removal of sperm, the overt physical sex acts began in 1985. An interesting aspect of the sexual contact is that it has been a regular event that happens on nearly the same day in March every other year.

Joel admits that he bruises very easily and has never gone willingly with his captors. The result of his struggles are marks, bruises, and cuts that seem to have no earthly explanation. He has documented this physical evidence well. He has shared his photographs with Estes and told him of the circumstances that produced the wounds.

One particular photo shows three distinct scratches down his arm. Joel said that during one of the abductions where physical sexual penetration took place, he grabbed the alien and she fought with him, producing the scratches. Estes asked if he had ever seen fingernails on the aliens. Joel said, "No, I have never seen any but it is obvious that they do."

Another photo is of a triangular wound on his penis that he claims he got when he struggled to remove the cone-shaped object that was placed on him to remove sperm. Joel's collection of photographs span a large number of abductions and related wounds.

Still another photo shows a number of symmetrical punctures from ear to ear, along the cheek line, and crossing under his chin. Joel explains that the puncture wounds were formed by a metal device that the aliens applied to his face. He has no idea what the device's purpose might have been.

Joel is also one of the few abductees who has had an object that is thought to be an alien implant removed and studied. The surgical procedure was documented photographically and the object was analyzed under a high-power electron microscope.

Joel has no idea how the object got into his leg. It had been there as long as he could remember. It was a small bump in the front of his lower left leg. "You could feel it with your fingers, it moved freely under the skin but caused me no pain."

The final analysis of the object showed it to be a shard of crystalline material with a composition not unlike glass.

Soon after the removal of the object Joel was abducted again. Joel recalls that the aliens were agitated and seemed to be looking for the missing implant. The resulting physical marks left by this abduction was a circular wound that resembled a burn on the back side of the leg where the implant was removed.

Estes asked, "Do you have any idea why any of this is happening to you?"

Joel's uneasy response was, "I have no idea." He added, "Some of my friends thought that they [the aliens] want my children, they want to abduct my earthly offspring, but that will never happen. I will not have any children."

"How can you be so sure?"

Joel replied, "It's a gender choice. I'm homosexual."

That candid, open response of Joel's sexuality led to even more questions. Estes asked if the alien contact affected his sexual performance with humans.

His response was quick and succinct "Yes! I don't want anyone to think that my alien contact made me a homosexual, it didn't. But it has had an adverse effect on my ability to have an ongoing relationship of any kind." At the time of the interview in 1993 Joel said that he had had no sexual contact with a human in eleven years.

Joel's memories of his abductions are a combination of conscious memory and hypnotic regression. He is a member of a support group and has said, "If it wasn't for this support group I wouldn't be here."

He has fought off his abductors in every way that he can and has suffered great mental and physical trauma at their hands. As a matter of fact, on two occasions Joel has had to be hospitalized after an abduction in shock and with a blood-pressure reading of 230 over 200. The price that he pays for the average abduction, if there is such a thing, is two or three days of deep depression.

It would seem that this horrific experience, so far out of his control would be something he would want to end. However, when Estes asked his final question, "Joel, do you want it to stop?" He paused for a moment and said, "No . . . not until I discover what it is."[4]

---

[4]Joel, interview by Russ Estes, videotape, Crystal Sky Productions.

4.

CONTACTEES VS. ABDUCTEES

As we look at the abduction phenomenon, we can't help but notice some glaring inconsistencies in the attitude of the UFO researchers. One of those incongruities is the fact that most researchers who truly believe that abduction is real also truly believe that contact, that is, meetings with the benevolent space brothers reported in the 1950s, is little more than bunk.

The contactee movement started many years before classic abductions began to appear. Or rather, to be completely accurate, we should say it was reported long before any of the abduction accounts were described. It started in the late 1940s and the early '50s with the likes of George van Tassel, "Professor" George Adamski, Robert Short, "Dr." Daniel Fry, Truman Bethurum, Ruth Norman, and many others. In most cases of contact, the alien entities communicated through the contactee. Not unlike the mediums in the occult world, the contactee "channeled" the entity through his or her human voice, and brought the messages to the masses.

George Adamski was the first to claim alien contact. His background, prior to his meeting with the benevolent space brothers is interesting. In the 1930s, for example, he established, in Laguna Beach, California, a religious order called The Royal Order of Tibet. He held a special license that allowed him to make wine for religious purposes. According to Curtis Peebles in *Watch the Skies,* Adamski later told two followers, "I made enough wine for all of Southern California. I was making a fortune."

The appeal of Adamski's religious order can be understood by remembering this was the time of Prohibition. The making, selling, and distribution of alcoholic beverages was against the law. Adamski had found a legal means for distributing his product. But Prohibition ended and, according to Adamski, if it hadn't been for that, he wouldn't have had to get into the "flying saucer crap."[1]

---

[1]George Adamski, *Behind the Flying Saucer Mystery* (New York: Paperback Library, 1967); *Inside the Flying Saucers* (New York: Paperback Library, 1967); Curtis Peebles, *Watch the Skies* (New York: Berkley, 1995), 112–20; Kevin D. Randle and Russ Estes, *Faces of the Visitors* (New York: Simon and Schuster, 1997), 89–91.

Others, such as George van Tassel, entered the picture just after Adamski. Van Tassel is responsible for the largest UFO-related gatherings ever held. His Giant Rock Airport hosted one convention where eighteen thousand people gathered. What makes it even more amazing is the location of Giant Rock. It is in the middle of the California desert, literally hundreds of miles from any real signs of civilization.

Almost all of the contacts were with the benevolent space brothers or light travelers who claim to visit us from Mars or Venus. Other space brothers later claimed to be from even more exotic locals such as Zeta Reticuli or the Pleiades. A common thread that binds them is membership in a confederation of planets. The confederations differ in number only, some claiming to be a group of thirty, thirty-three, or even three hundred planets.

Their visit to Earth is a precursor to the invitation for us to join them and to share the knowledge of the universe. Of course there are certain rules that we must follow. This is where the message given by the contactees, and later echoed by the abductees, becomes similar. First of all, we must become less violent. Second, we must be careful with atomic energy, specifically with the storage of nuclear weapons. And of course we must be kind to the planet and to each other.

Many contactees graduated from channeled contact to physical contact with the aliens. The description of these beings was almost always the same. The aliens are taller than humans, prettier than humans, with long flowing golden hair on both males and females, and they were bathed in brilliant light. The typical description was of very handsome, Nordic-looking humans.

Along with the physical contact came the obligatory invitation to ride on a flying saucer. Once on board, the contactee was shown things that went far beyond the technologies of the day. They might see the Earth from space, which is not a stunning show of technological muscle today, but in the early 1950s it was very impressive. Or they might take the contactee to the alien's home planet, which is a little more impressive even by today's standards. Almost always they would show them holographic images of the things to come on planet Earth. Not pretty sights of a utopian future but pictures of famine, atomic explosions, and the destruction of the world as we know it. The purpose of showing them the the bleak and horrible future was to give them a chance to spread the word among the heathens and save the planet. The contactee had been selected so that he or she could spread the word to the rest of us.

It should be pointed out that there has been an escalation of the contactee experience, just as there is with the abductees. Adamski's first contact was with Venusians he met in the desert. After Adamski's claim, Truman Bethurum said he entered the craft and spoke with the ship's captain. Dan Fry then said he flew in a saucer from New Mexico to New York, but did not leave the Earth's atmosphere. Orfeo Angelucci made it into space. Adamski then managed to board a ship and flew around the moon. Howard Menger, not to be out done by anyone, then claimed that he landed on the moon.

Each of the contactees expanded his or her tale to take the next logical step. Adamski might have met the benevolent space brothers first, but Dan Fry got the first ride and Howard Menger claims to have been the first man on the moon. Each tale of the contactee expanded on the last in an attempt to interest their readers and followers, who were becoming bored with the simple idea of alien contact and messages of doom.

These contacts, however, were never related as negative experiences. They were always positive and with a special, higher purpose. The message was, "You are not alone, be kind to the planet, learn to live together, and get ready to join the confederation of planets."

This is not to say that there weren't negative aspects to the contactee phenomenon. Adamski, among others, warned that there were "evil" aliens who were plotting with evil humans to take over the Earth. Some of these claims evolved into tales of Men in Black who harassed witnesses, stole evidence, and suppressed information to keep us all in the dark. And as so often happens, there isn't a single shred of evidence to convince us that these evil aliens are real. Governmental conspirators, sometimes working inside an alien conspiracy, have helped suppress the evidence.

The aliens often communicated through a human channel. When not communicating through the human's voice so that those around could "listen in," the aliens communicated telepathically with the contactee. All of the communication was done in perfect English. And "Dr." George H. Williamson, a witness to some of Adamski's contacts, sometimes used a Ouija board, and sometimes a more conventional shortwave radio, to communicate with the aliens.

In stark contrast to the willing communications and adventures of the contactees, the abductees are often taken against their will. Invasive medical and sexual experiments are performed on them, not by beautiful benevolent space brothers, but by small, horrific gray creatures. These three points are the real differences between the abductee and the contactee.

Most researchers see the abductee as the "hard evidence" of alien contact. They zero in on the medical/reproductive experiments and disregard almost everything else. They quiz the abductee on the experience and the procedures time and again, looking for the clue that will explain the whole situation to them. Those same researchers, claiming objectivity, never seem to notice the contradictory statements made by the abductees, nor do they investigate the situation in their search for the truth. They assume that the abductee wouldn't be lying to them and therefore the abductee is not lying to them.

But the fact remains that most abductees report that they were shown the same chilling stories of atomic or environmental destruction and a very bleak future for planet Earth if we don't change our ways. Many abductees feel that they were chosen because of their ability to spread the word. One abductee who was interviewed by us drew the analogy of feeling that she was a female version of Moses, chosen to lead the earthlings to the promised land in the very near future.

Some additional, though minor, differences between the contactee and the abductee are:

1. The contactee has vivid memories of the events. On the other hand, the abductee has little or no conscious memory of the events, just a sense of missing time, and it is through regressive hypnosis that memories of abduction take shape.

2. The contactee has little or no memory of any experiments conducted on him or her, either medical or sexual. The abductee almost always has repressed memories of experimentation and sexual reproduction.

3. The contactee places deep spiritual meaning in both the message and the messenger. Even though there are segments of the abductee community that feel that the events can be mind expanding, most still feel that it is not. It is more or less a religious experience as opposed to a scientific one.

So you can see that the split between the two is not as far as one may think. The contactees are not so different from the abductees. The common thread is the similarities:

Both have contact with alien creatures.
Both are taken to a spacecraft or facility.
Both are given a message to bring to our planet.
Both are shown apocalyptic images.
Both are returned relatively unharmed.

The abductee claims to have missing time, or blank spots during the abduction. Even though the contactee remembers the event as a real or conscious memory, it doesn't mean that they don't have missing time. As we examined the contactee phenomenon from start to finish, we have found that many contactees also have missing time. In one interview, the contactee was asked, "What happened between the time that you were on the spacecraft and the time that you realized that you were back at home"? The answer was, "No . . . I don't even remember if it was a physical or spiritual thing."

The contactee leaves the door open to many possibilities. The psychic experience is one answer to the question of faster than light travel. Many contactees claim that they have traveled psychically to other worlds. They tell of the "Brothers of the Light" and how they are transformed to light beings so they are able to free themselves from the bounds of Earth and visit other planets. The multidimensional aspect is also open to the contactee. Some of the Venusian space brothers have said that they live on Venus but not in a dimension that is recognizable to us. The multidimensional answer conveniently circumvents the scientific data that tells us that no life-forms live on Venus. The cloud-shrouded planet has a surface temperature hot enough to melt many metals.

The spiritual aspect of contact is the very thing that is most often ridiculed. The interesting thing is the people who are doing the ridiculing are the very same ones who claim that little gray creatures come through their walls while they sleep, and abduct them against their will taking them to a secret location, most often a spacecraft. They suggest they too are passed through the walls, that spouses and others in the house are "switched" off so that they will be unaware of the abduction, and that they communicate telepathically with the aliens.

Why does one story of contact bear more weight than another?

As mentioned, the tales that are told by both the abductee and the contactee bear a striking resemblance to one another in many ways. Even though we have looked at both aspects using generalities that have been used by the majority of both groups, it still leaves room for crossover. There are many abductees who feel that the aliens are not extraterrestrial and may possibly be extra-dimensional. There are also a number of abductees who feel that the abductions could be a positive event, even though they were taken against their will.

So if we conclude that both groups equally believe that their alien contact is real, then how can we, as researchers, take sides and choose which anomalous event is worth further study?

One of the arguments against the contactees is the that they have formed organizations that can be described as "cult-like." But were they not assigned the mission to pass the word? Aren't they supposed to form organizations so that we can stop our rush to atomic annihilation and the destruction of our planet?

Another reason that we have encountered in support of the abductee is the physical marks and trauma caused by these events. It seems to mean that someone has set up the criteria that states, "Trauma must be associated with the truth." Or the comment used quite often by a well-known abduction researcher, "Who would want to be a member of this club?"

But the real reason to accept the abductee over the contactee might be simpler. The contactee needed no intermediary. The contactee, after the experience, began to spread the word without the help of the UFO researcher. The contactee was the focus of the story, sought by the reporter and the fan alike. There was no need for a self-styled UFO investigator to translate the tale.

The abductee, however, seeks out the services of an "abduction" researcher. Often, without the assistance of hypnosis, administered by the investigator or one of his colleagues, there is no tale of abduction. Without the "inside" knowledge of the UFO researcher, without the understanding of the UFO researcher, the tale would remained buried in the subconscious of the abductee.

The real difference, then, is the focus of outside attention. The contactee is the focus of the tale. The contactee is the important part of the contact tale. With the abductee, the focus is on the researcher. The researcher is more important than the abductee because the researcher is the one who can answer all the questions. The researcher is the one contacted by the television shows and magazine reporters to tell the tale. The researcher is the one who sets up the interviews with those claiming abduction, but the spotlight remains on the researcher.

But either way you look at it, one tale of alien contact is as feasible as the other. It does become a rather silly "ship of fools" when you consider what is happening. Members of one group of people is saying, "My unbelievable story is more believable than your unbelievable story."

Another way of looking at it could be, never believe a story told with a smile. Or only believe a story told with fear and pain. Or it could be, always buy into it if the teller is at peace and happy. Or never listen to a fearful, panicked person.

As we consider all of the revelations that have been portrayed, a light shines brightly over our heads, "What is that one ingredient that separates the contactee from the abductee"? Well, the one factor that hasn't been mentioned in the past is the man in the middle, also known as the researcher/therapist.[2]

---

[2]Peeples, *Skies*, 112–28, 148–50; Randle and Estes, *Faces*, 79–116.

5.

THE SEXUAL COMPONENT OF ALIEN ABDUCTION

Most of the abduction researchers have not noticed a high percentage of victims reporting sexual activity by the alien beings. These ideas are not new and will be explored in depth later. The point here is that those conducting research have overlooked this aspect of the abduction phenomenon. For example, in one of his books Hopkins wrote, "I know of no case in which a female abductee has ever reported an act of intercourse."

John Mack, when asked about sexuality, including sexual dysfunction and homosexuality, said that the incidence among abductees was no higher than in the overall general population.

Other researchers have commented on the lack of any sexual connotations in the abduction phenomenon. Yes, they fully admit an interest in reproductive cycles, a collection of ova and sperm, or a search for genetic materials, but claim no overt sexual content to the abduction reports. Sexual activity, according to them, is nonexistent.

The only exception to the rule seems to be David Jacobs. In his book, *The Threat*, he made repeated mention of the sexual component to the abduction phenomenon. He detailed a number of reports of sexual activity between humans, aliens, and the various hybrids.

In the process of interviewing over a hundred people who claim to be abductees, it has become painfully clear that sexuality is a major part of abduction. A vast majority of the abductees felt that the experience went well beyond reproductive curiosity and smack dab into lusty sex. We've noted one who said, after an alleged encounter, that she knew the smell of sex. The aliens had been there for a single purpose.

There is a support group in the Midwest that is made up of abductees who are having sexual relationships with a reptilian alien race. All of the members feel that they are being raped. They also feel ashamed of the fact that they like the sex and that the quality of the orgasm is well beyond the feelings that they have with a human lover.

In fact, from the very beginning of the abduction phenomenon, no matter how far back we care to trace it, we can find a sexual component to it. During the great airship craze of the late nineteenth century, the *St. Louis Post-Dispatch* reported that a traveling salesman had discovered a landed airship outside of Springfield, Missouri. Inside was a naked couple who communicated with the man through hand signals and gestures. In the straight-laced Victorian times, that is about as racy as such a story could get.

The contactee movement produced little in the way of explicit sexual stories. There was always talk of beautiful aliens from Mars or Venus, and trips to those worlds, but the sexual component usually related to the men and women of Earth who made "contact," rather than beings from another world. The aliens were strangely asexual.

Even without the contactee claims, there were sexually explicit stories that mirror some of those told by the abductees of today. In the 1950s, John Stuart of Hamilton, New Zealand, entered the UFO arena, forming a UFO study group known as Flying Saucer Investigators. His organization consisted of himself and a younger woman identified as Doreen Wilkinson.

Apparently Stuart and Wilkinson spent many days together, researching and studying UFOs. They corresponded with American UFO researcher Gray Barker, who would eventually become a publisher of flying saucer books with an eye toward those of the contactees. One of the books he would publish was written by Stuart and titled *UFO Warning*. It was meant as a warning to others who might become interested in the UFO field. Barker would apparently also receive an audiotape that was, more or less, an earlier draft of the manuscript, and a report that covered much of the same ground as the manuscript he eventually published.

In the book, Stuart describes Doreen Wilkinson as a young and attractive young woman with a slim figure who eventually began to make sexual overtures toward him. These he ignored, believing for some inexplicable reason that her actions were evidence of a supernatural possession. Then, one evening, Wilkinson arrived at Stuart's house claiming that she had been raped, but the attackers, in this episode, were apparently Earth based.

But soon a phantasm appeared to them using telepathy to tell them that Wilkinson would be in danger if they did not stop their UFO investigations. Although Stuart, later, after Wilkinson had gone home, had a close encounter with a UFO, never ceased the work. It isn't clear if this warning about UFO investigation had anything to do with the alleged rape, or if it was a separate episode.

On a Friday that Stuart and Wilkinson were together not long after the warning and the sighting, Wilkinson went out to the store to buy some cigarettes. When she didn't return as quickly as Stuart believed she should have, he began to panic. Eventually Wilkinson did return, throwing open the door and rushing into Stuart's arms, screaming, "There's something out there."

Outside, they first encountered an overpowering odor that was so bad, Stuart wrote, that he nearly fainted. They also heard something moving and eventually came face to face with a creature that was eight feet tall. Stuart would later describe it as having a "head [that] was large and bulbous. No neck. A huge and ungainly body supported on ridiculously short legs. It had webbed feet. The arms were thin and not unlike stalks of bamboo. It had no hands, the long fingers jutting from the arms like stalks. Its eyes were about four inches across, red in color. There was no nose, just two holes, and the mouth was simply a straight slash across its appallingly lecherous face. The whole was a green lime in color, and it was possible to see red veins running through its ungainly form. The monster was definitely male."

According to the account, Stuart stood paralyzed as the creature approached them. The creature's attention was fixed on Doreen, and, according to what Stuart said later, the creature was communicating with her telepathically, much as the abductees of today sometimes suggest. Doreen told Stuart that it had sex on its mind.

Before any act could be accomplished, the creature stopped, retreated, and then disappeared. Wilkinson then fainted, and Stuart, no longer paralyzed, attempted to revive her with a glass of brandy. Wilkinson, after telling Stuart that she believed that she would be attacked again if she went home, did exactly that.

The chronology isn't quite clear after this point. Wilkinson said later that she was attacked in her room by an invisible, sandpaper-skinned alien, but it seems to have happened a couple of months after the day she ran into the house with cigarettes.

While telling Stuart this story, she removed her clothes. Stuart would later report that her body was covered with scratches, and on her ribs were two brown marks about the size of a dime. She told him that the marks had appeared after the thing had left. This was, after a fashion, the unexplained scarring that marks the modern world of abduction.

Wilkinson abandoned both UFO research and Stuart shortly after these events. That would be the last time he would see her, except for a short visit in 1955 when she would attribute the extraterrestrial (or supernatural) attack to nerves, and then accuse Stuart of not protecting her as she thought he should have. That was the end of their communication.

Stuart, however, hadn't finished with the alien creatures. Not long after the attack on Wilkinson, Stuart said that something had visited him as he sat at his desk early in the morning. This time he saw a creature that was vaguely human, which was male from the waist up and female from the waist down. The creature told him, telepathically, that his friend, Wilkinson, had seen too much and had to be silenced. With that, the thing began to flicker and the male and female parts of the body changed. That was it for Stuart. He quit UFO investigation.[1]

What is most disturbing about the account were not the sexual aspects of it but the foreshadowing of the abduction phenomenon. Though the description of the creature is at odds with the grays of today's stories it does include the large eyes, though they are not black, the two holes for a nose, and the slash for a mouth.

While this case also foreshadows the growing sexual content of the contacts with alien creatures, the first of the reported abductions was one of the more sexually explicit. As noted earlier, it was made on February 22, 1958, by the Brazilian farmer Antonio Villas-Boas, who reported that he had made love with a vaguely alien woman, twice. This case would become the model for abductions reported for the next twenty years.

Most UFO researchers in the late 1950s and early 1960s would ignore the Villas-Boas report simply because of the overwhelming sexual aspects of it. The explicit nature may have offended some of them. Or, it could be that most realized that the experiment performed, that is, the coupling of Villas-Boas and the woman from space, was not unlike a union between a chimpanzee and a human. No scientist would perform such an experiment in such a crude fashion. If interested in what would happen if a chimpanzee and a human mated, the experiment would be conducted at the genetic level with a mixing of the ova and sperm. It is unnecessary, extremely crude, and highly unethical to perform the experiment in a fashion similar to the one Villas-Boas reported.

Nonetheless, that was not the only such experiment reported. Ignoring the tabloid newspapers that often trumpet the union of humans and aliens, there are still these sort of reports. Hans Holzer, whose work is mainly in the paranormal field, reported that he had interviewed and hypnotized a young woman, Shane Kurz, who apparently was a victim of a sexual encounter not unlike that reported by Villas-Boas.

Under hypnosis, Kurz told Holzer that she had recalled hearing a voice in her bedroom and then seeing a bright light. Suddenly she was walking across a muddy field until a beam of light from an oval-shaped object drew her up, into the craft.

She found herself in a hospital-like room with a small creature who told her she was special, that she should lie on the table and she should remove her blouse. As she argued with the alien, a second creature approached from behind

---

[1]Gray Barker, *They Knew Too Much about Flying Saucers* (New York: University Books, 1967); Jerome Clark, *High Strangeness: UFOs from 1960 through 1979* (Detroit 1967: Omnigraphics, 1996), 437–50.

her. Although he looked like the first, he wore a long coat and she thought of him as the "doctor."

Eventually, she was taken to another room where she was stretched out on a table. Because she was a "good breed," according to the "doctor," a needle was pushed into her navel.

Another alien creature, this one wearing a scarf and who Kurz thought of as the leader, told her that she would have his baby. Although she resisted, the leader removed his clothes and then rubbed a warm, jelly-like substance on her chest and belly to stimulate her. He then entered her with a sex organ that was remarkably human in appearance. Finished, he said that she could go but that she would not remember the abduction.

She didn't consciously remember the abduction but, as has been reported so often, she did dream about it in the months that followed. Holzer's hypnosis merely provided a more elaborate account of her experiences.

Holzer's investigation didn't extend much beyond the hypnosis and some correspondence. He did, however, talk to the woman's mother and learn that on the night of the abduction, at least according to the mother, she had awakened to discover her daughter was not in her bed. The mother believed the daughter had gone to the bathroom and thought nothing more about it until the next morning, when she found her daughter sleeping on top of her bed, her feet muddy.[2]

The stage was now set for the abduction phenomenon that would follow. Villas-Boas, the first of the modern abduction cases, had established the sexual component of abduction. The Kurz tale, although ignored by the UFO community, certainly underscored the sexual nature of the accounts and moved it into the female arena.

As Hopkins began his abduction research, he did find a number of cases in which a male was involved in sexual intercourse with an alien female. A Wisconsin man, for example, claimed to have been mounted by a "hybrid-looking" woman while her shorter and obviously alien companions watched. The man was paralyzed and sad because he was certain that she got no sexual pleasure from the act.

In another of the cases reported by Hopkins, a truck driver from Wisconsin and identified only as Ed was abducted by the grays and encountered a naked female who was a head taller than her companions. As Ed lay naked and paralyzed, the woman somehow aroused him, brought him to orgasm and then left. Two grays, using spoons, collected the residue from his body.

---

[2]Hans Holzer, *The Ufonauts: New Facts on Extraterrestrial Landings* (New York: Fawcett Gold Medal, 1976).

From this point we begin to descend into new and adverse arenas, some of which have been studied in other forms by psychologists. Jerry Clark reported, "Sometimes the abductee may have an induced orgasm, in some instances including a hallucination of intercourse with a member of the opposite sex."

Clark also noted, "Another form of hallucination occurs when an 'abductee is made to believe that either her husband or loved one is with her,' even though her sexual partner is really an alien or hybrid."

While it can be argued that the grays, or the other alien beings who are participating in the abductions and the sexual experiments, are anything but handsome, it can also be argued that they are "attractive." They have superior intellects, and many women, when asked, suggest that intelligence is high on their list of attractive traits.

Even ignoring that point, what we have is a belief by the abductees that the alien creatures have traveled interstellar distances, overcoming a variety of scientific and technological problems, just so that they can abduct the victim, usually not once but many times. The victim is so attractive to the aliens, for whatever reason, that these beings have sought him or her out for their continued attention.

Another study, by Harold Leitenberg and Kris Henning, noted that "high guilt participants more frequently had the fantasy, 'I am so beautiful that men cannot resist me.' " They defined high guilt as resultant from engaging in sexual activity that was socially unacceptable, such as sleeping with a lover's best friend or with someone she hardly knows.

Jacobs explored the erotic dimensions of alien sex in his discussions of the Mindscan, or the staring into the alien's big black eyes. He wrote, "The alien can elicit a variety of feelings and he can make the abductee envision specific scenarios . . . One of the most common procedures is when the alien uses Mindscan to elicit sexual feelings that escalate unabated until the female abductee reaches a high sexual plateau or orgasm."

These sexual fantasies are not limited to women. Sexual fantasies in men, according to Arndt, Foehl, and Good, differ, though some of the same themes appear. Just over half the men surveyed reported a fantasy in which they were kidnaped by a woman and forced to do as she ordered. Over three quarters of the men reported fantasies in which a woman told him she wanted his body, or that he got a woman so excited that she screamed with pleasure. And nearly 90 percent reported they fantasized about a woman forcing her intentions on him.[3]

---

[3]William B. Arndt, John C. Foehl, and F. Elaine Good, "Specific Sexual Fantasy Themes: A Multidimensional Study," *Journal of Personality and Social Psychology*, 1985, 472–80.

The same comments that were made about the woman can be made here. A man finds himself so attractive and so important that the alien creatures search for him. He is selected from all those available in the human population. And although he claims, as do the women, that he wishes the abductions would stop, each of the abductions is a reinforcement of his own perceived worth and strengthens his own sense of self.

In fact, when we begin to examine those who claim alien abduction we often see a poor sense of self. This is underscored by those who claim that they have always felt to be on the outside of society and who have felt unattractive and unwanted. They are not part of the group, are the last to be invited to a party or picked for a team, and often sit home alone. Now, suddenly, they are important. Alien creatures have found something attractive and desirable in them. The gigantic hurtles of interstellar flight are overcome so that the victim can be taken on board the flying saucer for, more often than not, a sexual encounter.

This is an important aspect that has been ignored by those conducting abduction research. The sexual component of the abductions has been reduced to reproductive experimentation or genetic research, but when examined carefully, it is little more than sexual intercourse.

Leeza D.,[4] a lifetime abductee, has had conscious recollections of her alien contacts since childhood. She remembers at least five different types of aliens from a seven-foot-tall dark-hooded female to the three-to-five foot grays. During an interview in front of a video camera, Leeza was asked about the sexual aspect of her abductions. She said that they were charged with sexual energy and that the aliens often observed her during sexual activities. One particular abduction stood out as the most sexual of all. Leeza remembered laying on the all too familiar metal table surrounded by the aliens who had just removed a hybrid fetus that they had implanted into her. As she lay there, immobile on the table, one of the five-foot aliens mounted her, looked deep into her eyes, and what she heard him say was, "What you need is a good fuck!" A very earthly and very sexist statement if ever there was one. The alien then proceeded to, as she said, "Give me the most profound orgasm of my life!"

Leeza had been interviewed by many abduction researchers, but when asked if any of them had asked about the sexual aspect of abduction, she said, "No, they never seemed to have any interest beyond genetic and reproductive experiments."[5]

When we examine the tales of abduction carefully, we see that they are extremely sexual. If there was no sexual component to them, if the aliens were

---

[4]The name has been changed.

[5]Leeza D., interview by Russ Estes, Crystal Sky Productions.

actually collecting genetic material, then there would be no need for the crude experimentation. There would be no descriptions of activities that are clearly of a sexual nature. Genetic material can be collected in a large number of ways that involve no sexual activity and no invasive procedures.

In setting up our protocols for the interview process, we added questions on blatant sexuality, such as, "Was there any physical contact? Was there actual penetration? How has this affected your earthbound sex life? What was the gender of the alien beings? How has this experience affected your ability to maintain relationships? What do you remember about your family life as you were growing up?"

With few exceptions nearly all of the people interviewed claimed to have come from dysfunctional families. The majority claimed that human relationships were very difficult for them and that their sex life was impaired in some way due to the alien contact. Another interesting point is that nearly all of the people we interviewed claimed that they were either sexually penetrated or forced to sexually penetrate an alien creature. Women reported intense orgasms with the aliens. Some even claimed that they could no longer achieve orgasm with a human partner.

The men had totally the opposite reaction to alien sex. Most of them reported that a female alien mounted them. They couldn't understand how they achieved an erection under the stressful circumstances, but they did. All of them reported that the sex act was completed but without the pleasant sensations of orgasm.

When each person was interviewed about his or her alien contact we asked if they could recognize a gender. Almost all of them said no, at least not physically. Some of the abductees reported that they perceived a gender with one or more of the aliens that they had had the most contact with. Leeza D. had a long-term relationship with the seven-foot-tall leader whom she perceived was female. She also had a feeling that the head doctor type was male.

Jacobs, in *The Threat*, noted that the alien females have no physical attributes indicating their sex. They have no breasts or secondary sexual characteristics. Instead, Jacobs wrote, "The female aliens seem 'kinder,' 'gentler,' or more 'graceful' or 'feminine' in some ill-defined way." It suggests a way in which the abductees are drawing distinctions.

In the cases of the male abductees who were forced to penetrate an alien, all of the males claim that the aliens were female and emotionless. It seemed that they had breasts not unlike a human female's, but all of the other features were consistent with the description of all of the grays, which includes the absence of any sexual organs. It is also an interesting contradiction, since most female abductees report no such sexual characteristics.

Clark, in his massive UFO encyclopedia, noted the sexual nature of the growing abduction phenomenon. He wrote, "By the last decade of the twentieth century, sex had become an inescapable fact of the abduction phenomenon. Those who subscribed to the hypothesis that abductions are event-level ('real') experiences inferred that alien interest in human sexuality either stemmed from alien fascination with the emotional make-up of Homo sapiens or was associated with the generation of the widely reported hybrid entities, or both. Some of the sex clearly had no reproductive significance, as in cases in which abductees reportedly were compelled to masturbate. . . ."

Clark also noted that the tales, which were "outlandish enough to start with, have grown ever more fantastic." As an example, he wrote that John Mack reported that some of his witnesses became aliens and in one case, the man known as Joe on Earth became Orion when elsewhere. Joe then watched, while in his Orion body, as the smaller aliens undressed a human female and spread her legs. He then penetrated her briefly. We should note that about 40 percent of the men have reported a fantasy in which a woman is forced to have sex with him, according to our review of the psychological literature.

Clark, as well as others, have suggested that a "small but growing subset of stories concerns episodes of repeated rape, both in a UFO and in normal life circumstances, by men who may or may not be extraterrestrials'. . . ." And Jacobs has now turned that around until, according to him, the sexual aspect has become the driving force.

So when Hopkins reports that he knows of no cases in which a female human has engaged in sexual intercourse with alien beings, he is, at best uninformed. The number of cases that have been described in both private investigations and public reporting suggest that the cases of such intercourse are widespread, with the number growing.

A very interesting fact about the interviews we conducted is that when the people were asked if it would be all right to talk about the personal and sexual aspects of their abductions, each and every one of them said yes. Some even commented that nothing could be more personal than the abduction itself, and if anything that they could add would help others, they wanted to share it.

On one occasion we were interviewing a male abductee on camera. Also present were about eight other members of an abduction support group and the abduction researcher/hypnotherapist. The gentleman that we were interviewing was so open about his sexual preference, homosexuality, that the researcher who had been working with him for over two years was shocked. She was hearing things about his sexuality and sexual orientation of which she had no knowledge.

It seems that in the fervor to investigate the abduction phenomenon, the researchers have chosen to miss a very obvious point. It may be our very human and very prudish approach to sex that stops researchers from asking the direct questions concerning sex for the sake of sex. But miss it they do. Nearly every person that we interviewed was a long-term abductee who had been interviewed many times by many different people. We were told that the only interest other researchers had in the sexual aspect of abduction was in reproductive experiments. If any of the other researchers would have walked that highly personal path they might have found that a higher than average number of abductees were either homosexual, hypersexual, or asexual.

A very high percentage of both the male and female abductees that we interviewed openly stated a sexual preference of homosexuality or bisexuality. An equally high number were hypersexual and highly promiscuous in their human sex lives. Of the remaining abductees at least half of them claimed that they had no sex drive whatsoever. That leaves us with a low number of abductees who claim to have what would be considered a normal sex life. Of course, defining normal in any society would be a tough call.

None of the people who were interviewed said that the alien abduction had any affect on their sexual preference. All of the people who were interviewed said that it did have an impact on their ability, or their inability, to nurture human relationships.

Although they attempt to hide the explicit sexuality of the abductions in discussions of reproductive research and the gathering of genetic material, the fact is the sexuality is the main feature. When we look at these stories in an objective light, there is no reason for the alien beings to conduct their research in such a crude fashion. And once it has been conducted, there is no need to repeat it time and again.

Even Jacobs's implication that it is the development of the hybrids that has necessitated the continued abduction and sexual aspect fails to make sense. In todays' world, genetic material recovered from just two adults would be enough to create hundreds, if not thousands of hybrids. All of it could be done outside the human body, in a laboratory.

It would seem that the recovery of more of these tales has a motivation other than hybrid creation or genetic manipulation. It would seem, from a psychological perspective, that these tales perform some other function. It is an interesting fact that nearly all of the women who report the missing fetus syndrome, or who have been implanted with what Jacobs has termed the extrauterine gestation device are postmenopausal, have had hysterectomies, or are unable to conceive children.

To understand these tales, it is necessary to look beyond the abduction phenomenon and try to understand them in the terms of a psychological problem. While it is quite true that not all abductees suffer from psychological problems, it is also true that many do. Research by several different investigators has suggested that as many as half of the abductees are homosexual, and while homosexuality is not a psychological problem, the influences and opinions of society can certainly create them. Beyond that, however, as many as 90 percent of the abductees have some kind of sexual dysfunction. Their tales of rape and sexual activity on board the UFO are evidences of these sexual problems.

The sexual component of alien abduction has been virtually ignored by abduction researchers. If they address it at all, it is to deny that it exists. But exist it does, as seen by the simple expedient of reviewing the literature and acknowledging that there is a sexual component to it. And by acknowledging that, we take additional steps to understanding exactly what is happening to those thousands who claim to have been abducted. Once again, the scientific and objective evidence leads down a path that does not end with alien creatures. Instead, it leads us to common human problems that are ignored by the researchers.

PART 2

THE COMMON THREAD

6.

FOLKLORE AND UFO ABDUCTIONS

For those who accept the idea that alien abductions are real, there is a major stumbling block. Similar tales of abduction, though not of alien, meaning extraterrestrial, abduction, exist throughout all of human history. Folklore, or the stories, traditions, beliefs, and customs preserved among the common people, is filled with fables that match, in great detail, the reports of modern alien abduction. Folklore, then, might partially explain the abduction phenomenon as we understand it today.

Dr. Thomas "Eddie" Bullard, a folklorist at Indiana University, has written extensively on this topic, suggesting no such connection exists. Others, such as Jacques Vallee, writing in *Passport to Magonia*, have, in fact, linked UFO abduction to the folkloric traditions. The patterns of folklore from the stories of the gods of the ancient Greeks and Romans, to the incubi and succubi of the Middle Ages, suggest a strong thread tying UFO abduction to folklore tradition.

The stories of alien abduction also bear a striking resemblance to tales of fairy kidnapping, initiation ordeals, and journeys to the land of the dead. Bullard, in fact, realized this. In the *International UFO Reporter* he wrote, "The initiation vision of a Siberian shaman presents an especially vivid example: While lying in a comatose state for hours or days the candidate meets two friendly spirits who escort [float him into the craft?] him into the underworld. Unfriendly spirits dismember the candidate and reconstruct his body [medical examination?] with new powers added, including rock crystals [implants?], all with magical powers inserted into the head. The candidate spends part of his stay inside a domed cave with uniform but sourceless lighting [the examination room on the alien spacecraft], and returns to his home a changed man, with new abilities and a new vocation."

Bullard comments, "The two stories poles apart in time, location, language and culture still converge in many aspects in sequence and content. The abduction story is not so unique after all."

In his article, Bullard goes further, writing, "If one sort of story alone approached the mark so closely we could blame chance, but in fact the comparisons are far too widespread. . . . These same themes are prominent in fairy lore, where fairies take human mates, steal human children, leave changelings, and seek human midwives to deliver fairy babies; where magic spells make fairies look beautiful and their subterranean kingdom rich when in fact fairies are often ugly and their land poor. . . . Abduction reports share everything from small details to broad themes with several mythological religious and legendary narrative types. . . . The 'wise baby' and hybrid offspring accounts brought to light by Budd Hopkins in *Intruders* reintroduced the old motifs of the changeling and the child of fairy-human union to the modern narrative."

Bullard is acknowledging the many examples of the abduction phenomenon appearing in the folklore of the past. Again, this all parallels the tales told today, to a surprising, and for those who accept alien abduction, frightening degree. Nocturnal visitors who have some sort of "scientific" agenda, or at the very least, a perverted interest in human reproduction, have long been part of these ancient traditions.

It seems almost as if the ancient Romans and Greeks were telling of alien abduction in the modern sense when they spread their legends and told their epic tales. Their gods lived, away from the human race, on Mount Olympus and came down (from the sky) to breed with the humans. The gods, of course, possessed powers of which we could only dream. The mixing with the daughters of Earth produced hybrids that were half-god and half-mortal. Hercules had extraordinary strength because he was the product of such a union. He is not, by far, the only hybrid between humans and the gods.

It should also be remembered that many of these tales focused on the sexuality of the situation. The satyr, from Greek mythology, was a woodland god with pointed ears and goat's legs. The most prominent feature was a large, erect phallus. The satyrs pursued the wood nymphs with a single thought on their minds. They captured, abducted, and raped the nymphs, often taking them deep into the woods to perform their terrible, and sometimes inexplicable, rituals.

Pan was a fertility god who worked the high hills and attended the flocks. He was a shepherd who played the flute, which seemed to possess the ability to hypnotize his victims. The satyrs were the product of Pan's breeding with humans.

Both of these myths revolve around the theme of abduction and reproduction. The method used, rape, seems no less gross than the technique used by the modern aliens. In other words, the ancient Greeks and Romans possessed the genetic knowledge we might use to discuss modern alien abduction research. The method and the experimentation of the aliens is the same as that ascribed to Pan and the satyrs. No complex reproductive knowledge was needed, nor even the most primitive of scientific understanding.

By the seventh century, the belief in gods that could manifest themselves on Earth had disappeared. There were those, however, who claimed they could communicate with the supernatural and use that power for evil on Earth. And there were those who believed that God was in his heaven, but Satan, ruler of the underworld, could, and often did, walk the Earth. From 1140 Catholicism further defined Satan and his influence on the common man. By the twelfth century the clergy had clearly come to rule, and Satan, the witch, and the demon were clearly defined as enemies of God and the church. Beginning in 1230 with the establishment of the Inquisition, stories of demonic possession and witchcraft would cause misery and death for tens of thousands of people for the next several hundred years.

The witch was seen as the origin of this misery, and her eradication was of the utmost importance. The witch had become a symbol of forbidden sexuality, and her torture and death became a vehicle to absolve the righteous of their prurience. The thirteenth century furthered the concept through reaction to the idea of antinomian heresy; the church's belief that acts against the laws of Moses led to immoral behavior. From 1420 to 1487, scholars of the time, through persuasion, force, and torture, formally defined possession and witchcraft and established protocols that would last until the eighteenth century.

The earliest formal tool for the recognition of demonic possession was the *Malleus Maleficarum (The Witch's Hammer),* written by two fifteenth century Dominican monks, Father Jacob Sprenger and Father Heinrich Kramer. The authors put forth several criteria for possession that bear an uncanny resemblance to modern descriptions of alien abduction.

Those possessed by demons were able to float through the air, communicate telepathically, had been given knowledge of future events, and had increased intellectual abilities as a result of their possession. Witches were said to be able to pass through solid doors and walls, to fly, to steal and eat babies, and to take on the shapes of other beings to disguise themselves (shape shifting). They often engaged in sexual orgies.

It may well be that the idea of sexual activity with nighttime visitations was introduced to the general populace by the church. The practice of celibacy appeared in the Christian church about A.D. 400, but it did not originate with Christianity. It was probably introduced to the western world by wandering Buddhist monks, but once introduced, the idea took hold in the West.

Priests, as well as nuns, became celibate. This unnatural sexual repression led them to be plagued by erotic dreams. These powerful sexual feelings and fantasies were, of course, forbidden, and could not be admitted among laity or to the public. In an attempt to rationalize the sexual impulses they were feeling, priests and nuns claimed that the devil slipped succubi into their beds at night to seduce them. Although they awakened only after the demons were gone, evidence of the encounter often remained. Nocturnal emissions were *solid physical evidence* that the demons were collecting sperm for purposes of *interbreeding*.

Nuns began to claim that they had been raped in the night by incubi, and injected with sperm collected by the demons. It is in this way that the ideas of nocturnal sexual activity by a supernatural force and the collection and dissemination of sperm were introduced to mainstream culture, and the incubus experience became part of the lore of the witch.

Jacques du Clerc, a secular scribe of the era, chronicled the sexual encounters with demons. He wrote that the purpose of the visit was to transfer sperm. "At first he put it in the natural orifice and ejaculated the spoiled yellowing sperm, collected from nocturnal emission or elsewhere, and then in the anus, and in this manner inordinately abused her."

Once these tales had become popularized, many of the townspeople began to have similar experiences. Lauran Paine, an expert on the sexual stories of the era, described a story of the time. "An incubus who seduced women or a succubus who did the same with men, was essentially a lewd demon or imp who sought sexual relations with mortals. An incubus was usually seen as a handsome man, a succubus appeared most commonly as a petite woman or girl, abundantly endowed, pretty, and although often somewhat smaller than most women, very desirable."

Ludovico Maria Sinistrati, a Franciscan scholar and linguist, said of succubi, ". . . they merely have to choose ardent, robust men, whose semen is naturally very copious, and with whom the succubus has relations, and then the incubus copulates with women of like constitution, taking care that both shall enjoy more than normal orgasm. . . ."

St. Thomas Aquinas wrote, ". . . if sometimes children are born from intercourse with demons, this is not because of the semen emitted by them, or from the bodies they have assumed, but through the semen taken from some man for this purpose. . . ." Cesarius of Heisterback claimed that these devils collected human semen from which they fashioned earthly bodies for themselves.

As time passed, the act of having sexual activity with demons came to be seen as a definite sign of witchcraft. Women claiming this experience were arrested, and often executed. In 1591, Francoise Fontaine was charged with being a witch. Francoise's arrest was prompted by her fits of bizarre behavior. She had frequent episodes of hysteria, fainting spells, and seizures that would cause her body to become completely rigid. When this occurred, her arms would fly out to her sides and her body would form the shape of a cross. During these fits she appeared to be in a cataleptic trance.

A search was made for evidence of these nocturnal visits. Francoise was stripped and her body examined for any scars or marks that would indicate the presence of her demonic lover. It was believed that every witch had at least one mark on her body left by the devil.

During her trial, Francoise told vivid tales of copulation with the devil. Francoise told her inquisitors that the devil came down her chimney at night and sexually molested her. In prison the attacks continued, and other prisoners complained that they were kept awake by the sexual activity.

During her confessions, Francoise stated that the devil was a handsome man dressed entirely in black. He had powerful flashing eyes that would overpower her, and caused her to yield to him without protest. Francoise claimed that the demon had a huge black penis, very stiff and cold, and so thick that she experienced considerable pain during copulation. Once the demon was finished with her, he would sometimes have difficulty removing his penis, and "because she could not relax, as the devil had already done, they remained nipped together, like a dog and a bitch."

At other times the devil would place Francoise between his knees fully clothed and have intercourse with her. These episodes could last up to half an hour, and ended with the devil ejaculating "something very cold in her stomach."

Francoise's tales were so *sincere and convincing* that the prison provost Morel climbed on the roof of the prison to find evidence that the demon had been there. On the roof he found the tracks and scratch marks, clear *physical evidence that the visits were genuine.*

Francois's tale is unique only in the fact that she was not executed for her behavior but cured of her affliction by exorcism. Others were not so fortunate. For the next three hundred years, thousands of men and women would be tortured and executed for reminding the church of sexual pleasure.

As time passed more and more reports of nocturnal sexual visits emerged. As people became familiar with the reports, the stories began to take on a specific structure, *always containing certain elements occurring in a specific order*. Lauran Paine has said, "Confessions became more lurid, not less. They also became more numerous and more stereotyped, as though the demands were increasing in both obsession and fixation."

In 1645, a young girl named Ann Jeffries was found lying on the floor in a semiconscious state, apparently suffering from a convulsion. Later she *recalled being paralyzed, and surrounded by six little men* who swarmed around her and covered her with kisses. Suddenly she felt a sharp pricking sensation. She then found herself flying through the air to a palace filled with people. One of the men there seduced her. Suddenly an angry crowd burst in on them and she was once again whisked away back to the floor where she was found. After the event she reported further contact with the fairies. In 1646 she was arrested for witchcraft and imprisoned.

These tales show that the modern abduction phenomenon closely resembles ancient traditions (the most vivid and important details in italics). Those in the past didn't invent alien spacecraft to explain these aberrations, but did invent many different tales to explain what was happening. For those who accept the idea of alien abduction as real, the similarities between abduction and these folklore traditions must be very disquieting.

In fact, it can be suggested that a belief in fairies, trips into the underworld, or witchcraft, are no longer acceptable in our modern world. Science has shown that fairies don't exist, the underworld is a myth, and modern witches practice their craft as a religion but have no magical or supernatural powers. However, spacecraft do exist, travel into space by American astronauts is so common that the networks no longer cover the shuttle launches as live events, and science argues about the reality of alien visitation. Fairy abduction is a myth, but alien abduction could be real. We have changed the trappings so that the tale is acceptable to all of us in the modern world.

Bullard, responding to these criticisms, wrote in the July/August 1988 issue of the *International UFO Reporter*, ". . . but abductions have a more down-to-Earth side as well. They include multiple witnesses, emotional conviction, apparent physical evidence, and minute similarities of detail seemingly too insignificant to remember from readings from other cases or to dredge up from psychic depths [though we can't help commenting that many abduction researchers claim that everything ever seen or read is in the unconscious even in the face of evidence to the contrary]. These aspects lend the solidity of physical reality to the phenomenon."

But again, as we examine the folklore history, we find the same things can be said. Even physical evidence, in the form of footprints for example, have been reported in connection with these folklore stories. In fact, there is little to separate these old tales from their modern counterparts.

Bullard refutes this idea, claiming, "Abduction stories are unlike fantasies in a number of important ways. If the reports are more or less likely, they lack the strong individual element necessary to satisfy personal needs. Abductions offer little comfort or peace to many abductees. . . . The personal element seems slight in these narratives, while an impersonal core turns up again and again, a core that seems imposed from outside rather than generated from conscious needs, and provokes strong emotions that seem to defeat the usual purpose of fantasy or hoax."

Here Bullard is arguing that these tales do not provide the teller with any sort of comfort. Rather, they are horrifying. He is suggesting that imagination would provide tales of comfort for those telling them, therefore alien abduction must be grounded in reality.

However, Bullard overlooks the result of the tale. Those who claim abduction move to center stage. They are sought by investigators, researchers, others with similar tales, and even television talk show hosts. The group setting, an outgrowth of their tales, provides the very comfort, and support, that is missing from the tale itself.

Bullard also argues that folklore, as it spread across the land, evolved from teller to teller. This mutation is what makes a story a true folktale. While many of the elements will remain the same, each of those telling the tale will put his or her own spin on it. Because this "personalization" doesn't seem to exist in the abduction stories, it moves it out of the realm of the folktale. In other words, because it doesn't act like an old folktale, it is not a folk tale and therefore must be grounded in reality.

The skeptical community, however, argues that modern communications negate that premise. Stories that took decades to spread across Europe or Asia, and that relied on individuals to tell them, are now spread around the world in a matter of minutes. An abductee interviewed on CNN will be sharing the story with the entire world at the same time. Rather than an audience of fifty or a hundred, the audience can measure, easily, in the millions.

According to Bullard, the skeptics say that Steven Spielberg's film *Close Encounters of the Third Kind* is responsible for the small, big-eyed aliens that appear in the abduction phenomenon. It was only after that film appeared that the small aliens were reported, according to that particular argument. Bullard cites numerous cases of small, big-eyed aliens that predate the film, proving that it wasn't as influential as the skeptics believe and thereby negating their argument. Or so he thinks.

It turns out that both camps are right and both are wrong. *Close Encounters* certainly established the small aliens as the "norm." Whitley Steiber's *Communion* and later *Transformation* were even more influential, as were Budd Hopkins and his work. In other words, and as Bullard himself admits, once something is written and in this case filmed, it tends to eliminate the individuality of reports. We have been educated as to what the aliens are supposed to look like so that we don't have to rely on oral reports and visualize the aliens. The pictures are now easily and readily available.

The original description of the alien beings, as provided by Betty Hill after her abduction in 1961, failed to conform to the standard of today. She spoke, in her earliest interviews of small men with big noses. But today, she speaks of small men with no noticeable noses. This subtle change in the description suggests that modern communications, and the acceptance of a "standard," may have influenced the memory of the woman sometimes credited with reporting the first alien abduction.

Descriptions of small aliens in UFO reports do predate *Close Encounters*. While it seems reasonable to argue that the movie, books, and later television set the "grays" into the public consciousness, it is also reasonable to suggest that the media are not responsible for the creation of the small aliens. A literature search shows that small aliens were reported prior to those events, and even predate the Barney and Betty Hill abduction. Does this suggest there is something real to it?

Of course not. Once again we can retreat to the stories of the fairy. These were small creatures that came in the night to abduct people. If we had nothing else, we could point to that and suggest a long and fine tradition of little "aliens" abducting people.

In fact, if we examine the classic drawing of the gray, with its big, black eyes, almost featureless face, and teardrop-shaped head, we can find many additional precedents in our history. The Jolly Roger, flag of the pirate, is of a skull and crossbones. But the skull as represented on the flag is teardrop-shaped with two large black eyes. It is, in essence, the gray of alien abduction.

Taking that even farther, we notice that a skull, representing danger, has the same features as those of the gray. What is evident when a skull is examined is that it tapers to a pointed chin, there are no features on it, except for the two large, big holes of the eye sockets. Our history is filled with images of the gray, we just haven't recognized it as such.

No matter how far back we go into human history, we find the skull as an important icon. It is represented as white, sometimes gray, and certainly looking like the drawings of the classic gray made by abductees. But, the one feature that is most important and prominent are the eyes. The eye socket is six times bigger than a normal human eye, and because of the nature of the skull, looks like a black orb. The power of the human is fixed in the skull and the power of the skull is fixed in the eyes.

Throughout our history, we have represented a ghost as a white form with no facial features except big, black eyes. Those who report ghosts claim that the phantasm looks human and is sometimes recognizable as a long dead relative. Yet, when we begin to speak of ghosts, we think of the kid with a sheet over his head and two holes cut out for the eyes. It has become a another folklore icon for us.

To suggest, as have the skeptics, that the grays came from *Close Encounters* is to ignore the rich history of folklore and how both danger and ghosts have been portrayed throughout society. The skull is shown as a lightbulb with black eyes on it. If you superimpose that image on one of a gray, there is little difference. Those who are describing grays are merely reporting an image that had its origins at the very beginning of human history.

But let's return to the UFO phenomenon. For some reason many of the reports, from the very beginning, mentioned small aliens. Descriptions, as we would expect, were varied because there was no solid tradition to this point. Some of the early stories were ridiculous. In one, the alien was described as being two feet tall with a head as big as a basketball. In another, the aliens trying to grab a man were described as hairy beanbags.

Coral and Jim Lorenzen, directors of the now defunct Aerial Phenomena Research Organization (APRO), reported in various books published in the 1960s a number of occupant types. A chart included in *Flying Saucer Occupants* showed the majority of the reports were of small beings. It wasn't until 1957 that large creatures, that is, over forty inches tall, were reported with much frequency. The UFO literature was reporting small beings from its very inception.

The 1955 Kelly-Hopkinsville report contained descriptions of small creatures with big eyes. The drawings in the Project Blue Book files look nothing like the grays reported by abductees today, but it seems this is the very type of variety that Bullard and folklore tradition demands. And it is the type of variety that is ignored by those same researchers.

What we see, then, are reports of alien beings made prior to the 1977 film *Close Encounters*, but they are of the variety that Bullard suggests. There is no standardization. Yes, there are reports of beings that would be considered "gray" by today's standards, but those cases were spread thin among all the others.

Interestingly, if we look at the earliest reports of UFO occupants, we see the influence of science fiction books and movies. One of the most common reported uniforms in those early cases are the "diving suits." People who were reporting alien beings early on seemed to believe they wouldn't be able to breath in our atmosphere and they would require a self-contained environment. *Flying Saucer Occupants* shows that many of those early reports, especially from France and South America, were of beings in diving suits.

Later, as the contactee phenomenon began to spread, and movies were made showing that other planets held Earth-type atmospheres, the idea of diving suits gave way to silver jumpsuits. People reporting UFO occupants were reporting what they expected to see based on what they had seen in their own lives, meaning what they were seeing in the movies and on television or reading in books.

What we see in the survey of the literature then, is a diversity of UFO occupants until the information becomes set by media influence. There is little doubt that the media have influenced the descriptions of aliens. The fact that those descriptions existed before the spread of the media proves the point.

Bullard writes, "My training as a folklorist teaches me that this situation is unusual [that is, the stability of the abduction tales]. Narrators recombine story elements, replace them with new ones, or borrow from other stories in a ferment of creativity that no brevity, mundaneness, or close-knit structure of the narrative can hinder. The notion that only one version of a narrative exists and everyone repeats it as if reciting from rote memory is a fallacy. Folktales like that of Cinderella never circulated in the storybook version, but in as many variants as there were narrators."

But the difference here is that the stories of abduction have become well known. They are written down or filmed or played out in front of national and international television audiences. Those coming forward today are well schooled in what an abduction is supposed to be like. The training "films" are broadcast constantly. And the story of Cinderella is now set in the storybooks. Everyone knows how it is to be told.

What develops then, is an argument of semantics. Bullard asks, "Abduction reports may share some content with fairy kidnap or shamanic initiations, but do these reports act like folklore?" He then suggests the reasons that the abduction reports are not folklore. And that is the point. The reports are not folklore, but that doesn't mean they are, therefore, grounded in reality.

Bullard writes, "The abduction story in its most elaborate form consists of eight episodes—capture, examination, conference, tour, otherworldly journey, theophany, return, and aftermath. . . . Not every possible episode or event appears in every narrative, but whatever appears maintains the same relative position in 70 to 80 percent of 103 high-information, high-reliability cases. A single deviation accounts for most of the sequence differences."

From this it seems that Bullard is suggesting there is personal deviation, but it is not significant. He suggests that there are several components to the abduction report, but not all components are present in all the cases, and in 20 to 30 percent, there are some deviations. In other words, it seems that the individual variation that Bullard requires for a tale to be labeled as folklore exists in the tales being told today.

Martin S. Kottmeyer, writing the "The Cultural Background of UFO Abduction Reports" in the January 1990 issue of *Magonia,* said, "A relabeling of Bullard's elements should make the logic clearer: (i) character introduced, (ii) peril and conflict, (iii) explanation and insight, (iv) good will and attempt to impress, (v) excitement, (vi) climax, (vii) closure, and (viii) sequel."

Kottmeyer makes the point that these elements, now correctly relabeled, are the proper way to tell a story. Kottmeyer writes, "Most of Bullard's deviant cases involve the other worldly journey not staying in the place he deemed correct. To put it simply, Bullard's correct order is the right way to tell a story."

Again, we have been reduced to a discussion of semantics. That in no way translates into a reality of alien abduction. We are arguing not for the reality of alien abduction but for the reality of folklore. Bullard believes that these tales are not folklore because they are transmitted in a different fashion, they do not vary the way folklore does, and they all contain the same elements told in the same sequence.

But that does not make the abduction tale real. All it suggests is that abduction is not a folktale in the classic sense. The modern world, with its immediate communications, with its mass audience, and with its ability to produce millions of copies of a book and provide instant access to the data needed, has destroyed the traditional folktale.

So Bullard is probably right. Abduction tales do not conform to the folklore tradition. However, alien abduction tales do mimic much of the traditional folklore. This cannot be denied. To suggest otherwise is to miss the real point.

Contrary to what abduction researchers tell us, there is good reason to draw parallels between the two. Arguing against the parallels is to argue semantic differences and not reality.

7.

DOES POP CULTURE AFFECT OUR VIEWS?

It is clear from psychological research that traditional folklore has produced a number of tales that closely match the abduction scenario. Apparently there is something in the human mind that creates these tales on a regular basis. In today's environment it is not appropriate to speak of fairy abduction or initiation rites held in the underworld, but it is proper to speak of alien spacecraft. The question that must be asked today is if these tales of alien abduction are influenced by our pop culture including science fiction books and movies, so-called reality based programs that deal with abduction as if it were already established fact, or is it something that is unique and therefore outside the realm of pop culture?

If we examine modern society carefully, we can find the treads of alien abduction reproduced with the growth of science fiction and the movies. All elements of abduction existed before the first of the flying saucers was reported nationally in 1947. In fact, the first of the abduction reports took place during the great airship sightings of 1897. Those reports varied from modern abductions in a number of ways. First, most often, the pilots of the airship were human scientists who were on the verge of a great new scientific discovery. In only a few of the cases were nonhumans reported, but in those few cases, they predicted the current UFO trends.

It might be illustrative for us to look at some of those early predictors. For example, the tale of Alexander Hamilton, a rancher in Kansas, is the first to report a tale of cattle mutilation. Today, ranchers and researchers claim that an unknown agency, probably extraterrestrial, is killing and mutilating cattle around the world. According to these researchers, no one can explain how it is being done. It is a modern mystery, according to the mutologists—those who study mutilations—that has no simple solution.

Hamilton, in April 1897, reported that one of his cows had been stolen from a closed corral late at night. The entire account of the aerial cattle rustling was published in the *Yates Center Farmer's Advocate* on April 23. Hamilton and a number of his friends signed an affidavit attesting to the truthfulness of the story. According to that document, "Last Monday night [April 19] about half past ten we were awakened by a noise among the cattle. I arose thinking perhaps my bulldog was performing some pranks, but upon going to the door, saw to my utter amazement, an airship slowly descending over my cow lot about 40 rods from the house.

"Calling Gid Heslip, my tenant, and my son, Wall, we seized some axes and ran to the corral. Meanwhile the ship had been gently descending until it was not more than 30 feet about the ground and we came up to within 50 yards of it. . . . It was occupied by six of the strangest beings I ever saw. There were two men, a woman and three children. They were jabbering together but we could not understand. . . .

"When about 30 feet above us, it seemed to pause, and hover directly over a three-year-old heifer which was bawling and jumping, apparently fast in the fence. Going to her, we found a cable about half an inch in thickness . . . fastened in a slip knot around her neck, one end passing up to the vessel. . . . We tried to get it off but could not, so we cut the wire loose, and stood in amazement to see the ship, cow and all rise slowly and sail off. . . .

". . . Lank Thomas, who lives in Coffee Country about three or four miles west of LeRoy, had found the hide, legs, and head in his field that day. He, thinking someone had butchered a stolen beast and thrown the hide away, had brought it to town for identification but was greatly mystified in not being able to find a track of any kind on the soft ground. . . ."

The affidavit was signed by a number of men who claimed they had known Hamilton for years and "that for truth and veracity we have never heard his word questioned and that we do verily believe his statement to be true and correct."

During the mid-1960s, the Hamilton tale surfaced again and was repeated in a number of magazine articles and UFO books. Each time, the statement of Hamilton's friends was mentioned without question. Here was a tale that deviated from the airship stories in a number of ways, suggested an extraterrestrial explanation for the event, and involved the rustling of a cow. It proved the strangeness of some of the airship tales and suggested to those who wanted to believe that something otherworldly was happening.

Jerry Clark, who had reported the case seriously a number of times, did some additional investigation on his own in the 1970s. He learned that one of Hamilton's daughters still lived in Kansas and interviewed her about the story. Although she hadn't been born until after the event, she had heard her father talk about it on many occasions. She said that Hamilton belonged to a local liars club, as did all the men who signed the affidavit attesting to Hamilton's veracity.

Even worse for the true believers was a letter found by Eddie Bullard. In the May 7, 1897, edition of the *Atchison County Mail,* Hamilton wrote that he had fabricated the tale. There simply is no reason for it to continue to circulate.[1]

The story was a joke, invented by Hamilton, and told at the time the airship was reportedly flying cross-country. There was not a word of truth in it. Hamilton would have been delighted to learn that his joke had fooled another generation seventy years later.

The airship craze of 1897 also provided the first of the UFO crash retrieval stories to gain widespread publicity. According to many newspapers of the time, and later to dozens of UFO writers, the airship was seen over Aurora, Texas, early on the morning of April 17, 1897. It swooped low over the town, buzzed the town square, and continued on to the north. There it slammed into a windmill on the farm of Judge Proctor and exploded. The badly burned body of the dead pilot was recovered. T. J. Weems, a Signal Corps officer, said that the pilot, obviously not human, was probably from Mars.

In the wreckage, searchers found several documents covered with a strange writing. Some of these were somehow translated, telling the investigators that the airship weighed several tons and was made of aluminum. By noon that day, April 17, the debris had been collected, and later in the afternoon, the Martian was given a Christian burial.

It probably isn't necessary to point out that here we have all the elements of the modern UFO crash retrieval, including the mystery writing, the disappearance of the debris, and the recovery of an alien body. It wasn't covered up by the American government but disposed of by the local residents in their desire to provide the alien occupant with a proper burial. None of the debris has survived until today, so no analysis of it can be performed.

The truth of the Aurora case, however, is that it is another hoax, just like the Alexander Hamilton cow-napping. H. E. Hayden, a stringer for the *Dallas Morning News* had reported the Aurora crash to the newspaper. They printed it, just as they had printed several earlier airship stories. Hayden apparently saw an opportunity to put his hometown back on the map. He missed by more than seventy years.

The fact it was a hoax has been demonstrated time and again. T. J. Weems, the Signal Corps officer, was, in fact, the local blacksmith. Local residents of Aurora since 1897, interviewed by Kevin Randle in the early 1970s, claimed they knew nothing about the event. Members of the Wise County, Texas, his-

---

[1]Jerome Clark, "Airships: Part I," *International UFO Reporter*, Jan./Feb. 1991, 4–23; Jerome Clark, *The Emergence of a Phenomenon: UFOs from the Beginning through 1959* (Detroit: Omnigraphics, 1992), 192–3

torical society told Randle they wished the story was true, but there was just no reason to accept it.[2]

It can be argued, and certainly will be, that by the time the UFO phenomenon exploded in June and July 1947, the great airship stories were long forgotten. It wouldn't be until the 1960s that any of the stories of the airship would receive any sort of renewed publicity. But the point is, there was a wave of UFO sightings at the end of the nineteenth century that mirrored, almost perfectly, the events that began in late 1940s.

David Jacobs has argued that the UFO phenomenon sprang into existence in 1947. Thomas Bullard suggested that the Barney and Betty Hill abduction of 1961 had no cultural sources from which to draw. And Budd Hopkins has claimed that the beings reported by abductees are like no "traditional sci-fi gods and devils." In other words, each is arguing that UFOs and abductions must be real because there are no cultural sources from which the witnesses could draw the material. Without those sources of material, the witnesses must be relating real events rather than some sort of folklore history even though the airship scare of the late nineteenth century demonstrates that the fundamental assumptions of each are inaccurate.

It seems ridiculous to suggest that a phenomenon that has no evidence of its existence other than witness testimony must be real because there is nothing in the past that relates to it. Because there are no past traditions, how did each of these witnesses, who have never communicated, relate similar events if not reporting, accurately, something they have witnessed? This is the question posed by many UFO investigators and abduction researchers.

The answer is, of course, that the cultural precedents demanded by Hopkins, Jacobs, and Bullard do exist. Pop culture from the beginning of the twentieth century is filled with examples of alien beings and alien spacecraft that match, to an astonishing degree, the beings and craft being reported today by the abductees.

To completely understand the cultural influences we must examine the pop cultural world. At the turn of the century information moved at a slower pace, but it still had the impact it does today. For example, there were no radio stations that played the latest music. Instead, sheet music was sold. To sell it, without radio to play the songs, music stores hired piano players and singers. The music circulated through the culture much more slowly, but no less completely. A hit, on sheet music, might take weeks or months to move from one coast to the other, but the point is, it could and frequently did.

---

[2]Kevin D. Randle, "The Flight of the Great Airship," *True Flying Saucer and UFOs Quarterly*, Spring 1977, 58–9.

Think about that. Music would move from coast to coast. Musicians would hear it in one city and play it in the next. Vaudeville performers used the same popular music in their acts. Player pianos played it to audiences in all sorts of environments. Before long everyone in the country was singing the song, or playing it at home, all without records or radio, national broadcasts, or even MTV.

This demonstrates just how information can be passed from person to person without modern technology. It also suggests that arguments claiming that one person could not have heard a specific story because it had no national forum is wrong. The information, whether it is music at the turn of the century or information about abductions, can enter into a "collective consciousness." Simply, it moves from person to person until all have been exposed to it.

In *The Vanishing Hitchhiker* and *The Choking Doberman* Jan Harold Brunvald explores the concept of urban legends. These are tales that have no foundation in reality, yet spread through the culture with what seems to be a will of their own. One of the prominent urban legends is the tale of the choking doberman. A woman, though it can be a man, returns home to find the dog choking. The vet discovers human fingers lodged in the dog's throat. A search at home finds the unconscious thief, missing his fingers.

Attempts to verify the story fail. When the report circulated in Las Vegas in 1981, both the newspaper and the police tried to find the name of the woman but were unable to turn up a single shred of evidence that the tale was true. Similar stories circulated throughout the United States in the early 1980s.

What this demonstrates is that some tales are able to survive even when no evidence is presented. Procter and Gamble has worked for years to stop the urban legend that their corporate logo is inspired by members of the board who worship the devil. Although Procter and Gamble has access to national media, and has the money to buy time for advertising, the story of its devil worship continues to spread. It is an urban legend.

The introduction of movies, radio, and other mass media, however, has made it even easier to spread data, and provides more opportunities for all of us to be exposed to it. An abductee might claim no interest in science fiction, but that doesn't mean that he or she has not been exposed to the elements of science fiction.

One of the first movies made was the 1902 version of Jules Verne's *First Men in the Moon.* Since that time, Verne's work has been translated into dozens of films in dozens of versions. They have been broadcast on television for more than fifty years.

H. G. Wells was responsible for more than just adding science fiction to pop culture. His *War of the Worlds,* first published before the turn of the century, was responsible for one of the great "hoaxes" in American history. In 1938,

Orson Welles, in a radio program broadcast nationally, reported on an alien invasion launched by beings from the planet Mars. The panic that developed during that broadcast has been studied for years afterward.

Even those who hadn't heard the original radio broadcast learned about the aftereffects. Sociological studies have been done on the mob psychology that produced the panic. But more important, the event brought the concept of alien invasion into the homes of average American before the 1940s. They might not be reading science fiction, but they were seeing the results of science fiction spread across the front pages of their newspapers.

Science fiction has been an important part of pop culture since Hugo Grensback introduced it to American society in the 1920s. Grensback's idea was to sugarcoat science so that the young would be interested in it. He envisioned it as a way of teaching science to those who weren't interested in learning science. He wanted it to bubble through society, through our collective conscious.

Frederick Pohl, a science fiction writer of the first rank, has said that in 1945, when the detonation of the atomic bomb was announced, there were two thousand scientists who comprehended the situation. . . . And maybe five hundred thousand science fiction readers who understood it. Science fiction was teaching us all about science.

In the 1930s and 1940s there were many science fiction magazines. The covers of them featured full-color art designed to catch the eye. Scientists, looking like all-American heroes, monsters of all kinds, and women in scanty clothes and in peril were the themes of many. At the time, these were the pulp magazines, filled with action stories and exciting tales. Each month the newsstands had issues with new covers, all crying out for attention to convince us to buy the magazine.

One particular cover, from *Astounding Stories,* published in June 1935, is particularly important. It shows two alien beings with no hair, no noses, slit-like mouths, and large eyes. Through a door, one of the strange creatures is looking at a woman on an examination table. Her eyes are closed and she is covered by a sheet (a convention of the time), but it is clear that she is naked under the cloth. In the foreground another creature is restraining a man trying to break through to the woman.

This cover, printed in 1935, predicts many elements of the abduction phenomenon of forty years later. Although, the alien beings have pupils in large whites of the eyes, the similarities to the modern abductions is striking. To suggest that abductees of today could not have seen the cover of a science fiction magazine published sixty years earlier is to miss the point. It demonstrates that the idea of alien abduction is not something that developed recently, as aliens began abducting humans, but in fact had been announced in public long before anyone had heard of flying saucers and alien abduction.

The covers of these magazines, seen on newsstands for thirty years or more, have featured scantily clad women, sometimes in peril, and sometimes exploring strange new worlds. Many of these covers, which hinted at alien abduction, also hinted at the sexuality of that situation. This sexuality, though subtle, has always been a feature of the science fiction covers and is a main feature of modern alien abduction.

If we limit ourselves to a discussion of pop culture in the last fifty years or so and ignore the great airships, the science fiction magazines of the 1920s and 1930s, and pop culture that came before 1940, we still see the media loaded with everything that is found in the UFO sightings and abduction reports of today. It is important to understand that nothing the abductees report is new and specific to that phenomenon. There are cultural precedents as well as the folklore traditions.

The idea that the aliens are from a dying planet has been played out in everything from *Not of This Earth,* first released in 1956, to many of the most recent science fiction movies, including a 1994 remake of *Not of This Earth.* Interestingly, the alien depicted in the film is collecting blood in an aluminum briefcase and he always wears dark glasses to hide his eyes. Although he is not collecting genetic material, as has been suggested of the aliens reported by abductees, he is required to send humans to his home world as they attempt to end the plague destroying them. The obvious purpose is to gather human genetic material.

But that very problem is discussed in *The Night Caller,* made in 1965. In that movie the alien is sent to Earth to provide women for "genetic experiments" on his home world. The women are, of course, abducted by that alien.

During the 1960s, David Vincent fought the Invaders, who were billed as coming from a dying world. They were going to take over the Earth, making it their world.

The mini-series *V* again featured a race that was invading Earth, wanting to steal our resources. Other films, such as *This Island Earth,* contain alien scientists eventually abducting Earth scientists to help them defeat their enemies on their home world. *The 27th Day,* features potential alien invaders who provide several people with the power to destroy all human life on Earth so that the aliens can inherit it.

And each of these films suggest human abduction somewhere in the storyline. *The 27th Day* begins with five people abducted onto an alien ship where time slows almost to a standstill. The abductees are returned quickly, after being given their mission, and the weapons to wipe out the human race.

Peter Graves, a scientist working on atomic energy, is abducted from his jet as it crashes in *Killers from Space.* He returns to the base, confused, with a period of missing time and a huge scar on his chest. The one thing that stands out in the film is the huge eyes of the aliens. Although not the jet black orbs of the modern abduction tales, these eyes haunt Graves as he tries to remember exactly what has happened to him. And Graves remembers nothing of the encounter until he undergoes a chemical regression aided by sodium amytal.

To take the *Killers from Space* theme even a step further, in 1975 Randle attended a UFO conference in Fort Smith, Arkansas. A man there claimed to have been abducted while waiting in his car at a railroad crossing. Under hypnosis, arranged by the conference organizer, Bill Pitts, he told a story of being subjected to a medical examination of some kind. He said that while lying on the table, surrounded by aliens, he could see a huge screen near him. It was a display of his internal organs, including his beating heart. And it is a scene right out of *Killers from Space*. Randle recognized the scene as soon as he heard it.

In *Earth vs. the Flying Saucers,* Hugh Marlowe meets the aliens on their ship, in which time stands still. This movie also features the abduction of an Army general in which his mind is probed and his thoughts are taken to be stored in the "Infinitely Indexed Memory Bank."

Small aliens, some with large heads, pointy chins, and huge eyes are featured in many of the 1950s science fiction films. *Invasion of the Saucermen* is only one of these. Of course the aliens also have a pronounced network of blood vessels and large pointed ears.

The implants claimed by some as proof that abductions are real have also been featured in science fiction movies. Tiny probes, pushed into the back of the neck to monitor the victims, are found in 1953's *Invader's from Mars*. In fact, there are several scenes in the movie that mirror the stories told by modern abductees.

Implants, delivered by "flying flapjacks," are also a feature of *It Conquered the World*. Once the implant is made, the person is then part of the alien invasion. He or she is controlled by the alien.

And for those who find these examples interesting but not persuasive there is *Mars Needs Women*. Overlooking the obvious, which is, of course, the abduction of women for reproductive purposes, there are the costumes worn by the Martians. These include a tight-fitting helmet, not unlike those worn by skin divers. Over this was a small, round radio with a short antenna sticking up. This exact costume was reproduced by Herbert Schirmer after his abduction was reported to the Condon Committee in 1968. The contamination by the movie is unmistakable.

What we find by searching the science fiction movies of the 1950s and 1960s are dozens of examples of aliens invading from a dying planet, abducting people for reproductive purposes, and implanting small devices into them for a variety of reasons. To suggest, as Budd Hopkins has, that there is no similarity to the "traditional sci-fi gods and devils" is ridiculous. The similarity to many of the alien beings and abduction situations in science fiction is overwhelming.

What we have demonstrated here is that all the elements of the abduction phenomenon have already appeared in dozens of science fiction stories. These films might have been poorly attended when first released to the theaters but have been replayed time and again on late-night television. Even those who claim no interest in science fiction movies have had the opportunity to see them on the late shows. It cannot be suggested that these films have had no influence on the abduction phenomenon, for even if a specific witness could prove he or she had never seen any of these movies, there are dozens of others who have. There is no denying that this aspect of pop culture has had an influence on our view of the aliens and their motivations, and therefore on the reporting of stories of alien abduction.

And even if the witness could somehow prove that he or she had not watched such films on late-night television, there would be other arenas for exposure. Again, we slip into a look at pop culture in the 1950s and 1960s. While a specific abductee might have avoided films with flying saucers and aliens in them, he or she would have attended movies. We all did, whether it was the Friday night date, or the kid's matinee on Saturday afternoon. One of the many features of the era's theater presentations was the trailers, or the previews of coming attractions. So even if the abductee didn't go to the science fiction movies of the era, he or she would have seen the previews for them. The abductee might have avoided seeing the whole film, but would have seen pieces of it while at another movie.

Or, to take it a step further. How many families made it an outing to attend the drive-in theater on a Friday or Saturday night? It didn't matter so much which films were showing, but that the family was going out together. Many of the drive-in movies were the "B" films, those made to support the main attraction. These were black-and-white science fiction films made cheaply. Many of them were of alien invasions, monsters from outer space, and as we have noted, included many of the elements of the abduction phenomenon of today.

And often, at those Friday night movies, or Saturday matinees, a chapter of a serial was shown. These films featured everything from Flash Gordon and Buck Rogers to Superman and tales of the Lost Continent of Atlantis. Robots, spaceships, and evil aliens were the norm. Trips through the solar system and to planets far away were taken. Many times the main film program was what people attended to see, but the "boring" shorts were shown first, including a serial.

Sometimes the "next exciting chapter" was the reason for going to the movies the next week. So maybe the abductee didn't go to see the science fiction movies but did see the serial. Maybe that was the reason to attend, but it was so long ago that he or she has forgotten. To suggest no exposure to the pop culture is to suggest someone who never left the house, never looked at a magazine, watched television, read the newspaper, or listened to the radio. Such a suggestion is ridiculous.

In today's environment, the influence is even more obvious. NBC broadcast the story of Barney and Betty Hill to a national audience in October 1975. If nothing else, it focused the alien abduction in the minds of many viewers. After that, millions knew that aliens were smaller than humans and they had big eyes.

Bullard opens his massive study of the abduction phenomenon by reporting on the Hill case. Prior to the release in 1966 of *The Interrupted Journey,* John Fuller's book about the Hills, there had been no discussion, in this country, of alien abduction. The Antonio Villas-Boas case, known to few even inside the UFO community, would not be known to Betty Hill. Yet, without that prompting, Betty Hill tells a tale of alien abduction that is similar to that related by Villas-Boas. The question that plagues the researchers, including Bullard, is where did she get the idea?

The answer, however, might be relatively simple to understand. Before the events in 1961, Betty Hill had discussed UFOs with her sister, who claimed to have seen one. Just after her experience, she read Donald Keyhoe's *The Flying Saucer Conspiracy*. In fact, within days of her sighting, Betty Hill wrote to Keyhoe at NICAP headquarters to relate her tale to him. She was wondering if there were any additional books, with "any information more up to date than this book [*The Flying Saucer Conspiracy*]."

Reading Betty Hill's letter to NICAP carefully, we find that Barney believed he saw in the flying saucer "many figures scurrying about as though they were making some hurried type of preparation. One figure was observing us from the windows. . . . At this point, my husband became shocked and got back in the car, in a hysterical condition, laughing and repeating that they were going to capture us."

What's this? The idea of alien abduction originated with Barney Hill, as he became hysterical. He was telling Betty that it was his belief that the aliens were going to capture them. In other words, it means the idea of an abduction was borne of the stress of the situation and the panicked reaction of Barney Hill to what he believed was an alien spacecraft. To answer Bullard's question of where Betty Hill got the idea, she got it from Barney.

Bullard believes that the Hills didn't possess the knowledge to construct the nightmare of alien abduction. And he might be right. We have, however, just been provided with a clue about how the idea originated. The question is, are there other facts that add to this? Barney Hill's hysterical reaction certainly isn't enough to add the details of small alien creatures. The answer to this can be found in Keyhoe's *The Flying Saucer Conspiracy*.

At the time of the Hill abduction, there were few public reports of alien creatures. It was not a topic discussed much in UFO circles. Keyhoe cites a dozen or so of these cases, ignoring the majority of them. He does, however, treat with respect the case of a pilot in Hawaii who claimed, "I actually saw him," meaning the creature from the craft. Keyhoe seems to be suggesting that the story, while wildly extreme, at that time, has an undercurrent of authenticity.

More important, however, Keyhoe writes of UFO reports from Venezuela that seem to have contributed to Betty Hill's nightmare. In his book, Keyhoe reports on two men who sight a bright light on a nearby road. Hovering over the ground is a round craft with a brilliant glow on the underside. According to Keyhoe, four little men came from it and tried to drag Jesus Gomez to it in an apparent abduction that failed.

In another report, Keyhoe mentions Jesus Pas, who was found unconscious after being attacked by a hairy dwarf. And finally, he reports on Jose Parra, who was "paralyzed" by a bright light after he had seen six hairy little dwarfs near a saucer. Keyhoe had written, then, of small beings associated with flying saucers, and Betty Hill was exposed to this within days of her own experience.

Betty wrote to Keyhoe, "At this time we are searching for any clue that might be helpful to my husband, in recalling whatever it was he saw that caused him to panic. His mind has completely blacked out at this point. Every attempt to recall, leaves him very frightened."

What is interesting about the Hill case is that Fuller, when he printed Betty's letter in *The Interrupted Journey*, left out a single, key sentence. "We are considering the possibility of a competent psychiatrist who uses hypnotism."[3]

All of this, from Keyhoe's writings about nasty, hairy dwarfs who are attempting to kidnap humans to the idea that the aliens are conducting some kind of experimentation, was introduced prior to 1961. The elements for the abduction scenario as outlined by the Hills were abundant throughout the media. If Bullard wonders where Betty Hill got the idea, a study of the case will provide an answer for it. There is no denying that pop culture had supplied the various elements. Betty Hill may have pulled them together into a single, neat package.

There is one point that is made, repeatedly, about the Hill abduction. The procedure that Betty Hill claimed was performed on her, that is, the needle in the navel, was not used by human doctors at the time of the encounter. According to many, this bizarre procedure was not introduced until after the Hill abduction. It seems to be of a predictive nature, and if not in use, how did Betty Hill invent it?

---

[3]Mark Rodeghier, "Hypnosis and the Hill Abduction Case," *International UFO Reporter*, Mar./Apr. 1994, 4–6, 23–4.

This, however, is not correct. Amniocentesis, as a medical procedure, has been around since the nineteenth century. In those days, the needle was pushed into the belly to draw off the amniotic fluid when there was too much pressure during pregnancy. It was also used, in the late 1950s, as a testing procedure for women when there was a suspicion of blood group incompatibility. After 1966 amniocentesis began to be used in genetic screening. The point, however, is that medical procedures requiring a needle pressed into the belly did exist in the medical literature. Betty Hill invented nothing, and the procedure itself was not predictive.

Once again, we find all elements of the Hill abduction in the pop culture. Of course, an exact match is not possible, but then Bullard in his work on folklore explains the imperfection. As the story is told, each teller adds and subtracts elements so that the story fits his or her personality. At the time of the Hill abduction, the phenomenon hadn't been as well publicized as it is today. The differences, then, are easily explainable in the formation and spread of folklore.

Martin S. Kottmeyer, writing in *Magonia,* presents a good argument for the introduction of elements from pop culture. For example, Barney Hill talked of "wraparound eyes" when he described the aliens to his psychiatrist, an element of extreme rarity in science fiction films. But Kottmeyer found the exception. He wrote, "They appeared on the alien episode of an old TV series, 'The Outer Limits' entitled the 'The Bellero Shield.' A person familiar with Barney's sketch in *The Interrupted Journey* and the sketch done in collaboration with the artist David Baker will find a 'frisson' of 'déjà vu' creeping up his spine when seeing this episode. The resemblance is much abetted by an absence of ears, hair, and nose on both aliens. Could it be by chance? Consider this: Barney first described and drew the wraparound eyes during the hypnosis session dated 22 February 1964. 'The Bellero Shield' was first broadcast 10 February 1964. Only twelve days separated the two instances. If the identification is admitted, the commonness of wraparound eyes in the abduction literature falls to cultural forces."

Betty Hill was eventually asked about this by UFO researchers. She claimed that neither she nor Barney ever watched *The Outer Limits.* It seems ridiculous to believe that she would be able to recall if her husband watched a television show some thirty years earlier. It could simply have been the only time that he ever watched it. The coincidence between the airing of "The Bellero Shield" and Barney's description some twelve days later is amazing.

It should also be noted that *The Twilight Zone* aired a program that dealt with alien abduction and had gray-faced beings with big black eyes. That episode, "Hocus Pocus and Frisby," was first broadcast on April 13, 1962. The gray aliens as described by Betty Hill in the years after her abduction bear a striking resemblance to the aliens seen by the character Frisby.

The situation of April 1961 is slightly different than we have been led to believe. The Hill abduction didn't spring into existence in a cultural vacuum but in a society where information was shared nationally on television and through the movies, not to mention magazines and books. Betty's interest in UFOs predated her experience because of her sister's UFO sighting, and Barney's fear of capture while driving on a lonely stretch of highway in New Hampshire created the scenario. As the days passed, Betty Hill dreamed of the incident, writing about it in her diary. When interviewed by interested UFO researchers, she always told about her dreams, with Barney sitting in the room with her. The rest of it came together almost naturally.

It is important to note that the Hills' psychiatrist, Dr. Benjamin Simon, never believed the story told under hypnosis. He didn't accept the abduction as real. He believed it to be a confabulation, a fact often forgotten by UFO researchers.

Kottmeyer, writing his article, makes another point that has often been lost on researchers. The original UFO sighting, the one that alerted the public to the existence of the flying discs, was not of a flying saucer but of a crescent-shaped object. Kenneth Arnold, the Boise, Idaho, businessman who made the report, said that the objects moved like saucers skipping across a pond, not that they were saucer-shaped.

Bill Bequette, a Washington State newspaper reporter, invented the term flying saucers, though the shape was not of a saucer. The shape, as drawn by Arnold in the report submitted to Project Sign (later Project Blue Book) was of a heel-shaped object. Kottmeyer writes, "The public, however, did not know that. No drawing accompanied the article. People started looking for flying saucers and that is exactly what they found."

Kottmeyer then makes a mistake in his analysis that is not important to his overall point. He wrote, "They reported flat, circular objects that look like flying saucers sound like they should look like. Equally important: no one reported objects like the drawing in Arnold's report to the Air Force. The implications of this journalistic error are staggering in the extreme. Not only does it unambiguously point to a cultural origin of the whole flying saucer phenomenon, it erects a first-order paradox into any attempt to interpret the phenomenon in extraterrestrial terms. Why should extraterrestrials redesign their craft to conform to Bequette's error?"

There were, however, at least two other instances in which heel-shaped craft were reported. One was the Roswell crash on July 4, and the second was a photograph taken over Phoenix, Arizona, on July 7. The lion's share of objects reported, however, were disc-shaped.

Those two aberrations do not negate the point made by Kottmeyer. He was suggesting that people were looking for saucers and that is what they found. Thousands of them. Few reported spheres, cigar-shaped craft, and a variety of other shapes, but the majority of reports were of flying saucers.

It probably should also be pointed out that at that time, the summer of 1947, space flight and space travel were on the minds of many. World War II had ended only two years earlier, and people were beginning to think of expansion into space. The newspapers, in their attempts to explain the flying saucers, reported that some scientists suspected the craft were from other planets. They didn't think in terms of interstellar but interplanetary travel, naming Venus and Mars as the two most likely candidates.

The idea, then, that flying saucers, or UFOs as they would be later known, are alien craft was spread from the first moment that they were reported. The step, from alien craft in the sky to alien craft on the ground was not a large one. In fact, there were a dozen or so reports of crashed flying saucers during those first few weeks, and there were a few reports of the creatures from them.

The idea that flying saucers were from outer space was advanced almost from the very beginning. It was one of many ideas that were proposed, but it is the one that stuck. Other suggestions, of experimental aircraft and rockets, war hysteria, or unknown natural phenomena, have all slipped away over the years. Many believed then, as they believe now, that flying saucers are extraterrestrial.

What we have is a well-ingrained theory that states aliens are abducting humans, fueled by speculation from science fiction movies and the popular press. All the ideas have been discussed, in the movies, on the radio, on television, and in books. All elements of the abduction phenomenon have been well publicized long before the first of the abductions is reported. Contrary to what the UFO researchers might want to believe, we can find all the elements of abduction in pop culture. We may have to search several sources, but there is no denying that the elements were all present before Betty Hill made her astonishing report.

If alien abductions are real, and even if we find precedents in pop culture and in folklore traditions, the abduction experience itself should be unique. We should find nothing similar to it in our society. It turns out that such is not the case. Alien abduction is not unique. There is another phenomenon that has grown out of pop culture, whose traditions and traits mimic UFO abduction almost step for step. It is a phenomenon reported, essentially, by the same kinds of people, investigated by the same kinds of people, and it provides us with clues about the reality of claims of alien abduction.

Satanic ritual abuse has all the same characteristics, including a group of clinicians who claim it to be as real as any other area of human experience. By examining it carefully, we begin to see what is happening in our society. Maybe the answer isn't in the stars but in ourselves.

8.

A FURTHER MECHANISM FOR THE CREATION OF ABDUCTION SCENARIOS

All of us dream. Some more than others. Even though many people do not remember their dreams, the average person has about four dreams every night. A dream is a specific state of consciousness wherein the mind generates stories and fantasies. Psychological researchers believe that dreams are an artifact of the brain's filing system. During the dream, information gathered during the day is edited, categorized, and stored in long-term memory.

Dreams have always been a source of fascination. Early man did not understand dreaming, and often failed to distinguish dreams from reality. He believed that he was being visited by gods in the night. Many cultures believe that during sleep or fainting, the soul separates from the body. During this sojourn, it has been said that the dreamer will see beings that differ from those of his or her conscious experience. Frazer wrote, "The soul of the sleeper is supposed to wander away from his body and actually visit the places, see the persons, and perform the acts of which he dreams." Today, most societies differentiate dream images from those of wakefulness, but many still place great significance on the dream experience. Many psychotherapists today see dream analysis as an important tool for cure, but unlike their predecessors, they do not suggest that dreams are actual experiences.

However, abduction researcher John Mack has recently said, "The word 'dream' or the idea of dreams provides a good example of how a familiar term has to be looked at more carefully, even redefined." Mack is suggesting that we turn back the clock several centuries and redefine dreams as memories of actual experiences.

Of course, if you redefine a dream this way, it is no longer a dream. Dreams, according to abduction researchers, are actual experiences that we have been mistaking as fantasy. A careful reading of many books on abduction shows that the concepts of dream, fantasy, and actual memory become blurred again and again. In a person with faulty boundaries to begin with, this process is not difficult.

Through hypnosis, abductees are taught to believe that their dreams are memories of actual experiences. Their criterion for this conclusion is that "the dream was so vivid." Abductees are also encouraged to believe that the reason they are having so many dreams about aliens is because they were actually abducted. They don't realize that thinking about aliens and abductions in the daytime increases the likelihood that they will dream of aliens and abductions at night. In fact, we all dream about the things we experience and think about during the day, a phenomenon called "day residue" by dream researchers. A majority of abductees have sleep disturbances, which in many cases increases dream recall and causes dream material to become more bizarre. Many abductees also have vivid nightmares.

Dreaming is a fascinating phenomenon. During sleep the body goes through a cycle consisting of alternate periods of light sleep, deep sleep, and another type of sleep called rapid eye movement, or REM, sleep. We do much of our dreaming during REM sleep, which occurs just about every ninety minutes. During REM sleep, hands and feet may twitch, and eyes dart around under the lids. While we dream, muscles are relaxed, and the body becomes detached from the mind.

As previously mentioned, research suggests that information gathered during the day is categorized and stored in long-term memory during REM sleep. In 1978, Howard Rothwarg at Montefiore Hospital in New York conducted a fascinating study that helped explain how what we do while awake is incorporated into our dreams at night. During the day, nine college students wore goggles that colored everything red. The subjects soon became acclimated to a "goggle colored," red world. On the day that they started wearing the lenses, all of their visual input was "tagged" with color.

Rothwarg believed that information processed during REM sleep would include the colored goggle world input. Each night the subjects slept in a sleep lab, where their EEG's and eye movements were monitored. They were awakened during REM sleep and asked to report their dreams. Like most of us, the subjects experienced four REM periods each night.

As predicted, on the first night, red goggle material began to enter their dreams, but only during the first REM period. In that first dream, about half of the scenes contained red information. But on succeeding nights, red information crept into later and later dream periods. By the fourth or fifth night, all REM periods contained goggle-colored material. As the red color moved into later dreams, it also increased in the first dream period, so that by the fifth night the first dream period contained 83 percent red scenes.

This study showed that material from dreams later in the night comes from earlier memories, and suggests that past material is mixed with recent material during sleep. Based on these findings, it would not be surprising that a member of an abduction support group would have frequent dreams about abductions. As the days go by, and abductees spend more and more time thinking and talking about abductions, it is inevitable that abduction dreams become more frequent.

For most of us, a dream is immediately recognized as such. But before the age of five, we cannot tell the difference between reality, dreams, and fantasy. We see our dreams as real and become terrified of the monsters we see on television. By the time we reach school age, however, we form mental boundaries between reality, dreams, and fantasies. At this point scary television shows and monsters no longer frighten us. As adults we are even able to catalog memories of dreams as separate from memories of reality. However, for some these distinctions remain unclear. Some people never form boundaries between dreams and reality. Rather than a clear boundary, reality has a ruffled edge.

When boundaries fail to form, it becomes difficult for the person to tell what is real and what is not. To the person with boundary problems, REM dreams become reminiscences, and what has been imagined is interpreted as actual experience. The boundary impaired make statements like, "I don't know if I dreamed this or if it really happened."

Often, they attempt to make the distinction by categorizing the "intensity" of the dream. They say, "The dream was so vivid, it must have been real," even though there is no evidence that vivid dreams are any more real than uninteresting and less vivid ones.

Ernest Hartmann, who has spent many years researching this problem, has described these people as having "thin boundaries." A thin-boundary person lacks the normal boundaries between reality, fantasy, and dreaming. Because of this they are not able to tell reality from fantasy. These people are prone to nightmares, and also prone to paranormal and unusual experiences.

It is clear that many abductees fit the description of those with severe boundary problems. A few of them actually cut themselves repeatedly in an attempt to feel the boundary of their skin. Some of these people say that cutting themselves is the only way they can feel that they exist. This is surely a severe boundary problem.

Therapists have talked about boundary problems for years. Many theorists feel that boundary problems originate from an "enmeshed family," in which personal boundaries are often ignored and violated. In these families, everyone is intrusive. Personal space and personal property are constantly invaded. A person in this type of family is never given a solid sense of identity and grows up with confusion between inside and outside. The classic statement of an enmeshed family is, "I'm cold, put your sweater on."

One lady entered therapy with the opening statement, "I don't know if I'm me or my mother." Many people with poor boundaries cannot tell the difference between their dreams and their actual memories. Most of these people also have sleep disorders. This set of symptoms *is not a mental disorder,* it is an artifact of character structure.

People living in families like this have experienced boundary violations since they were born. They often find it difficult to know who they are, what they feel, or what is real. Because they have such a poor sense of self, they often have trouble establishing and maintaining relationships.

The garbage pail, pop-psych term for this today is *codependence* (a term that originally meant something very different). Thousands of people with boundary problems find themselves in long-term therapy or in "recovery" support groups. Most of them never leave. Instead of getting well, they re-enmesh themselves into the group. The label of codependent, abuse victim, or abductee becomes a new way to identify oneself. But to maintain their membership, and their identity, these people must continue to produce material that fits the belief system of the therapist and the group. Not surprisingly, much of the material comes from their dreams.

It is important to note that alien abduction dreams revolve around themes of boundary violation. The dreams of abduction and medical experiments become both a metaphor for the early violations and a vehicle to express the feeling of helpless violation. For an "abductee" with poor boundaries, dreams may become the vehicle that leads him or her into a life of endless therapy and constantly increasing dreams of abduction. As the dreams inevitably increase in frequency, the abductee becomes more and more convinced that his or her experiences are real, and that they are escalating. Soon the dreams become indistinguishable from actual experience; fantasy thrusts its way into reality. It is no accident that Budd Hopkins called his book about abduction *Intruders*.

In the first of the American abductions to receive national publicity, the problem between dreams and reality can be readily seen. Betty Hill, after the horrifying experience along the New Hampshire highway, became obsessed with UFOs. Although she had no real interest in flying saucers prior to this event, she had discussed the topic with her sister, who claimed to have seen one. After her own experience, Betty Hill began reading books about UFOs, contacted Major Don Keyhoe at NICAP, and became the subject of a long-term, ongoing UFO investigation.

What is important, however, is that both Barney and Betty Hill remembered, consciously, seeing the UFO along the highway. As they drove, they argued over the identity of the object. They stopped so they could see it better, and at one point Barney left the car with a pair of binoculars and returned convinced he and his wife would be captured.

They arrived home worn out and convinced that the trip had taken much longer than it should have. There was a period of "missing" time. Over the next few days, Betty Hill had a number of disquieting dreams concerning alien creatures and abduction. According to Fuller, writing in *The Interrupted Journey*, "Some ten days after the sighting, Betty began having a series of vivid dreams. They continued for five successive nights. Never in her memory has she recalled dreams of such detail and intensity. They dominated her waking life during that week and continued to plague her afterward. . . . When eventually she did mention rather casually that she was having a series of nightmares, Barney was sympathetic. . . ."

Fuller continued, "Realizing that Barney was attempting to put the UFO event out his mind, Betty refrained from discussing the nightmares with him. But she began telling a few close friends, one of whom was a fellow social worker, who urged her to write down her dreams."

Fuller then revealed what Betty Hill was writing about from her dreams. "Her dreams were unusual in subject matter and detail. They revealed that she encountered a strange road block on a lonely New Hampshire road as a group of men approached the car. The men were dressed alike. As soon as they reached the car, she slipped into unconsciousness. She awoke to find herself and Barney being taken aboard a wholly strange craft, where she was given a complete physical examination by intelligent, humanoid beings. . . . They were assured, in the dream, that no harm would come to them and that they would be released without any conscious memory of the strange happening."

What this does is reinforce the triggering mechanism for a belief in alien abduction. Betty Hill certainly reported that. There was no conscious memory of the abduction and the information came to her through dreams. Once again, we find a number of researchers accepting the dreams as reality. Although there is no reason to believe that these specific dreams reflected reality, that point has not been questioned.

Later, under hypnosis with Dr. Benjamin Simon, both Barney and Betty, with sufficient emotion to impress researchers, told the tale of abduction. But it should be clear to everyone, especially those who read the introduction that Simon provided to *The Interrupted Journey,* that he was interested only in treating the psychological problems of his patients. The introduction gives no indication that he believed the tales, first described in Betty Hill's dreams, were true.

Validation of Betty Hill's dreams came from Benjamin Simon. His work with her, and her husband, suggested to her that something real had happened. Here was an authority, a medical doctor, who was listening to her tales but not trying to convince her that they weren't real. Instead, he was attempting to treat what he, Simon, believed to be a psychological problem. His course of treatment did not require that he accept the dreams as reality, though many have since done that.

Of course, it can be argued that because Betty Hill was interested in finding out what had happened, and because she was deeply concerned with it, that anxiety manifested itself in her dreams. But that in no way translates into a dream being grounded in reality. We again we crossed the boundary between the real world around us and the unreal, fantasy world of our dreams.

It is necessary to explore, at this point, the Allagash abduction. It, like the Hill case, is multiple witness. Four young men, on a fishing and camping trip along the Allagash Waterway had, at first, what they believed were two sightings of a UFO. Under hypnotic regression later, it became clear that they had all been abducted, and they all told a similar story.

Raymond Fowler, author of *The Watchers* (in which he details his own abduction), met Jim Weiner prior to speaking at a PSI symposium held by the Universal-Unitarian Church in May 1988. Weiner had attended the conference to search for someone who was knowledgeable about UFO abductions on the advice of his doctor.

According to Fowler in *The Allagash Abductions*, Weiner had been plagued by nightmares for a number of years. The dreams centered around strange creatures looking at him, temporary paralysis, and a manipulation of the area around his genitals. Weiner told Fowler that he'd seen a UFO close at hand, and that his brother and friends had also been having similar nightmares. Importantly, Weiner said that he had read Whitley Strieber's *Communion* and believed that Strieber's experiences mirrored his own. He also mentioned a period of missing time, the additional witnesses to the abduction, and the nightmares about the alien creatures.

After a meeting with Weiner, Fowler decided that the case demanded an investigation. He arranged for Weiner to undergo hypnotic regression in a number of sessions. While under hypnosis, Weiner described their first sighting of the UFO, which Fowler believes might have been Jupiter. Two nights later, an object appeared overhead, coming closer to the four men in their canoe. It hovered above them, played a light across them and the boat, and then disappeared. Weiner told Fowler they had built a large fire on the shore as a beacon for themselves, but when they returned to the beach after the UFO had disappeared, the fire had burned down to glowing coals, suggesting to Weiner that an hour had passed for which they could not account.

During that first hypnotic regression session, Weiner merely related the story of the sighting. Weiner expressed his surprise that the fire had burned faster than they thought it should have. Fowler asked, "If there is time unaccounted for, why can't you remember?"

Weiner replied, "I don't know. I, I'd swear I remembered everything that happened."

Although Fowler claimed that he, along with his fellow investigators, takes great pains to refrain from leading questions, Fowler's demand to know why Weiner couldn't remember anything else is, in fact, a leading question. It told Weiner there was something else that happened and he should be able to remember it.

At that point, Fowler, and the hypnotherapist Tony Constantino induced

Wiener into a deeper state of hypnosis and began probing again. Constantino then said, "You sense there is something you should remember?"

Again, this is leading. The hypnotist is telling the subject there is something more to remember. The hints have been planted by the conversations held earlier as well as Weiner's introduction to alien abduction by reading Strieber's work.

The questioning continued. Weiner finally described events on board the craft but refused to describe the faces of the alien beings. It was clear to the men conducting the research that much information was being suppressed by Weiner.

At the conclusion of the first session, Fowler was "stuck by the emotions that Jim manifested when reliving the experience." He also wondered if "the aliens programmed them not to see their faces or whether the sight of them caused self-imposed amnesia."

Fowler also told Weiner that they would talk more about UFO abductions once the investigation was over. Fowler advised him not to talk to his fellow abductees because they didn't want to taint the investigation. This warning issued after eight years of discussion among the four men and the fact that Wiener had already read Strieber's work was a bit late. Contamination had already taken place.

Fowler, after the second session with Jim Weiner, in which he described the faces of the creatures for the first time, wrote, "I remembered Jim telling me about a series of lifelike nightmares involving bedroom visitations by alien creatures and paranormal happenings."

According to Fowler, at the end of that session, "Out of curiosity, I snapped a quick question at Jim. . . . When did you see the creatures again?"

"In Texas."

"What year was that?"

"I think it was 1980."

Is it necessary to point out that Fowler's quick question, snapped at a witness under hypnosis, was leading. Fowler, in his own words, was asking Weiner when he saw the aliens again. He was telling the subject that he had seen them again and Weiner, obligingly, provided an answer. The question obviously suggested to Weiner that he had been abducted more than once.

What we end up with here is another series of abductions that began as vivid and realistic nightmares. The four men on the camping trip saw what they believed to be a flying saucer, and they discussed it among themselves for years. At least one of them read Strieber's book, which contained the descriptions of the alien creatures and the details of an alien abduction.

What is interesting is that when Weiner was asked to tell the researchers how tall the aliens were, he gave the standard of five feet. Yet when he drew the encounter after hypnosis, he showed an alien creature that was taller than he while both were standing. It means that Weiner is either shorter than five feet tall, or that the aliens were taller than five feet.

It should probably be pointed out that the drawings of the aliens made by each of the men do nothing to corroborate one another. While generally similar, that can easily be accounted for by the information circulating throughout the media. By 1988, when Fowler began his investigations, the covers of a number of books had already published the stylized alien abductor. In fact, the Avon paperback edition of *Communion* was published in February 1988, which means it was available in bookstores by the middle of January. The cover shows an alien head with a pointed chin and big, dark eyes. What should have surprised Fowler is that the drawings of the four men didn't agree more closely.

Once again we have a witness, Weiner, who is provided with the validation for his memories by the UFO investigators. What began as nightmares suddenly became real memories buried in the subconscious. While the conscious refused to acknowledge the events, the subconscious was forcing them to the surface in the form of nightmares. In each case, we are crossing the boundary between dreams and reality.

The problems of Leah Haley, described in detail later, reflect much the same thing. She began with strange dreams, but friends and family convinced her they were repressed memories of "real" events. She sought professional help, and the professionals, Budd Hopkins and John Carpenter, pushed her along the alien abduction road with books, articles, and videotapes. Carpenter added to the confusion by confirming for her that the theories of alien abduction told by others were accurate. According to Carpenter, the dreams of Leah Haley were "real."

This idea, that dreams can trigger a belief that an abduction has taken place, is not confined to Betty Hill, John Carpenter's research with Leah Haley, or the Allagash affair. It was driven home even in a more direct fashion at the UFO conference hosted by Bill Pitt and held in Fort Smith, Arkansas, in 1975. A man, who claimed to have been abducted, agreed to undergo both hypnotic regression and a polygraph examination.

The story the man told prior to the hypnotic regression was that in November 1953, while working for the railroad, he had seen a UFO. He had been driving on back, county roads, and was on an incline leading to a railroad crossing when the engine of his truck died. Remember, he was telling this tale in 1975, two years before *Close Encounters of the Third Kind* was released.

A bright light flashed on the other side of the tracks and he leaped from his truck before losing consciousness. When he awoke, sometime later, he was apparently strapped to some kind of table. Near him was a mirror-like device in which he could see his body's organs. He could see his heart beating and his lung expanding. As noted earlier, such descriptions sound suspiciously like a scene from *Killers from Space.*

He described the tiny, alien creatures as having small chests and being covered with a greenish, rough skin. There were no features on the face, but he noticed two holes on the side of the head. There was strip of stiff hair on the top of the head, reminding the witness of a Mohawk Indian. The creature's hands had only three digits, a thick thumb and two fingers.

After twenty minutes or so, the man lost consciousness again and woke up next to his truck. The bright light was gone, so he climbed into the cab. The engine, apparently stalled by the UFO, started easily. He rushed back to the railroad yard, but when he returned to the site, with friends, there was nothing to be seen.

A police sergeant conducted the polygraph examination. In three tests, the sergeant received a reaction on the critical questions. Although the sergeant was careful, the witness, according to him, failed all three of the examinations. In other words, the UFO witness had not passed any of the polygraph tests.

The hypnotic regression test turned out no better. After an exhausting two-hour session, the investigator concluded, "He may have had an experience at one time, but it is now so buried in dreams, projections and sublimations that we might never get to the real experience, *if there was one* [emphasis added]. There is no way that we could get to the facts."

An important aside is the reaction of the conference audience to this tale. There were three men on the stage. There were questions directed at both the police sergeant and the hypnotist, but those were asked in a way to suggest that no one believed them. The audience was trying to learn what sort of mistakes those two men might have made that would color their conclusions. It was clear that the audience believed the abductee, who had failed both the polygraph and hypnotic regression tests. They were more interested in a alleged tale of alien abduction than they were in the truth as outlined by disinterested investigators.

Once again, we have an event that is wrapped in the world of dreams. Here was a man who claimed to remember his abduction without the aid of hypnosis but who apparently got the tale from his dreams and the pop culture around him. We must not forget that important connection to the abduction phenomenon.

It might be claimed, and sometimes has been, that these tales harm no one. Betty Hill's abduction, while frightening to her, didn't radically alter her life. The man in Fort Smith had an interesting tale to tell, but he did not become obsessed with it. The difference is what happened to Leah Haley. She is the woman who had gone to John Carpenter with her possible tale of abduction as detailed in her dreams. He validated it, drawing more information from her. Like the other abductees, her dreams became reality. It filled her life to the exclusion of all else. She became obsessed with the dreams and details of alien abduction. Her life would be radically altered by these beliefs.

Still another example of abductions from dreams came from Karla Turner, who said her abduction experiences also began with a series of dreams. As she described it, "On April 21 I dreamed I saw my husband sitting with a group of friends in a round environment. . . . On the twenty-second, I dreamed that the worldwide disaster or catastrophe had occurred.. . . .On the twenty-fourth, I began reading *Communion.*" On April 25, the day after reading Strieber's book, she dreamed she saw a UFO land. As she approached it, the ship exploded and she concluded, "I knew the government was responsible."

Turner said that this is the first time she had ever had a UFO dream. Although she had spent time in therapy, and routinely logged her dreams, she discounts the phenomenon of day residue. On April 27, she bought and read *Missing Time,* by Budd Hopkins.

Although she had no formal training, Turner began hypnotizing her husband, who produced some memories that resembled the abduction material she had read. For example, "He remembered once when he was thirteen waking up to see a strange woman, dark eyed with white wispy hair approach him in unfamiliar surroundings. She got on top of him, and engaged in sex. . . ." This is, in fact, a classic example of sleep paralysis, a phenomenon to be discussed at length later.

Alarmed by these and other memories, the couple decided to seek professional help and contacted a UFO organization. Although Turner was seeking help for her husband, the UFO people were more interested in her. "They insisted I talk about any unusual events or recurrent dreams I'd had." She told them of a childhood dream fragment where she was standing next to a large insect.

She was introduced to an investigator named Barbara Bartholic. Bartholic gave Turner and her husband her version of the abduction experience, including her belief that the aliens are crossbreeding with humans. Shaken by what they had been told, they returned home with a feeling that they were being followed.

Turner's research took her to other people. A friend, Megan, described a dream she had when she was ten or eleven. While she was taking a nap, she awoke to see a monkey outside the window. In another incident, she awakened to see visions on her bedroom wall. Although her experiences are quite common, and have nothing to do with alien abduction, Turner thought otherwise. "I didn't want to frighten Megan by telling her how much these things sounded like screen memories, protective disguises of events too frightening to face."

Continuing her research, Turner met a woman named Ellen who had been unsuccessful in having a baby. She'd had a series of miscarriages and finally decided to use a surrogate mother. Although Ellen was not involved in abductions, Bartholic asked Ellen if she had had any nightmares. Ellen described a nightmare where a strange woman was trying to steal her baby from the surrogate. After what Ellen had been through, it is not surprising that she would have a dream about her fears, but Barbara redefined the dream as an abduction experience.

As we look at the work completed by Turner, Carpenter, and the others who develop abduction scenarios, including the way they were recovered from memory, a pattern begins to emerge. Not the pattern that abduction researchers see, but one that requires nothing more than open eyes to understand. While abduction researchers will look only for the corroboration of a tale from one witness to the next, we see the contamination from one researcher to the next. They overlook common and well-established psychological phenomena such as sleep paralysis because they don't fit into their belief structure. Classic examples of it are turned into abductions. Dreams that have no basis in reality are turned into screened memories, or repressed memories of actual abduction. Validation is offered by the "experts" and the arguments then become that these people couldn't be imagining these events. The all match too closely.

Yet here, in these cases, that is exactly what is happening. Barney and Betty Hill scared themselves by convincing themselves they were being paced by a flying saucer. They stopped repeatedly on their long drive home, arriving after being awake all night. Their trip took longer than it was supposed to, but *only because they stopped repeatedly* to study the flying saucer. Neither had any memory of the abduction until Betty had her dreams.

Episodes of sleep paralysis, which are often accompanied by the manifestation of some kind of creature, are sometimes incorporated into these fantasies. But rather than examine the established psychological literature, these tales are turned into reality by those who want to believe in alien abduction. There was

THE DEFINITION OF SLEEP PARALYSIS

It turns out that there is no single, simple definition of what sleep paralysis is. Sleep researcher K. Fukuda wrote, "Sleep paralysis has usually been described in terms of its association with narcolepsy." And Jorge Conesa wrote that when sleep paralysis occurs without the additional symptoms, it is known as isolated (idiopathic) sleep paralysis.

A precise definition of sleep paralysis was written by R. J. Campbell for his dictionary of psychiatric terms. "A benign neurologic phenomenon, more probably due to some temporary dysfunction of the reticular activating sys-

nothing presented in any of the accounts, whether it was Barney and Betty Hill, Leah Haley, Karla Turner, or the Allagash four, to suggest the abductions were real events in the real world. The stories came out of dreams, discussed by those having them. The triggering mechanism that abduction investigators search for is provided in the interpretation of those dreams. Without the dreams in these cases, there would be no tales of abduction.

In fact, psychologist Edith Fiore sees this herself. In the preface of her book *Encounters* Fiore sets the stage for many to begin manufacturing abduction memories and to interpret dreams as reality. "As you read *Encounters* it would be helpful for you to keep a notepad nearby and record anything that you suspect as being evidence of an encounter. . . . Mark the margin or highlight the material and record the page number so you can check back later when you start exploring your own possible contacts. You also may start remembering seeing lights in the sky, or recall a dream you had totally forgotten. This book could bring closer to the surface of your conscious mind experiences that have been deeply buried for years."

Fiore herself suspects that she was abducted. Her experience began in a dream. "In 1979 I joined a UFO study group and became intensely interested in extraterrestrial visitations. . . . It was during this period that I awoke one morning, feeling particularly good, and remembered a fragment of a dream. It was vague and fuzzy, but I did recall being in a 'room' in a UFO." She shared her dream with Dr. James Harder, a civil engineering professor in California, who told her, "Dreams are very commonly the tip of the iceberg of a meeting with ETs."

Harder hypnotized Fiore, who quickly recovered an abduction memory. However she was not convinced of its reality, and made a very astute observation. "I had read a great deal on the subject and done so many regressions that I could not separate fact from fiction."

tem consisting of brief episodes of inability to move and/or speak when awakening or less commonly, when falling asleep. There is no accompanying disturbance of consciousness, and the subject has complete recall for the episode. The incidence of the phenomenon is highest in younger age groups (children and young adults) and much higher in males (80%) than females. It occurs in narcolepsy. . . . The terms by which the phenomenon has been known are nocturnal hemiplegia, nocturnal paralysis, sleep numbness, delayed psychomotor awakening, cataplexy of awakening and post dormital chalastic fits."

So another part of the question has been answered. There are those who remember their abduction without the aid of hypnosis, but many of those "conscious" memories first surfaced in dreams. These witnesses are later convinced that the dreams are real and not fantasy.

And once the "authority figure" validates the claim, rational thought is ended.

PART 3

THE RESEARCHERS

9.

MARSHALL HERFF APPLEWHITE, AKA BO, DO

We should begin by drawing a heavy dark line between the cult status of those following Marshall Applewhite and those who conduct abduction research or claim to be the victims of the aliens. And we should also note that the line between the contactee phenomenon and that of alien abduction is very fine. What we have is a dichotomy of beliefs and evidence that is sometimes nearly impossible to detect. That is not to say that those who practice and conduct abduction research should be lumped in with those who establish cults that use alien contact as the source of their inspiration. It is merely to note that there are some similarities, but more important, some obvious differences.

The definition of a cult is neither positive nor negative. It is simply the zealous devotion to a person, ideal, or thing. It can have a religious significance and many cults do, but that is not a requirement. Instead, there is just a belief structure that might not be grounded in reality and an obsessive devotion to that belief and those who hold it.

Many of those in the UFO community, especially the contactee community, have been involved in cults. The followers of George Adamski, George van Tassel, and other such contactees held an almost religious devotion to their leaders. They believed everything their leaders said, often provided them with money and expensive gifts, and listened as they spun tales of trips to other planets and other worlds.

As noted earlier, much of the contactee movement collapsed as our space program revealed the true nature of Mars and Venus. When both worlds were found to be barren wastes with no traces of advanced civilization, many followers abandoned the contactees. There were always some who held on, but by the 1960s, the contactee movement, for the most part, was finished.

One small segment of it was revitalized in the mid-1970s when "The Two" began their missionary work on behalf of the Earth. Their messages and missions were found in the old teachings of the contactee movement, but that made little difference to their expanding audiences. There are always more people wanting to go along for the ride, and hundreds did.

"The Two" were finally identified as Marshall Herff Applewhite and Bonnie Lu Nettles. They had met in 1972 at a Houston hospital where Applewhite was being treated for psychological problems and Nettles worked as a nurse. Both were highly educated and Applewhite had once taught music at the University of St. Thomas. He had also been a director of music at the St. Mark's Episcopal Church. They left Houston together in 1973.[1]

In July 1974, The Two arrived at the International UFO Bureau headquarters in Oklahoma City to speak to its director, Hayden Hewes. They believed that Hewes could help them spread their message to the rest of the world. They told Hewes that they would prove they were "real" because they would, one day, be assassinated and after three days would return to life in full view of the news media.

In the beginning The Two were just representatives of the alien creatures who were on Earth to lead the human race to the next evolutionary level. Again, this would all be proven after their assassination and resurrection in front of the media.

Within months of that first meeting a letter from Human Individual Metamorphosis (HIM, which would eventually evolve into Heaven's Gate) began to circulate announcing that the prophets from the next kingdom, as important as Jesus, were now present on Earth. They were circulating throughout the United States to spread the word.

In early 1975, Applewhite and Nettles, now calling themselves Bo and Peep, made what many consider their first public appearance. In Los Angeles, at the home of psychic Joan Culpepper, they told all present of what they believed. Although they refused to provide any sort of personal history, and their insights were no more instructive than those of the contactees of a couple of decades earlier, twenty-four of those who listened that night decided to join Bo and Peep. They began to wander through the West, holding meetings in several locations in California and Colorado.

At some point that spring, they arrived in Iowa and camped on Kevin Randle's doorstep. As they had done with Hewes months before, they selected him because he had been teaching a class about UFOs at the local community college. They wanted his help in setting up and publicizing a lecture.

Randle, however, was reluctant. He listened as they explained their philosophy about UFOs and their mission. Their message to Randle was no different than that he had heard on a number of occasions, or read about in the books

---

[1]Jerome Clark, *High Strangeness: UFOs from 1960 through 1979* (Detroit: Omnigraphics, 1996), 477–80; Michael Miley, "Return of 'The Two,' " *UFO,* May/June 1996, 26–31; Brad Steiger, *Gods of Aquarius: UFOs and the Transformation of Man* (New York: Harcourt Brace Jovanovich, 1976); Brad Steiger and Hayden Hewes, *Inside Heaven's Gate* (New York: Signet, 1997).

about and by contactees. Bo and Peep had no proof of their claims, other than their word that they were " 'sent' from that kingdom by the 'Father' to bear the same truth that was Jesus'. This is like a repeat performance, except this time by two (a man and a woman), to restate the truth Jesus bore, restore its accurate meaning, and again show that any individual who seeks that kingdom will find it through the same process."

Although the three talked through the afternoon, Randle didn't offer any assistance to them. On their own, they did set up a lecture at a local college. Randle and about two hundred others attended, but the audience was hostile, demanding evidence, logic, and proof. Those who lectured, disciples of Bo and Peep, were not able to provide persuasive arguments or the physical evidence being demanded by the audience. When the lecture ended, they left, but no one joined them.

But other people in other parts of the country certainly did. At a lecture held on September 14, 1975, in Waldport, Oregon, twenty people, of the audience of three hundred, disappeared after the meeting. The media became interested at that point because no one seemed to know who the mysterious couple was or where they were going. The UFO angle, mentioned during the lectures, was played up by the media. One member of the audience told reporters that the message delivered had been vague, but the implication seemed to be that they might leave on a UFO.

Robert Balch and David Taylor, sociologists from the University of Montana, joined the group some six weeks after the Oregon meeting. By the time they arrived, however, Bo and Peep were gone. They had learned their group have been infiltrated by two men who were looking for a friend who had disappeared in Oregon. Bo and Peep claimed they feared assassination before their mission was complete.

Followers saw no inconsistency in their fear of assassination and believed that Bo and Peep would be martyred, rise from the dead three days later, and ascend into heaven. Within days of the deaths of Bo and Peep, the devout followers, those who had divested themselves of their worldly possessions, their attachments to human emotions, and the material world, would be carried away in a flying saucer.

Before they left the group of some 150 to 200 believers, Bo and Peep broke them into smaller groups to send them out with the "word." These smaller groups, made up of fourteen to twenty people, were equally divided between men and women, with each member assigned a partner, typically of the opposite sex. They had now established the core of their "movement."

Interestingly, the members of the groups were forbidden to engage in romantic, sexual, or friendly relations. They were to be together twenty-four hours a day and were to become familiar with the human qualities of the other. But they were not to see each other as sexual partners. They were not to communicate on a personal level.

Nick Cooke, a former member of the group that would eventually be known as Heaven's Gate and who was interviewed by Brad Steiger in 1997, noted that he still supported the beliefs of Applewhite about abstaining from sex. Newspaper stories mentioned the fact that many of the male members had been castrated. This fit with the asexual path that had been preached by Bo and Peep from the very beginning.

The small groups, these "families," wandered through the United States, living on food that they found through begging and slept outside each night. There was little communication among the groups, but periodically they would stop to hold meetings in an attempt to recruit new members. Some of the older members dropped out, but they were always replaced with other people eager to join. Balch and Taylor reported that in one small Arizona town of five hundred, eight residents joined the group. Mainly, however, it was a repeat of what Randle had witnessed in Iowa. The audience made sport of The Two and their disciples.

Hewes, because he had been the first of the UFO researchers to be visited, along with colleague Brad Steiger, was selected by Bo and Peep to help publish their manifesto. This was shortly after the media flap around the Waldport Oregon disappearances. Steiger, in an updated version of that book, wrote, "On the next occasion when Bo and Peep appeared in Hewes' office, it was to announce that we [Hewes and Steiger] had been chosen to present their teachings, their creed, their extraterrestrial manifesto, to the world."

The timing was perfect. Steiger noted, ". . . as destiny would decree it, one of my editors called to say that there was publishing interest in the mysterious Two who were luring men and women away from their families to join a UFO cult. If anyone could track Bo and Peep down, my editor expressed confidently, it would be me with my numerous contacts in the paranormal-UFO field."

Steiger traveled to Oklahoma City to meet with The Two. He noted that the media attitude had turned negative. The cult status of their group was being underscored. They were breaking up traditional families, as one spouse would join and the other wouldn't. And to make it worse, the orthodox religious community was outraged by what they considered the blasphemous teachings of Bo and Peep. The Two were claiming to have divine guidance and were also claiming to be the Second Coming. Their mission, they claimed, was the same as that of Jesus.

But The Two were not satisfied with the book that Steiger and Hewes wrote. To their credit, both Steiger and Hewes wanted to provide an opposing point of view. They didn't want to just regurgitate the ramblings of The Two. The alternative information that Steiger wanted to add did not sit well with either Bo or Peep.

According to Steiger and Hewes in *Inside Heaven's Gate*, "Although we played fair by presenting a large portion of their teachings exactly as they were written in their original manuscript, editors deemed much of the text repetitious and deleted a number of pages of their manifesto."

Just before the publication, Hewes was visited by The Two and several of their followers. They claimed that the final document was not what they had hoped for, so the demonstration, that is, their deaths by assassination and their subsequent resurrections, would not be carried out on the publication of the book as planned. That surprised few in the UFO community.

In October 1976, while Steiger was in Washington, D.C., to do a television show, he met with other members of the group. Steiger made it clear that the confrontation was nonviolent. It was just another attempt by the group to explain their disappointment with the book. Apparently membership was down and they were going to close the doors to the next level. Fewer than one hundred of the Earth's five billion population would be saved when the time came.

It was about this time that Bo and Peep retreated to a mountain camp near Laramie, Wyoming. Followers were provided with uniforms and training to prepare them for their trip into space. Within months, they had moved to Salt Lake City.

In 1979, Bo and Peep sent out an emissary. Paul Groll said that the group divided its time between the Wyoming Rockies and north Texas. Followers were to monitor one another looking for violations of the rules set down by Bo and Peep. Other than the words "yes," "no," and "I don't know," the members communicated only through written messages. When the flying saucer landed to pick them up, they would move to a realm where they would live forever in bodies with neither hair nor teeth.

Steiger reported that in 1985 he had "learned that Bonnie ('Peep,' then called 'Ti') had died of cancer and that Marshall, ('Bo,' now 'Do') had carried on their mission of informing Earth's humans that salvation hovered overhead in a spaceship." Bo, or Do, apparently did not explain why Peep, or Ti, had not been resurrected.

Do, however, carried on the work that he had started with Ti. The group was renamed the Total Overcomers, and only occasionally gave public testimony. They continued to warn of the imminent destruction of earthly life and of the plans of evil extraterrestrials.

Steiger reported that in May 1993 he noted an advertisement in *USA Today* that declared Earth's civilization was about to be recycled. He thought that it sounded like Applewhite. He wrote for information about the group but received no reply. In *Inside Heaven's Gate,* he wrote, "I was undoubtedly still persona non grata after the publication of *UFO Missionaries Extraordinary* [the title that had been given to The Two's manifesto]."

On March 26, 1997, Applewhite and thirty-eight of his followers finally launched themselves into the next realm. It was their last, narrow window of opportunity, apparently triggered by discovery of and then discussions about comet Hale-Bopp. For some reason, Applewhite believed that the comet itself, or an object alleged to be following in its trail, was the spaceship for which he had waited. He shed his "earthly" container so that he could be "beamed" aboard the alien craft, as did thirty-eight followers in a mass suicide.

The question that can be, and should be, asked is how does this all relate to alien abductions? Here was a man who was deluded into believing that he had some special knowledge about the alien presence. He didn't claim to be an abductee, only that he had been in communication with the aliens who have been visiting. . . . Or rather, claimed that he was one of those aliens who had been sent here on a mission to save us all.

We should not, however, confuse Applewhite and his "teachings" for anything more than what they were. Applewhite was not divinely inspired, nor was he in communication with alien intelligence. In a world that has become cynical, a belief in alien beings is easier to support than one that relies on fairies, devils, demons, and angels.

Applewhite knew no more about UFOs than the average man on the street. The connection to UFOs was made for the publicity it garnered. Without that specific link The Two, Bo and Beep, Do and Ti, and Heaven's Gate would have been nothing more than so many other small cults circulating in today's environment. Without the UFO connection, the news media would have ignored him until March 1997, when he inspired thirty-eight of his followers to join him in a mass suicide. Without the UFO connection, Applewhite would have been ignored completely.

10.

DR. RICHARD BOYLAN

It has been argued by many that the similarities of abduction told by witnesses in widely separated parts of this country and around the world suggest there is something real happening. How could abductees who don't know one another, who claim never to have read a book about abductions or watched a television program about it, or who claimed no interest in UFOs in general, tell stories that are so similar if the abduction experience isn't real? Doesn't the lack of diversity of tales told by the witnesses suggest the abductions are being accomplished by alien beings from outer space?

Part of the answer to that question lies at the feet of the hypnotherapists, psychologists, and UFO investigators who have been conducting the research.

Dr. Richard Boylan, onetime licensed psychologist in the state of California, and one of the founding members of Academy of Clinical Close Encounter Therapists (ACCET), which taught therapists how to treat abductees, has been accused of implanting such memories. Patients who came to him for treatment for other psychological problems, including satanic ritual abuse, were eventually convinced they had experienced alien abduction instead. In two cases reviewed by the Board of Psychology, Department of Consumer Affairs, for the State of California, evidence showed that Boylan's personal beliefs were imposed upon his patients.

In the board's "Findings of Fact" it was reported that Boylan had treated a woman identified only as D. W. She had been a member of Incest Survivors Anonymous (ISA) and was diagnosed as suffering from a variety of conditions, including post-traumatic stress disorder, dissociative disorder, depressive disorder, and a number of substance abuse problems that were all in remission when she first visited Boylan.[1]

---

[1]The "Findings of Fact" is a public record of the testimony given during the licensing review by the State of California. On August 4, 1995, the Board of Psychology, Department of Consumer Affairs, State of California revoked the license of Dr. Richard Boylan. The documents are available from the state California and is marked as No. W-14, and OAH No. N-9404129. Other data for the chapter is from Boylan's Internet postings and letters.

During the early sessions with D. W. Boylan advocated the use of hypnosis to retrieve memories of abuse, but she was reluctant to try it. At their fifth session, however, Boylan convinced her of the benefits of hypnosis. At that session, using hypnotic regression, they uncovered sexual abuse at age four by D. W.'s mother and at age five by her father. According to the "Findings of Fact," more hypnosis produced more memories of sexual abuse and then the possibility of ritualistic abuse. To this point in her treatment, there had been no mention of UFOs or alien abductions by either Boylan or D. W.

It was D. W. who eventually brought up the subject of alien abductions, but only because she had heard a radio talk show that discussed hypnosis and reports of UFOs. At the first session after listening to that program, she asked Boylan about it, thinking that these people had been abused as children and were confused about the memories. Boylan, however, took the question seriously, suggesting that he thought there might be something real to to these reports of alien abduction.

Boylan then gave her *Encounters,* a book by Dr. Edith Fiore that describes alien abductions in a positive light. He also told her that he was about to tour various southwestern sites where aliens or their craft have been reported by many people. Boylan told her he believed there was a government cover-up concerning these events, and even mentioned Area 51, which supposedly houses captured and recovered wreckage of crashed UFOs. Government scientists and military officers were working there to understand the dynamics of the captured alien spaceships. Boylan also said that during the trip he was *afraid that he would be abducted by aliens* (emphasis added).[2]

The idea of alien abduction was now firmly implanted in the mind of D. W. Boylan didn't bring the subject up himself initially, but his belief in alien abduction and other theories surrounding the government conspiracy made these concepts seem all the more realistic to her. They were no longer theories to be ignored but something that was happening in the real world outside his office.

Not long after Boylan's return from his southwestern trip, D. W. told him in a private session that she remembered a "daydream-type image or fleeting memory from childhood of a strange-looking man, who might have been a molester." Boylan asked for a description, but D. W. said that the man wasn't clear to her. She remembered that he wasn't tall, had sharp features and narrow eyes. She tried to draw the man for Boylan, who apparently thought that she was attempting to describe an alien being for him. She had told Boylan the man

---

[2]C. D. B. Bryan, *Close Encounters of the Fourth Kind: Alien Abduction, UFOs, and the Conference at M.I.T.* (New York: Alfred A. Knopf, 1995), 129–31, 162, 165–82, 196–7, 237–49.

had narrow eyes. Large, black, almond-shaped eyes with no pupil are the most dramatic of the alien features, according to many of the researchers.

Under hypnosis, Boylan had her again describe the man. As Boylan questioned her closely, she "started getting an image of that like an extraterrestrial face and this really bright light." When she came out of the hypnosis, Boylan was smiling. "He was leaning back in his chair and smiling. And I remember feeling really weird. I was really confused and scared," she told the California licensing board.

Boylan told her that he was starting a support group at his home for people who had had extraterrestrial experiences. Boylan said that he believed the invitation he extended to her might have been premature, but he wanted D. W. to attend some of the sessions. This group might help her understand what had happened to her.

Shortly after the hypnosis session, D. W. attended her first alien abduction support group meeting. Interestingly, she recognized one of the members from her Incest Survivors Anonymous group. During the meeting Boylan played a tape that told the group about the extraterrestrials and who was likely to become a victim of alien abduction.

For over six months, she continued attending the meetings of the abduction group and continued her weekly or biweekly treatments with Boylan. With alien abductions in mind, she told Boylan of a nightmare she had experienced repeatedly since childhood. She dreamed that she was floating down a hallway and she saw a monster. She was aware now that those in the abduction group had had similar experiences.

Under hypnosis she provided more details of her dream for Boylan. According to the "Findings of Fact," she said, "Then the monster became an ET, and then there were other extraterrestrials. And then I was, like, walked out of my home where I saw some brilliant lights out on the front lawn."

The change was now complete. According to D. W., she began to identify with the abduction group and her belief in childhood molestation and ritualistic abuse had been pushed to the side. She believed that she might have had an abduction experience. The "Findings of Fact" reported that notes from her therapy sessions "show increased ET references, along with continued references to ritualistic abuse."

In the fall of 1992, Boylan told his abductions group that there was a UFO conference to be held in Las Vegas. He would be speaking there, presenting information from his research. He invited the members of the group to attend with him. He added that if they were willing to share their experiences with those attending the conference, their expenses would be paid. A number of the group agreed to go.

The history of D. W. now paralleled that of many abductees. She had gone from no conscious memory of any experience of alien abduction to attending a UFO conference to tell others of her experiences. Her memories were based on dreams and guidance under hypnosis by Boylan. She had arrived at his office for therapy because she believed she had been molested as a child and now she believed she was an abductee, taken from her home by alien beings. The process of change was subtle but the guidance was real. A dream, or nightmare, was transformed by Boylan into a real alien abduction. The coaching was very refined, almost nonexistent, but very real. Careful reading of the "Findings of Fact" demonstrates this.

Before leaving for the Las Vegas UFO conference, Boylan had members of the support group draw "visual representations" of the aliens. D. W. gave Boylan a drawing from "her experience," and in the last therapy session before the conference was told that she could attend, all her expenses paid.

Boylan offered D. W. an expenses-paid trip to Las Vegas. Before going to the conference, Boylan had D. W. provide a drawing that was sufficiently accurate, in this case that matched Boylan's beliefs. She already had been provided with all the information she needed to complete the task, from the books and articles Boylan supplied, to his theories about extraterrestrial abduction at the group sessions, and finally to the stories told by the other members of the abduction group.

Although somewhat irrelevant, it must also be noted that the last session prior to the Las Vegas conference was held at Boylan's home. This was to be a planning session for the trip, and that part of the planning would take place in Boylan's hot tub. He made it clear that suits were not optional because he "believed residual detergent in the suits left soap scum in the tub."

The trip to Las Vegas was made by a few chosen members. Two women, D. W. and another, shared one room, and Boylan, with a male patient, occupied another. Boylan and his group attended a number of sessions at the conference. Later, Boylan and D. W. made a trip out to the top-secret Air Force installation at Area 51, which is about a hundred miles north of Las Vegas. They became lost and confused during the drive. They were unable to get back to Las Vegas until early the next morning. At the hotel, they discovered that the man and woman who had not accompanied them on the expedition were already asleep in separate beds in one of the two rooms. Boylan invited D. W. to share his room with him.

According to the "Findings of Fact," "after the lights were out, D. W. began moaning in pain, apparently some form of gastric distress as a result of fast food eaten on the Area 51 drive. Respondent [Boylan] offered to give her a massage to help relieve the pains. Both D. W. and respondent [Boylan] were nude, although respondent was covered by the sheet. As D. W. came over to respondent's bed, she stated she did not want any sexual relationship. Respondent [Boylan] advised her that he did not want one because he did not want to risk his license, his marriage or his therapeutic relationship with D. W. He then gave her an abdominal massage. Afterward, he turned over and went to sleep, expecting D. W. to return to her bed. She did not and slept next to respondent until the alarm went off at 6:00 A.M."

It was also in the fall of 1992 that Boylan told his abduction group that he planned to write a book about their experiences. He and his wife would edit the completed manuscript, and the abductees would write the individual chapters about their experiences. D. W., now believing herself to be an abductee, wrote her chapter and submitted it. At a therapy session, Boylan returned the edited copy to her and asked for a revision.

In February 1993, D. W. stopped her therapy with Boylan and dropped out of the abduction group. She didn't submit the final version of her chapter, and when the book was published, her experiences were not included.

D. W. wasn't the only Boylan patient who showed up with one set of symptoms and developed a new set. D. S., who said she had been molested by a teacher at thirteen and raped at eighteen, also complained of chronic fatigue immune deficiency syndrome (CFIDS). She wanted to reduce stress in her life and be able to put behind her the sexual abuse and rape, difficulties from a car accident, and other personal problems. Boylan agreed to see her on a twice-weekly basis to help her cope with her problems and relieve her stress.

With D. W., Boylan didn't mention the alien abductions. D. W. learned of the phenomenon on a radio talk show and then asked Boylan about it, believing he would reject the notion as ridiculous. With D. S., however, it was Boylan who introduced the extraterrestrials into the sessions.

D. S., on February 21, 1991, told Boylan of "a recurring dream involving small figures which looked like monks at the end of a hall. In the dream, she tried to turn on lights, but none worked. She got angry because she could not see the figures and hit one of the 'monks.' After that, they all disappeared. Then the dream would repeat. D. S. had a second dream about a man, dressed in black, whose face she could not see. . . ."

About a year later, Boylan brought up the subject of extraterrestrials and dwarflike beings. She reminded him of her dream. Boylan then gave her an article, "The UFO Experience," that had appeared in the *Atlantic Monthly.* Boylan suggested that she read the article. He wanted to know if any of the events described were similar to her experiences.

D. S. thought of the monks as part of a dream, but Boylan told her it wasn't a dream. It was a real visitation. The monks she described were a type of alien that Boylan labeled "Jawas." Boylan also showed her photographs of drawings of aliens that he said were "Grays."

Boylan began to suggest that hypnosis would help her learn if her dream about monks was a real visitation. He told her that the aliens could cloud the minds of their victims, making them believe that the visitation was nothing more than a dream, that it was a pleasant experience, or that it never happened. The only way to learn the truth about the alien visitation was through the use of hypnosis.

When D. S. said she didn't want to be hypnotized, Boylan suggested that she attend his close encounters group sessions. Although she didn't keep her first appointments to attend, she finally did "because he kept bringing it up."

The first session she attended concerned Area 51, the government coverup, and other aspects of the UFO phenomenon. Boylan, during the discussion, suggested that his telephone was probably tapped (undoubtedly by government agents) because he was becoming vocal about alien abductions. Boylan, feeding the paranoia of his patients in the abduction group, suggested that others were being spied on as well. It was all part of the government plan to keep the information buried.

Not long after the group session, Boylan began to talk about publishing a book on alien abductions to "get the message out." D. S. told Boylan that she had nothing to contribute to such a book. Boylan reassured her, "Just write what you have. It's enough." Boylan provided her with a table of contents and a copy of the chapter he was writing about his experiences.

In the therapy sessions, Boylan told her about the benefits of writing the book, and laid out the risks as he saw them. By entering the "public eye" with his book, he would be calling more attention to himself and risked "silencing by the secret government."

In the "Findings of Fact" D. S. apprised the Board of Psychology, "He told me that they [the 'secret' government] have the ability to make a person die very quickly looking like it was natural causes. They have the ability to use a little dart that doesn't leave a trace. They can make a person die of cancer within a few weeks. That he would probably be the target because he was the leader."

It was also in the fall of 1992 that D. S. began to complain of difficulties eating and sleeping, and of a persistent ringing in her ears. Boylan told her that maybe the extraterrestrials were "doing a tune-up on her."

In early 1993, D. S. reported that she had had another dream. In this dream, she was awakened by a roar, got out of bed, and walked into her living room, where a pink light was coming through the skylight. She felt joy and thought that she saw Gandhi. According to her, after the dream, she felt peace and slept well the rest of the night. She awoke feeling rested and good about the dream.

Boylan suggested he use hypnosis to help her explore the dream further. Because she felt so good about it, she agreed, even when Boylan said that there was a strong possibility that extraterrestrials were probably responsible for it.

There is nothing like setting up the subject before beginning the session to make sure the results are the ones wanted. Boylan, in telling her about the strong possibility of ET involvement, was "priming" her for the hypnosis. Instead of making such a suggestion, Boylan, if he felt hypnosis necessary, should have confined his remarks to the facts as presented or detailed the capabilities of hypnosis. By suggesting the extraterrestrial intervention, Boylan had contaminated the session before it even began. It could also be claimed that Boylan had told her what he expected to find in the session. It was not a very subtle form of coaching.

Once Boylan had induced the trance, he moved her quickly through the dream to the point in the narrative where the pink light was coming through the skylight.

Here Boylan began bombarding her with questions. "Was it a ship? Was it metal? What kind of metal?" It was Boylan who introduced the idea of a ship into the session by suggesting the idea of a metallic ship hovering over the skylight. His pressure for details, and his questions, reinforced his ideas and led D. S. into areas that she had never mentioned to him. Although she'd had no initial belief in any sort of physical ship, Boylan kept pressing her for details of both the exterior and the interior of it.

D. S. later said that her original dream never contained a ship or metallic walls as Boylan suggested repeatedly. She told the board investigators that what had been for her a comforting and peaceful dream was altered under hypnosis by Boylan. It became an ugly, frightening nightmare, including the elements of terror introduced by Boylan.

D. S. told the investigators, "[T]he dream turned into [me] being put on some kind of a table and probed with some kind of probe that hurt really bad, and me feeling very angry about, 'Why are you hurting me?' And that there was something wrong with me and 'Why don't you fix it?' and then one of the—after this, like escorting me back in my room and I couldn't move, and that was it." She also mentioned, "They did something. Removed something."

Analyzing the information after the session, Boylan told D. S. that the aliens sounded "like they were reptilians or amphibs." He asked her to draw one, but she couldn't do it. Boylan then sketched for her one of the reptilian aliens, but D. S. didn't think it looked quite right. It might be pointed out that Boylan, having apparently failed to gain the data he wanted, supplied her with the information. He had contaminated his witness by providing her with a drawing of what he believed to be an accurate representation of the aliens, rather than letting her come to her own conclusions about it.

In Boylan's close encounters group, D. S. had learned that the aliens often return. She told Boylan that she was scared because she believed the aliens could return at any time and there was nothing she could do to stop them. D. S. told the board investigators that Boylan didn't seem particularly concerned about her fear.

What had originally been a dream of getting out of bed, seeing a light, and then Gandhi had, under Boylan's hypnosis, changed into a "classic" alien abduction, complete with a physical examination on a table. According to D. S., none of that was in the original dream. She did say, however, that in 1991, she had undergone surgery for the removal of an ovary. She had a bad reaction to the anesthesia, according to her, and had been scared and angry at that time.

It seems reasonable to believe that Boylan, in his search for alien abduction, was able to retrieve part of the regular, earthbound surgery memory and combine it into a space spectacular that fit into his personal agenda. D. S., in attempting to answer his questions, used the earlier experience as a basis for her answers to Boylan.

To make matters worse, at least for D. S., Boylan began another support group. D. S. had told Boylan that she was part Cherokee and was very interested in her Indian heritage. Boylan told her that many people who had abduction experiences identified with the Indian traditions because of a need for spiritual belief. The two worked together because of the "concept of interrelationship and the living earth." Boylan was forming a side group from his close encounters patients who had an interest in the spiritual aspects of their experiences.

In February 1993, D. S. was told about the first meeting of the side group. Boylan, using the trappings of Indian culture, conducted the session near the American River. D. S. said that she felt participation in the makeshift ritual was disrespectful to her heritage. When asked, by Boylan, to make a statement about her beliefs, she said she had nothing to add. According to her, Boylan "became angry with her for failing to participate."

A few days later, at her normal session, she ceased her relation with Boylan. She told the investigators that she no longer trusted him and didn't believe he was a caring person. She had been upset with his treatment of the Indian culture.

Boylan made a termination diagnosis: "Factitious Disorder with Psychological Symptoms (Psychological Munchausen Syndrome); Personality Disorder NOS (Addition to Victim Identity Syndrome)."

In their final report on Boylan, the California board wrote, "Respondent [Boylan] was not fueled by evil motive. He believes in extraterrestrial life and believes he has had ET experiences. In 1989, he had three patients who presented him with stories of ET contact. It was those contacts that inspired his interest in researching ET issues. He formed his CE-IV group so that experiencers would have others with him to share, so they would not feel isolated."

There are two other points that impact on all of this. First it should be pointed out that Boylan does have impressive academic credentials. He was graduated from a small private college in California in 1966, received a master's degree in educational administration in 1977, a second in social work, and in 1984 received a Ph.D. in anthropological psychology. What is usually left out of the picture is that Boylan, according to the "Findings of Fact," was also a Catholic priest. It is the Catholic church that seems to have the most intense interest in the paranormal and the supernatural. It is the Catholic church that brings the realm of the supernormal down to Earth, and it is Catholic priests who perform exorcisms.

Second, as noted earlier, Boylan told D. W. that he was going to make a trip into the Southwest, swinging by the various sites of extraterrestrial activity including the areas of Dulce, New Mexico, Area 51 in Nevada, and even stopping by the Very Large Array Radio Observatory west of Socorro, New Mexico. During that trip, according to D. W., Boylan thought he would be abducted by aliens.

Boylan says his prediction proved to be accurate. On U.S. Highway 180, north of Deming, New Mexico, as he headed for the Gila National Forest, Boylan drove into a cloud of smoke. According to him, this was the last thing he consciously remembered until he exited the smoke and continued to his campsite. He thought nothing of the odd experience as he camped for the night.

The next morning, he became more concerned, wondering about the strange nature of the smoke that didn't seem to be either smoke or fog and the peculiar "scoop" marks he found on his foot. Thinking about the night's drive, he discovered there was an hour of missing time. The classic clues that had led others into a belief they had been abducted by aliens were now all available for Boylan.

He had no real memory of the encounter, but through hypnosis, conducted on several occasions by three different hypnotists, Boylan remembered what had happened. He learned that he had driven in the smoke until it became so thick that he couldn't see the road or the white lines painted on it. He stopped dead on the highway.

Unable to determine anything, he climbed out of the car, walked to the side of the road, and was suddenly paralyzed. Two beings appeared from the smoke, grabbed him by the upper arms, and "guided" him a short distance to their ship.

Boylan described the landed, "metallic" vehicle. Remember, in his questioning of D. S., he kept asking her about the ship in her dream, wanting her to tell him that it was metallic. Now he had confirmation of that point based on his own, alleged firsthand observations.

But Boylan's story is better than most of the abduction accounts. Not only was his ship shaped slightly differently, being a thick disc with no dome on it, but it was embedded in the ground. Yes, Boylan was claiming that he was taken aboard a crashed flying saucer.

Boylan was held in a room on the ship. The two beings put him in a chair and then left. Boylan realized that he had no idea what the creatures looked like, so when they returned, he made the effort to see them. Their faces were oval, without noses, but did contain big, black oval-shaped eyes, as demanded by abduction researchers.

He was removed from the one room, taken to another, where he was put into a chair that reclined. He wasn't laid completely flat. His feet seemed to be held by some kind of a force field. There was an intense pressure as something was forced up his nose. Once the object, or the assumed object, was implanted, Boylan's feet were released. He was returned to his car.

So Boylan, the former Catholic priest, who had treated forty or more people who thought they had been abducted by aliens, who believed that the government knows the truth and is hiding it, became an abductee complete with nasal implant. Not only is he a researcher but he's also a victim.

Boylan's interest in UFO abductions was not limited to his patients, the book he wrote using them, or his own experiences as an abductee. With Dr. Leo Sprinkle, he founded the Academy of Clinical Close Encounter Therapists so that they could develop a program for the treatment, by qualified therapists, of those who claimed alien abduction. Both Sprinkle and Boylan (both of whom are convinced they have been abducted) believe that the abductions were real events and not psychological problems. The trauma of the abduction, or the manipulation of the subject by the aliens, often resulted in psychological problems for the experiencer. Although mental health professionals and abduction researchers disagree about the positive or negative nature of the event, many of them believe their patients are telling the truth.

Sprinkle, the president of ACCET, who has been using hypnosis to recover suppressed or forgotten memories for thirty years or more, believes the experiences to be real. Sprinkle, for example, was involved in an investiation that one UFO researcher said is the best UFO case on record, that of police officer Herbert Schirmer in Ashland, Nebraska, who reported a flying saucer but had no conscious memory of alien abduction. The sighting, as described in chapter 1, took place in December 1967, while the Air Force–sponsored University of Colorado UFO study was still under way. Because of that, investigation into the report was made by the Colorado researchers.

Sprinkle, as detailed earlier, was brought in by the Colorado scientists and used hypnotic regression on Schirmer after it was discovered that twenty minutes seemed to be missing from Schirmer's police log. Under the questioning by Sprinkle, Schirmer remembered that the craft had landed just off the road and that alien creatures had gotten out. Although Schirmer's description of them doesn't match precisely the grays reported by abductee's today, it is close. He mentioned the gray-white skin, small creatures about four and a half to five feet tall, and strange eyes.

Like Boylan, however, Sprinkle has demonstrated that he can take an innocent dream and turn it into an alien abduction. Both Boylan and Sprinkle have been actively, though unconsciously in many cases, imposing their own personal opinions and beliefs on those who have come to them for help.

In *Mute Evidence,* by Daniel Kagan and Ian Summers, Sprinkle is interviewed. Kagan wanted to try an experiment with Sprinkle, telling him that he, Kagan, never had a "conscious or unconscious feeling of any contact with a UFO or space aliens." Kagan told Sprinkle, "I'm a good hypnotic subject. Why don't you put me under, and let's see if I come up with any UFO images?"

Sprinkle agreed. Kagan was hypnotized and Sprinkle had him review his memories for anything that would fit into the UFO abduction scenario. Kagan did remember a recurring dream that was pleasant. Kagan thought of it because there were objects flying through the air. These were balloons, blimps, and even heavy structures that should not float in oil, including a "gigantic oil tanker, maybe a thousand feet long. I can see the rusted hull."

That dream went no further, but another reoccurring dream "suggested itself to him. . . . [H]e had the sensation of being suspended high in the air over one end of an immensely high and long suspension bridge."

Sprinkle suggested that Kagan try to cross the bridge but even in the car he imagined, a "hopped-up, red convertible, something like a fuel-injected 1962 Chevy," he couldn't seem to get across the bridge. Kagan then saw the car was two thousand feet above the road and Kagan thought he would feel safer in an airplane and knew what airplane it would be.

But before Kagan could describe the little yellow Piper Cub of his youth, Sprinkle interjected, "Imagine yourself in a spacecraft."

In *Mute Evidence* Kagan and Summers reported, "Kagan felt his mind breaking into two distinct parts. One part shuffled through his memory of visual images to try to find a spacecraft interior to fulfill Sprinkle's request. The other segment of his consciousness kept saying, 'Here it is. He's steering your mental images. He's taking this opportunity to try to implant the image of a flying saucer into your recall.' "

According to Kagan and Summers, ". . . Sprinkle had just demonstrated how much he had probably been responsible for the UFO imagery reported by so many of his hypnotic subjects. It meant that none of Sprinkle's oral histories could be taken seriously, because his role as hypnotist could have been the single most powerful factor in introducing UFO images into the subjects' memories."

Of course, it could be argued that Sprinkle's strong beliefs are the result of long work in the field. After all, it was more than thirty years ago that Sprinkle had used hypnosis on Herbert Schirmer. Besides, according to Ralph Blum's *Beyond Earth: Man's Contact with UFOs,* Warren Smith, who investigated the Schirmer case, told Blum he had found the landing site, "an unplowed sloping field, not far from the highway. There were three-pointed marks where the tripod landing gear had sunk deep into the earth. 'Patches of grass had been swirled and twisted into odd patterns as though the vegetation had been under powerful centrifugal pressure.' "

That would be a nice piece of corroboration for the Schirmer case, and for the idea of alien abduction, if it were true. It would be the physical evidence that so many of the skeptics demand. However, Smith seems to be the only person to have seen it, and although he claims to be a professional photograper with his many other traits, he failed to photograph the evidence. According to the Air Force–commissioned Scientific Study of Unidentified Flying Objects (the so-called Condon Report), "The site was checked for radioactivity, no evidence of which was found. No other evidence that an unusual object had landed on or hovered over the site was found."

What is most important is that no one, other than Smith, and including Schirmer or any of the other private UFO researchers, ever reported the physical traces. No pictures of it were taken. While it might be argued that the Condon scientists wouldn't have seen the craft had it still been there, other independent researchers failed to see the evidence reported by Smith when they searched for it. Given that, it is only fair to conclude that the physical evidence, in the form of landing gear marks and "swirled" grasses, probably never existed.

So what we see is that two of the founders of an organization designed to teach therapists how to deal with victims of alien abduction do, in fact, implant those very memories. Dr. Sprinkle is subtle in his methods and might not realize exactly what he is doing. Boylan, however, is much less subtle in his attempts to implant his theories in the minds of his patients.

Kagan and Summers, in their investigation, speculated that Sprinkle had a disproportionate number of abductees because of the nature of his work. People were sent to him because they believed they might have had a UFO experience. They were primed when they arrived. Sprinkle, in his search of their memories under hypnosis, asked questions that would elicit abduction memory.[3]

But Boylan, with an agenda, injected an element of extraterrestrial terror into the lives of his patients, priming them with books and magazine articles on abduction. He suggested that the aliens often returned, said that his life was probably in danger (implying that his patients' lives were also in danger), and that the government was out to get him. He talked of the government coverup

---

[3]Clark, *UFO's in the 1980s* (Detroit: Apogee, 1990), 198–9; Daniel Kagan and Ian Summers, *Mute Evidence* (New York: Bantam, 1983); R. Leo Sprinkle, "Hypnotic and Psychic Implications in the Investigation of UFO Reports, in Coral and Jim Lorenzen, *Encounters with UFO Occupants* (New York: Berkley, 1976), 256–329.

and, on a trip to explore the "secret" bases for himself, claims to have been abducted.

And now these two are training others in their techniques. They are trying to convince other therapists that alien abduction is real. They are providing other therapists with their expertise based on what they believe to be the truth.

11.

JOHN CARPENTER

John Carpenter is a psychiatric social worker who has been appointed MUFON's Director of Abduction Research. He is one of the few of those conducting abduction research who has been trained as a mental health professional and who is employed in that capacity. He has a B.A. in psychology from DePauw University, a master's in social work from Washington University in St. Louis, and was trained in clinical hypnosis at the Menninger Clinic in Kansas in 1980. He belongs to several professional organizations, including the American Group Psychotherapy Association, and works for the Center for Neuropsychiatry as a licensed clinical social worker in Springfield, Missouri. He has received more than 150 referrals from others interested in alien abduction and has used hypnosis in about 100 of those cases.[1]

His training and experience, then, would seem to place him at the top of the list of those conducting research into alien abduction. His interest in UFOs of more than thirty yeas should provide Carpenter with a keen insight into alien abduction. But his paper delivered at the Abduction Study Conference held at MIT in 1992, "Resolution of Phobias from Recall of Abductions in Alien Discussions," reflects what some could conclude is a lack of scientific research and understanding of scientific protocols.

Carpenter, in his paper, defined a phobia as "a psychological problem that reaches deeply into a person's emotional feelings." This is not the standard definition as outlined in various texts and must be one of Carpenter's own creation.

The *Diagnostic and Statistical Manual IV,* published by the American Psychiatric Association, defines a phobia as, "a marked or persistent fear that is excessive or unreasonable, cued by the presence or anticipation of a specific object or situation." The only feeling that is provoked by a phobia is fear, and it's not necessary to "reach deeply into a person's emotional feelings" to find it.

Carpenter followed his definition by writing, "A phobia neither develops

---

[1]John Carpenter, "The Significance of Multiple Participant Abductions," in *MUFON 1996 International UFO Symposium Proceedings* (Seguin, TX: MUFON, 1996).

nor vanishes without a significant emotional experience attached." He offered no proof that his statement is accurate and, in fact, a search of the psychological literature failed to reveal scientific evidence to support his claim. Many of the people treated for phobias have no idea where or when their phobia developed, and are cured of it without experiencing any deep emotional trauma.

Carpenter also noted, "Successful resolution of a phobia comes from a direct focus on the genuine source of its existence." Carpenter apparently achieved this result through the use of hypnotic regression techniques. Neither the statement by Carpenter nor the technique used by Carpenter has any independent scientific support for the resolution of phobias.[2]

Dr. William Cone ran a literature search through the National Library of Medicine looking for any research showing that hypnosis used to recover memories was an effective treatment for phobias. In a search that spanned nearly thirty years, from 1966 to 1995, he found not a single article that recommended the technique being advocated by Carpenter in his paper at the conference.

Further, Cone searched his personal library and nine textbooks on phobias failed to mention any recommendation for Carpenter's interesting technique. *The Comprehensive Textbook of Psychiatry* stated, "Therapists soon recognized that, despite progress in uncovering unconscious conflicts, patients frequently failed to lose their phobic symptoms . . . a growing body of evidence has made it evident that phobic patients are not helped by analytic techniques."

What the scientific and medical literature suggests, then, is a treatment that is at odds with what Carpenter wrote. Carpenter's proposed treatment suggests a lack of understanding of the mechanisms responsible for the induction of phobias, or for the proper and successful treatments of them.

Carpenter seemed to believe, "When phobias can be successfully treated during the recall of UFO/abduction experiences, this strongly suggests that the individual perceived UFO-related events as quite real." Although the meaning here isn't clear, it may mean that the person experiencing the abductions believe they are real and that the successful treatment of the phobia validates that belief. While it might be true that the abductees perceive the experience as reality, that certainly doesn't mean they have any basis in reality. This seems to be a subtle distinction that is all too often overlooked by Carpenter, as well as many of the other abduction researchers.

Carpenter, in his paper, described a hypnotic regression session of a client

---

[2]C. D. B. Bryan, *Close Encounters of the Fourth Kind: Alien Abduction, UFOs, and the Conference at M.I.T.* (New York: Alfred A. Knopf, 1995), 25–6, 30–1, 76–89; John Carpenter, "Resolutions of Phobias from UFO Data," in *Proceedings of the 1992 Abduction Study Conference at MIT.*

who was afraid of dolls. During the session, she recalled being abducted and then shown small human/alien "hybrids," which other abductees had described as dolls. This, according to Carpenter, was the origin of her phobia. After recovering the memory, the phobia, according to Carpenter, disappeared.

Carpenter then wrote, "The significance of her intense and unexpected emotional reaction adds credence to the belief that this is the genuine source of her phobic fear [sic]."

While it can be questioned that the hypnotic regression and the retrieval of the memory were actually the reason the phobia disappeared, it should also be noted that the fact a person displays emotion while recounting the memory is often taken as evidence that the experience must be based in reality. We have found no scientific evidence to back up this claim and there is a body of clinical evidence to suggest it is false. Emotional response under hypnotic regression is no indicator of the validity of the memory or the event.

Disturbing experiences are usually recounted with much emotion, but the amount, or intensity, of that emotion has nothing to do with the reality of the experience. Michael Yapko reported that a man diagnosed as suffering from post-traumatic stress disorder often had emotionally charged flashbacks of his horrific combat experiences. Later, it was discovered that the man had never been in combat.

William Sargent, who spent years treating soldiers with post-traumatic stress disorder, corroborates the view that emotion is irrelevant, writing, "However real and vivid personal and apparently remembered experiences may seem, this is no evidence of their reality, if they are brought to the surface under conditions of stress and in states of abnormal brain activity and heightened suggestibility [that is, hypnotic regression]. And the overwhelmingly vivid and convincing nature of so many experiences reported in the same states of brain activity induced by drugs, sex, hellfire preaching, mob oratory or other mind bending agencies provides no evidence of their truth. . . ."

Carpenter, apparently unaware of the scientific research completed in this arena, finished his section of the MIT conference report, writing, "It is important to realize that the resolution of such phobic fears cannot be accomplished by the telling of fantasy."

Again, he appears to be at odds with the psychological and scientific evidence. Past life therapists, those who search our alleged past lives for our psychological problems in today's world, have been treating phobias this way for years. When an MIT conference attendee asked how past life therapists treat phobias through what is most likely fantasy in the minds of those reporting them, Carpenter answered, "I'd really have to study the symptoms to see that it's a genuine phobia."

According to Carpenter, then, past life therapists can't cure "real" phobias through the exploration of the client's past lives. But this makes no logical sense. Does this mean that people who claim to have lived before suffer from "*psuedophobias*" which are only cured by reliving "psuedo-lives"? Isn't this the logical extension of Carpenter's claim?

The point here is not to endorse past life therapy as a viable method of treatment but to draw a parallel. If Carpenter's premise, that the reduction of a phobia through a hypnotic regression of an alien abduction experience is evidence of the reality of that experience, then the reduction of a phobia through a past life regression would also be evidence of a reality of experience.

Carpenter seemed to understand the box he had found himself in when he said that he would have to study the symptoms to see if it was a geniune phobia. But that wasn't the point. The question was raised to show that a similar argument, one that the conference attendees were not willing to concede, could be made about past life regressions.

The real point, and the one that might have been lost, was that the reduction of the phobia through a hypnotic reliving of an alien abduction did not underscore the reality of the memories. The reduction of the phobia by a hypnotic reliving of the abduction experience did not prove the reality of that experience.

Or, to examine it another way, we can look at another's work. Sargent found that hypnosis, used to assist his patients relive an event, often helped. But rather than insisting that the recovered memory used to cure phobia responses be the actual memory of the origin of the phobia, Sargent found that "the release of great anger or fear could be more effectively produced around incidents which were entirely *imaginary and had never happened at all,* and such abreactions of imaginary events could have remarkably beneficial effects."

What he was saying, and what his research had found, was that the memory that reduced or extinguished the phobia did not have to be a real memory. An imaginary event could be, and often was, sufficient to extinguish the phobia, which is contrary to Carpenter's conclusion.

Robert Schaeffer, who attended Carpenter's presentation at MIT, and who seemed to understand this, asked if a powerful fantasy would produce the same catharsis. At this point John Mack took center stage, saying, "A phobia has a core experience. It's not something a person elaborates totally from within. It has an actual event at some point early in the person's life. It's a powerful magnet for later events and displacements that occur. But there is an experiential source. It doesn't just come from the imagination."

Carpenter, not missing a beat, responded to this incomprehensible double-speak by saying, "Thank you, John. I agree one hundred percent."

Scientifically speaking, however, the statement by Mack, and endorsed by Carpenter, made no sense. It is true that phobias are caused by something. But if we follow what Mack said carefully, that the phobia "is not something a person elaborates totally from within," the only thing this could mean is that it is something that is partially elaborated from within. This is, in fact, one definition of fantasy.

Mack also said that phobias develop early in a person's life, but phobias, according to David Sheahan, author of *The Anxiety Disease,* begin, on average, at age twenty-six or twenty-seven. Others, however, can begin earlier. The fear of animals often starts about age seven, fear of blood, or the sight of blood, about nine, fear of the dentist about twelve, claustrophobia about twenty, and agoraphobia about twenty-five.

We have found no objective scientific data to support Carpenter's premise, in his presentation and paper. His conclusions, drawn on his own work are badly skewed. In fact, by reviewing his sessions with one of his abductees, as told in her own book about those sessions, we can see that Carpenter, though he denies it, leads his patients to the point where they believe they have been abducted by alien creatures.

Leah Haley eventually found her way to Carpenter to ask for his assistance in understanding what had happened, and was happening to her. When Haley began her journey into the world of alien abduction, she was a happily married, intelligent, certified public accountant who was pursuing a doctoral degree. Haley's trek began with a seemingly innocuous but very vivid dream. On page twelve of her book, *Lost Was the Key,* she wrote, "I thought about a vivid dream I once had about being on a spaceship. . . . I wondered if I should tell Seth [her brother] about the dream. The dream had always haunted me because it seemed so real."

One of the points made by many abduction researchers is that there must be some sort of triggering mechanism for the belief in alien abduction. Sometimes it is little more than a "feeling of dread on a certain stretch of highway," but others times it is something a little more concrete. Here we have a woman who has had some vivid dreams, dreams that she said had haunted her. But it wasn't until she had conversations with her brother, and her husband, that she began to suspect there was something more to the dreams. Her brother and her husband told her that there was something real about alien abduction, which set the ball in motion. Carpenter, however, is right there to pick it up and run with it, eventually suggesting that her dreams were not just dreams but were the actual memories of an alien abduction.

Haley had always accepted the dream as a dream, but her brother Seth suggested it might be something more. Seth was a staunch believer in abductions, having read *Intruders* and watched several abductees on the television talk shows relate their experiences. Through a series of questions and suggestions, he set out to convince Haley that her experience was more than a dream. He redefined a urinary tract infection she once had as a symptom of abduction. He asked her to search her body for any unusual marks or scars. Without any formal investigation, without any real expertise in alien abduction, other than having read a book and watched a few television shows, he redefined an ear problem from an allergic reaction she had to duplicating machines as pain caused by an alien implant in her ear.

Haley, at first, rejected his suggestions and insisted that her experience was just a dream. But her brother countered by saying, "That's what all the other people said." The seed was planted. After sleeping on it, Haley awakened the next day unable to shake the conversation from her mind. When she discussed her thoughts with her husband, he told her he believed her, and believed that abduction was possible because he had seen the movie *Communion* in a hotel, and her experience sounded similar to that portrayed in the movie.

Prompted by her brother and her husband, she wrote her first letter to Budd Hopkins. When she got no reply she sent a second letter. This time she received an *information kit for abductees* (emphasis added). The kit contained accounts of experiences of other abductees. One account reminded her of another dream, and this began to build, in her mind, a confusion between the world of dreams and the world of reality.

"Suddenly I remembered a strange dream I had in October. . . . Maybe these dreams hadn't been dreams at all. And if they weren't dreams, that meant that aliens were coming to get me."

Filled with dread and desperation, "trembling and my heart beating," she made a call to abduction researcher John Carpenter. She set up an appointment. By this time, it appeared that her ability to tell reality from fantasy was beginning to crumble. "The more I thought about the prospect of aliens visiting me, the more confused I got. One minute I thought the prospect of aliens actually coming anywhere near me was absurd. The next minute I questioned why my dreams were similar to those of other people."

It's important to remember here that her state of mind at this time is one of heightened suggestibility. It is also important to remember that she had a cadre of friends and relatives handing her everything she needed to "understand" abduction, not to mention the "abduction package" supplied by Hopkins. Everyone to whom she turned seemed to believe that her dreams and anxieties were the result not of the fantasy of the dream world but because she had been abducted by alien creatures. No one was stopping to say maybe we should reevaluate this from a logical position.

Not surprisingly, while waiting to begin hypnosis with Carpenter she began to have more dreams about aliens and abductions. She described the new dreams in her next letter to Carpenter. Her anxiety continued to mount.

During this waiting period, she befriended a woman named Judith in one of her graduate classes. She told Judith about her experiences. Just like her family members, Judith accepted the idea of abduction and further reinforced her belief that her dreams were real experiences. Dreams, with the help of friends and family, in fact, practically everyone to whom she spoke, had moved from fantasy into reality.

Soon the anxiety and uncertainty she was experiencing began to take its toll. Two months after she had begun to discuss her dreams with her brother, she developed feelings that she was being followed. She said, "I couldn't prove the cars were following me. I just had a gut feeling they were. I felt as if I were being watched most of the time."

When she finally met with Carpenter, she was primed with expectation. She was desperate for an explanation for her distress, she was primed with a belief in the reality of alien abduction, and she was certain that Carpenter was the expert who could help her. It is clear that she had high expectations of him and had a high respect for him as an expert. She wrote, "I kept waiting for him to say it was OK to call him John. I was taught at an early age not to call an *authority figure* [emphasis added] by his first name until he grants permission."

Haley and Carpenter began by discussing the dreams she had written about to him. Haley described the dreams in detail, but said she had decided that one dream was an actual experience and the other was "just a dream." She made the judgment based on the fact that in the first dream she experienced being "beamed" up by aliens, which seemed plausible to her. But the second dream contained a scene in a conference room where she was talking to other people. The second dream was seen as only a dream because it seemed implausible to her that she would go to a conference.

She then recounted a dream she had shortly after her hysterectomy. In this dream she was having sex with "a spirit of some kind." She told Carpenter, "I don't remember what the spirit looked like or what the experience felt like. . . . And I remembered being thankful that I couldn't get pregnant."

Haley then began her first hypnotic session with Carpenter. Under hypnosis she recovered memories of encountering a strange, alien being in the woods when she was a child. Even though she was confident of Carpenter's abilities, she did not accept the memories as real. "If the encounter had actually taken place, why hadn't I remembered it long ago?" she asked. She disagreed with Carpenter's interpretation. "I denied being hypnotized, and stated the story I told resulted from my husband planting it in my brain." This is a very plausible explanation and quite perceptive on Haley's part, but Carpenter's reaction clearly showed he had his own beliefs and expectations about the session. More important, he communicated those to her.

She wrote about that as well. " 'Right,' John said *in a way that let me know he did not accept my explanation* [emphasis added]. 'We'll tell you all your details are right on the money. Grace [his assistant] and I are sitting here predicting what you are going to say before you say it.' "

That paragraph written by Haley is extremely important in understanding the abduction phenomenon. Haley realized that Carpenter, the authority figure, did not accept her explanation, which is, of course, leading the witness. But more importantly, Carpenter and his assistant are predicting what Haley is going to say. Carpenter's own expectations are being borne out in the session. Again, it seems that the abduction phenomenon is an outgrowth of what the researcher expects rather than what the abductee experienced. It is interesting that Haley was able to pick up this so early in her sessions. And it is important to note that she did pick it up. It is a symptom of what is wrong with abduction research.

Though reluctant to believe she was hypnotized, she eventually accepted Carpenter's stance. As her sense of psychological boundaries began to crumble, she wrote, "Today I found out that as a child I had been abducted by aliens . . ."

Her own beliefs, her own world, and her own memories were being changed as she continued her sessions with Carpenter. She was aware of the process, noting it in her own writings, but was unable to stop it. Psychologist Stephen Ceci states that there are three prerequisites for implanting false memories in a subject:

1. Shatter the persons' confidence.

2. Elicit one confabulation from them.

3. Develop it into a story.

Once this has been accomplished the person will accept the new memory without question. This is exactly what was beginning to happen to Haley. She was frightened and confused. Several people whom she respected and trusted were telling her she was an abductee, and that her dreams were her actual memories of real experiences. Despite her protests, and despite her awareness that, at least in the beginning, these "memories" were dreams, Carpenter continued weaving her dream fragments into his reality of alien abduction. As he did so, he was unaware that he was destroying her ability to tell her dreams from reality.

Although Carpenter has written many articles stating that he takes great precautions not to lead his subjects, it seems from Haley's account that he has a clear agenda. In Haley's second session, for example, she explored the origin of some bruises on her body. "Let your mind drift," said Carpenter, "and find out what touched you to leave bruises," suggesting that the bruises were caused by being touched rather than falling or bumping into something.

"I keep seeing the creature," she replied.

Later Carpenter suggested, "It all has to do with not remembering. When we talk about bruises, you see the creature. They didn't want you to remember how you got the bruises." Of course, no one asked how Carpenter knew what the aliens did or didn't want or even how he knew they were present at the time the bruises were created. And no one pointed out that his discussion of alien desires is just another form of leading the witness. He was subtly, and probably unconsciously, planting there ideas in Haley's memory.

Carpenter then addressed the ear problem that Haley had told him about earlier. "When did you have those ear problems?" he asked.

Haley said that they started when she was in Florence, the place where she got the bruises.

"Bingo!" Carpenter exclaimed, obviously delighted. "And you said this was an allergy to copy machines, right? I never heard of such a thing before. The doctor you asked had never heard of that."

The statements made here by Carpenter are certainly subtle forms of leading his subject. He was now telling Haley that her allergy couldn't be caused by the copy machines as she believed, and therefore there had to be another cause for it. It is also clear from his comments what Carpenter believed that cause to be.

Haley, still somewhat skeptical, said, "It was apparent John believed the pain I experienced in my ear was the result of the alien's procedure. I could tell I had failed in my attempts to convince him the problem resulted from copy machine ink."

It is important to note, once again, that Haley understood what Carpenter wanted her to say. She could tell when her efforts to present a rational and mundane cause for her reaction were rejected by Carpenter. He might say that he takes pains to avoid leading his subjects, but Haley appears to be picking up on the nonverbal cues given by Carpenter. Leading a witness is much more than just the asking of leading questions.

Carpenter seemed to be telling her that because he had never heard of this type of allergy, it simply doesn't exist. There is no indication that he ever researched this assumption in the medical literature or that he consulted a doctor to learn if such conditions exist. Instead, he seemingly rejected it because it didn't fit with his views of reality. In actuality, there have been hundreds of reported cases of allergic reactions to copy machine ink.

But more important is that Carpenter has allowed his personal beliefs to influence his subject. In her own words, as she recounted the sessions, she explained how her vivid dreams were turned into real experiences. And she was able to repeatedly determine what Carpenter was trying to accomplish and what answers he wanted to hear. Carpenter's leading of Haley to the point he wants to reach is blatant and undeniable.

Carpenter then told Haley that it was through dreaming that she would recover the rest of her memories. "It may leak out in dreams again," he said, 'cause that's a point in time when we're at rest, where our defenses are down and what really is inside with our feelings and our thoughts comes to the surface."

Keep in mind that we are not aware of any scientific evidence to suggest that what Carpenter said is true. Several decades of dream research have failed to establish that dreams represent real events buried in our unconscious.

To further drive home his view that the dreams are actually real experiences, Carpenter said, "If it was a nice little dream, it would sound much different than this stuff. 'Cause this stuff is really, in a lot of ways, boring. There's a real lack of creativity and diversity in these images." In fact, the majority of our dreams are mundane, uncreative, and uninteresting, which is one reason why we soon forget them."

But Carpenter continued, telling Haley what his research had suggested to him. He told her that he believed that alien abduction was real because of the number of "independent" reports that seemed to corroborate one another. He told Haley, "Especially when you get thousands of people who report the same exact thing, down to the skin color and texture and those eyes."

This is another statement by Carpenter that is not backed by any identified hard or objective evidence. No one has yet to produce these "thousands of people" of whom Carpenter speaks. No one has done any definitive study showing that the similarity of thousands of dreams in any way corroborates the tales of alien abduction. But he continued his technique, verifying that what Haley had recovered under hypnosis was an actual abduction experience.

He told her, "They [other witnesses] always talk about the eyes, over and over. And that's the main thing you always focus on—those eyes."

Carpenter sent Haley back to her hotel room after the session with two videotapes showing other abductees detailing their experiences. Haley said of this, "I was having a hard time accepting the idea I had actually experienced the three alien encounters that had surfaced in the hypnosis sessions. Furthermore, I still refused to believe that I had really been hypnotized. I thought my mind was playing tricks on me or that I was mentally ill."

She quite astutely pointed out, "I sensed that John wanted me to see the videotapes so that I would accept the reality of the abduction experiences and admit I had indeed been hypnotized." This, too, is a form of leading the witness. Handing them information about the abduction phenomenon that suggests that abduction is real is a way of leading them to the conclusion that the researcher wants to find. It also "primes" them with the information about abduction that the researcher believes to be accurate. Is it necessary to point out here that Carpenter, before he had completed his investigation, had supplied Haley with the details of other victims' abduction? Is it necessary to point out that we now have an answer to the question that Carpenter had asked Haley earlier? How could the abductees give details that matched one another though they had never met and didn't read the abduction literature? Hadn't Carpenter just supplied the necessary information?

After Haley watched the videotapes, Carpenter introduced her to another abductee, named Paula. At this point he took what he believed was a great precaution. He told Paula not to tell Haley "too much" to avoid "contamination." This precaution was taken after he had just handed Haley two videotapes of other people's abduction experiences and introduced Haley to a woman Carpenter believed to be an abductee. Paula's very presence was a source of contamination. And remember, Hopkins had already sent her an "abduction package." The mere fact that Paula claimed to have been abducted would have an effect on Haley. Even with this precaution in place, Haley said, "She managed to tell me enough to convince me she had been abducted by aliens." And this would validate the topic for Haley and underscore the veracity of the information being provided by Carpenter.

Before she returned home, Carpenter set the stage for further abduction memories. "Carpenter told me not to be surprised if memories of prior experiences started creeping into my consciousness. He explained that hypnotic regression sometimes triggers memories that have been buried."

And once again, it should be noted that there is no scientific evidence to support this statement. There is no evidence that experiences become "deeply buried" in the unconscious, or that hypnosis "triggers" them, bringing them to the surface. However, Haley respected and believed Carpenter, and produced the results he predicted. "I immediately started having flashbacks related to my prior alien encounters," she said.

As always happens with dreams, the more she began to think about her alleged abductions during the day, the more the material appeared in her dreams at night. Because she had been told by Carpenter that the memories would leak out as dreams, she could no longer tell dreams from reality, and began to conclude each dream was further evidence of abduction. After having two vivid dreams of aliens, she said, ". . . I could not deny the feeling that portions of both dreams were more real than dreamlike. I decided if aliens truly did exist, then I probably was abducted while I was asleep those two nights."

But the process left Haley confused. She said of her experiences with Carpenter, "I was annoyed that instead of answering questions for me, the hypnosis sessions were bringing up more questions."

As time went by the dreams became more and more frequent. Throughout the rest of her book, Haley described a long series of abduction dreams. By this time, however, because she had been told by Carpenter that real memories "leak through" in dreams, she accepted every dream she had as evidence of a real experience. In this regard, she said, "I began to get indications that my dream of having sex with a spirit was not a dream at all."

As she became more obsessed with the dreams, she continued to deteriorate, and became paranoid. The "mental flashes" she was experiencing began to happen more often. The boundaries between dream, fantasy, and reality had become so blurred that she accepted them all as remnants of actual experiences.

This blurring of boundaries is typical of all recovered memory therapy. Consider, for example, the experience of Jane, a therapy victim discussed in Richard Ofshe's book *Making Monsters.* "As the visual images became stronger over the course of therapy, Jane began to experience waking hallucinations and often had trouble distinguishing between reality and what she would picture in her mind . . . the images that she believed came from her past became more bizarre. In one scene she saw herself at the age of one, being forced to watch her father rape a dog."

As Haley's confusion grew, she began to become paranoid. In the early stages of paranoia, the victim may merely be suspicious, angry, and indignant. She may feel persecuted by only a few individuals. But as the disease progresses, the power of the imaginary pseudocommunity grows. Eventually, she begins to find hidden meaning in every conversation. Strangers on the street are talking about her. Newspaper stories, radio, and television programs contain hidden messages directed at her. These delusions are called ideas of reference. This experience results in the person believing that conversations of others have personal meaning to the unfortunate victim.

Haley described this process quite well, when she recounted her experience in a hotel, ". . . the sound of a man's voice in the next room awakened me. The voice sounded vaguely familiar, so I put my ear to the wall and strained to hear what he was saying."

What's important here is the fact that she was now sufficiently paranoid to put her ear to the wall just to hear a "vaguely familiar" voice in the next hotel room. All of us have heard noises in hotel rooms, but it's very unusual to become so caught up in them that one must put an ear to the wall to listen. When the process of paranoia takes hold, it takes on an obsessional quality. Describing her obsessional state, Haley said, "I listened for what seemed like a very long time, but I was able to make out only a very few words dispersed throughout the conversation."

This isn't surprising, since it's hard to hear clearly through a wall. But the few words she thought she heard took on great significance. A phrase that sounded to her like, "we got most of it," immediately became important to Haley. Although this was only her interpretation of what she heard, and even if it were correct, could mean any number of things, she quickly concluded, "the man had succeeded in recording my last two hypnosis sessions."

As often happens in paranoia, Haley developed an inner core of grandiosity. These beliefs are a reaction to feelings of being powerless to stop her life from being monitored. Paranoia results in the conviction that the victim has been singled out by her persecutors because she has some special significance. This feeling of importance is known clinically as paranoid illumination. Many paranoids eventually become convinced that they are a new messiah and that they have a vitally important message to share with the rest of the human race. There is always some startling truth that will be revealed at any moment.

Leah Haley described this facet of the disorder when she announced, "While I was reading my bible, a silent voice told me I am needed to help spread the word about the existence and visitations of the aliens, through both speaking and writing. I was told the creatures will be visiting Earth more frequently, and humans need to be prepared." When she protested against this role, she was told that like any messiah, she would "be ridiculed and persecuted." But, the inner voice also told her, "This is necessary. Someone must inform the public. We need your help with this."

It is tragic that Carpenter seemed oblivious the fact that his patient was deteriorating. During his treatment her life was disrupted severely, causing her to drop out of school. "I told Mike [her husband] that I could no longer pursue my doctorate," she lamented. "Dealing with them [the flashes and dreams] had begun to require so much of my time and energy that a life dedicated to concentration and study was now out of the question."

Tragically, Haley appears to have become another victim of this absurd pseudo-science. Haley was transformed from an intelligent, high-functioning person into a dyed-in-the-wool abductee. This eventually led her to see her abductions as the central focus of her life, and eventually cost her several friends, her doctorate, her job, and her marriage. Haley's experience is a prime example of how a normal, healthy person can be led into a world of paranoia and isolation by a misguided and misinformed therapist.

All her fears were confirmed by Carpenter during another hypnotic regression session. Carpenter began by telling her, ". . . there was a time when you dreamed that something had some kind of relations with you. You can slowly sink back into that dream and what you remember about the images in the dream."

Carpenter was again clearly implying that dreams are real memories of actual experiences. During this session, Haley remembered being raped by a reptilian alien. Although she described the experience as rape, she said that she couldn't move during the experience, and went back to sleep when it was over. Falling asleep is highly unlikely to happen after a rape, but it fits perfectly with a classic description of a sleep paralysis episode. But instead of sleep paralysis, Haley became another victim of alien rape.

So here we have a researcher, working hard to avoid leading his subjects, and then providing videotapes and introducing them to other abductees. He suggests to his subjects that dreams are the doorway to reality, and if the subject dreams of an abduction, it must be the subconscious trying to remember that abduction.

Like the others, Carpenter appears to have his own agenda. He believes that people are being abducted by alien beings. But the consequences for some of his subjects is extreme. The damage to Leah Haley's life can't be underestimated. She lost her job, her husband, and dropped out of graduate school. Had she got the help she needed that might not have happened. Instead she was sucked into the quagmire of alien abduction.

12.

DR. JAMES HARDER—THE INTERRUPTED JOURNEY CONTINUES

Although it was Barney and Betty Hill who began the interrupted journey in 1961, there are many others who continued it. Pat Roach, a divorced mother of five, living in a small house in American Fork, Utah, is one of those who took her own "interrupted" journey. She wasn't traveling when it happened, but home, asleep, when the intruders arrived to change her life radically.

Roach wrote to the editors of *Saga* (a men's magazine of the 1960s and '70s), telling them that she had read the abduction-related article in the Spring 1975 issue (the Dionisio Llanca abduction) and saying "I think I know how entire families can disappear." She then wrote, "We had a visit from someone about 11:00 at night in the middle of October 1973."

She wrote that there had been stories of a prowler in the neighborhood but that he seemed quite harmless. It seemed that he would unlock doors or gates and leave them unlocked. He took only food, and the few witnesses who saw him said that he was "dressed, 'like for Holloween.' "

Roach then explained what had happened that night in October. ". . . I lay on my living room couch and my four year old son lay beside me dragging a blanket along. I fell asleep and when I awoke the entire house was in commotion. The cat was screaming. My son was across the length of the living room staring at the space between the bookcase and drapes hysterically saying, 'Skeleton skeleton.'

After she quieted her son, she heard a noise outside that sounded as if someone were dragging the branches of a tree across the side of the house. Something shook the windows. Although Roach wrote that she wasn't terribly frightened, she couldn't bring herself to look outside for the prowler.

The next morning, when she inspected the fence around the empty field next to the house, she discovered the middle strand of barbed wire had been broken. Standing there, near the fence, she told her oldest daughter, "They must have made us forget." Her youngest daughter, Debbie, who was six at the time, then said, "They didn't make me forget."

Debbie, according to Roach's letter, said that two men had walked her out of her room the same way they had walked her sister out. "She thought she floated out rather than walked. . . . She did say that she was afraid they wouldn't bring her back. She said there was a man in the corner of the living room and he smiled at her."

Apparently Debbie and the man had a conversation. "She said there were no lips on his mouth and he didn't talk with his mouth but with his 'head.' . . . She said in the spaceship they told her she wouldn't be sick anymore. She said the spacemen looked like Indians but with shorter hair. There was an 'Indian' girl with a long dress in the spaceship seated at some controls."

Debbie said that she had seen "a lot of children from our neighborhood in the ship. There seemed to be a few from each family on the block. She said one child was lying on the examining machine and another was standing in a small room off the large entrance room. She said they told her to tell no one but her family about the incident."

Roach, in her letter, wrote, "When I tried to think if I could remember anything about the night it was very hazy. All I could remember was a bright light coming into the living room. I remember walking up steps like that of an airplane with a solid grey steel wall to the side."

In the text of the letter, Roach also reported that family members had been moved. Of Bonnie and Debbie, her daughters, she wrote, "She [Bonnie] woke in bed and Debbie, the six year old, was gone. She awoke again and their places were switched in the bed."

Reporting about what her daughter had told her, Roach wrote, "She also said Ken was across the room covered by a blanket and I was on the couch. He would never voluntarily leave my side so someone had moved and covered him."

Next to the Roach house was an empty field. Although partially hidden from the rest of the neighborhood by trees, the side next to the street was open and would allow those living across the street to see anything in the field. Roach wrote, "She [Debbie] said there was a spaceship parked in the field. It was saucer shaped with port holes on the sides. She said as she walked up the steps entering the ship she heard a 'beep beep' and didn't remember anything [sic] except pressure on the top of her left arm. She said she returned through the fence and as she did cut her chin on the wire. She did have a cut in the morning that hadn't been there the night before. As they took her through the dining room she noticed the clock said 1:00 A.M."

She finished the letter, writing, "It was hard to believe although I knew 'something' had happened that night so I placed Bonnie and Debra in separate rooms and told them to draw a picture of the 'spacemen.' The drawings were just alike except the triangle at the top of the suit was reversed in Debbie's drawing."

Saga editor Martin Singer passed the letter along to Kevin Randle, suggesting that he contact the woman. Singer found the story fascinating, especially because of the mention of others besides Roach's children being abducted. He wanted to follow up, as long as it didn't cost the magazine too much money.

Since Roach had suggested hypnosis, Randle called APRO headquarters and asked for help. Coral Lorenzen had one suggestion, but the psychologist was unavailable. She then suggested Dr. James Harder, APRO's director of research. James Harder was not a psychologist but a civil engineer who had been trained in the use of hypnosis and had been involved in a number of other such investigations.

On July 8, 1975, Harder and Randle visited Roach at her home in Lehi, Utah. Harder discussed the case with her for a short time, told her about hypnosis, and suggested a session to put her at ease. He wouldn't ask any questions about the sighting. He would just put her under so that she would experience hypnosis with no pressure.

Two hours later, with Roach relaxed, the first of the three hypnotic regression sessions began. Harder put her under and then told her to get the "feeling of concentration, going back in time, get that feeling that you had that day, that you were going to bed . . . tell me, tell me that . . . you've got the feeling of being on air. . . . What was the feeling you had."

"I'm surprised by. . . . It was a bright light. . . ."

"Did you go to the window?"

"No. I was in the living room and I was on the couch. . . . I sleep there occasionally. . . . You know two figures were standing over me."

Harder said, "It's all right, no one is going to hurt you."

"I wasn't really afraid. Maybe I felt they must know what they were doing."

"What happened then? You can get that memory back. Just feel yourself being there."

"I was lying down, you know, and they're bright. They're skinny. Whatever they were, they're skinny and they look like they've dressed up in all white. People that would be in the service or something?"

"What gave you that idea?" asked Harder, his voice low and quiet.

"Their uniforms."

"They seemed to have a uniform?"

"Yes."

"Did it seem to have any buttons or anything?" Harder asked, probing for information.

"No."

"Did they talk to you at that time?"

"No."

Harder continued to probe, telling her to get the feeling of being there and then trying to determine who was present in the room. Once he had established that some of Roach's children, but not all of them, were with her in the living room, he wanted to know what happened next.

But Roach claimed she couldn't remember. She mentioned that one of the men was in the corner. "He's standing by us. . . . I don't remember what happened."

Harder told her that she could remember and she said, "They have a machine that they carry. They're very businesslike, and they hurt my arms because I don't want to go anywhere. . . . They seemed to grasp me on my upper arms. . . . I don't remember going out the door. . . . I see bright room, big bright room. . . . They're standing around."

Harder had her describe what she could see around her. She was in a big, round room and she could see stars. "It looked like a lot of technology. It's all machines and buttons and on the wall." Finally she said, "That's all I want to remember."

Harder wasn't finished. He asked, "Is there something that you think would be frightening to remember?"

"Yes."

"Now you got out of this perfectly safe, nothing happened to you that hurt you, but it might have been a very frightening experience at that time. Is there something you think that would really . . ."

"No, I don't remember being examined but I know I was and that's what bothers me."

"You think you have been physically examined?"

"Yes."

"Probed?" asked Harder. "Somebody touched you?"

"Yes."

"Did you feel there there was a machine involved in this"

"And people."

Harder pressed on. "Did you get the impression that you were up on a table?"

"Yes."

"Were your clothes on? Did they take your clothes off?"

"I don't remember."

"It might be hard for you to remember," said Harder. "Did they tell you that you wouldn't remember this? Did you get that impression that you wouldn't remember?"

Roach responded, saying, "They really didn't talk to me."

Harder asked her more about the creatures surrounding her, trying to learn what he could about their attempts, or the lack of attempts to communicate with Roach. She told him that she didn't like their attitudes. She found them to be coldhearted and cold-blooded. According to Roach, they were interested in gathering their data but cared nothing for her emotional state or her feelings.

Harder then tried to get a description of the beings. He asked her, "Can you remember what the face looks like?"

"I remember the big eyes."

"And do you remember a pupil in the eyes, a round pupil, or was it a slitted pupil like a cat?"

"It doesn't matter . . . let me think. 'Cause they looked at me closer in my face."

"Did they?" asked Harder. "How big would you say their eyes were? The size of a quarter?"

"They were big."

"A fifty-cent piece?"

"No. Quarter."

"Was it round?"

"No . . . Oval. It had a big pupil. It was a round pupil."

"Was it black?"

"Yes."

"What about the nose? Do you remember anything about the nose?"

"Don't remember a nose."

"Was there anything that looked like holes there?"

"No," said Roach.

"What about a mouth?"

"A fish."

"It looked like a fish?" said Harder. "Does that mean it didn't have any lips?"

"Yes."

"Did the mouth look wide?"

"It looked long from side to side."

Harder continued his questioning of Roach. He wanted to know how tall the beings were, suggesting three feet and then four feet. He wanted to know how their arms related, proportionally, to the bodies. He then said, "Remember their hands. What they looked like."

"They have those funny hands like Bonnie said but they're orange."

"Orange color? Did they seem to have fingers?"

"Didn't look like fingers."

"Did they move their hands ever?"

"Yeah."

"Did they open their hands ever?"

"Yes. They opened . . . it was almost like a clasp."

"Like there were two fingers, or three?"

"I wouldn't call them fingers, they were big . . ."

Harder worked to reinforce the session, saying, "That's all right. You can remember it . . . I can understand that you didn't like them. Did they seem to have feet that looked like ours. You really didn't have a chance to see them?"

Harder continued for a few minutes more, asking about the appearance of the creatures and trying to learn what he could about how they were dressed. He asked specific questions about the belts the aliens wore and if their clothing was the same color above the belt as it was below.

Then, switching tracks, he went back to a comment that Roach had made earlier, asking her, "When they put this pan on you, did you get any, was it a little pan?"

"They didn't put the pan on me, they took something out of it and wiped me with it."

"Wiped you, just like a piece of cloth? Did it hurt?"

"I don't remember."

"It probably didn't hurt you. They probably were just taking a little skin sample or something superficial. Cells or something."

"I don't know."

"You can really remember, you just don't want to remember."

"I don't want to."

Harder, again telling Roach she could remember, said, "I can imagine, you were worried about your children. Your children may remember what happened and then afterwards you may want to. You will want to remember what happened to the children so that you can reassure them, probably. So it would be a good idea if you remembered what happened to you, if you can possibly do that without its bothering you too much."

"After I left that room, I wasn't with the children."

"I see," said Harder. "But they may be worried a little bit about what happened to them and you'll want to make sure it isn't too frightening. You don't want to upset them unnecessarily."

"No."

"I want to ask you one question, and you don't have to answer it. Did they put a needle in your stomach or anything like that? You can just answer with your fingers, you don't have to say."

"I'd rather say, I don't remember anything like that."

"You don't remember any blood samples that they took?"

"Nothing. They hooked me up to a machine. Checked everything, examined me from top to bottom. They put needles in me in places."

"Do you remember what places?"

"No."

"Perhaps in your arms or legs?"

"They put needles everywhere it seemed like."

"Was it Chinese acupuncture do you suppose?" asked Harder.

"I don't know."

Harder couldn't learn any more about the needles or the probing. He wanted to know if she had watched them work or if she had kept her eyes closed. He asked her about leaving the craft. He said, "Did they carry you?"

"Yes, more or less. I don't know how it was. I wasn't really walking."

"Did you feel yourself floating? Did they take you back to the living room?"

Roach told him that she was angry with them. She said that she had no desire to communicate with them. Harder suggested that they might have put the thought, the anger, in her mind. It was clear that Roach didn't want to discuss the abduction any further.

Harder, however, said, "It would be very helpful for me to know, as a scientist, what kinds of things that they are looking for. That would be very helpful if you could remember that . . . if it wouldn't be too much trouble."

"They wanted to know how our minds work."

"That's very interesting," said Harder.

"They want . . . to give them certain information that they don't understand yet."

"What kind of information?"

"How we think, how we feel, our emotions. They don't know about us."

"That's very interesting," said Harder. "Do you think that makes them interesting?"

"No . . . I don't like what they want."

"You thought you were being intruded upon."

"Yes. They didn't care, because they don't have an understanding of emotions like ours. Maybe they're trying to understand our emotions. I may be wrong . . ."

"You know, Pat," said Harder, "you're one of the more intelligent people that have been in touch with this thing."

That ended the first session. Roach had awakened at that point. Harder, with Randle present, would conduct two more regression sessions, but all were contaminated by the first. Harder made no suggestion that Roach would be unable to recall what had been discussed. He believed that she should be aware of everything that had transpired, and this would help him learn more in the additional sessions.[1]

In fact, after the end of that session, Harder and Randle asked additional questions. She remembered a few more details about what had happened. She now believed that Kent, her youngest son, had been on the craft. That was a detail she hadn't known before the regression.

---

[1]Kevin D. Randle, "The Family That Was Kidnapped by Ufonauts," *UFO Report*, June 1976, 21–3, 50–2, 54; Kevin D. Randle, *The October Scenario* (Iowa City, IA: Middle Coast Publishing, 1988), 17–30; Pat Roach, interview by Kevin D. Randle and James H. Harder, July 8 and 9, 1975, tape recording.

That same afternoon, July 8, Harder and Randle interviewed the oldest daughter, Bonnie, to learn if she would corroborate what her mother had told us. In the letter to *Saga*, Roach had made it clear that her children had more memories of the situation than she did.

The session with Bonnie was a disaster. She seemed apprehensive about hypnosis but Harder did manage to induce a light trance. The distractions proved to be too great and no progress was made. Bonnie woke quickly.

A second attempt met with the same results. Although Harder could induce the hypnotic state, it wouldn't hold as the probing moved to the abduction. The first question would destroy the mood, and Bonnie would sit up, blinking. Bonnie provided no information at that time.

On the morning of July 9, Roach was ready to try again. She was sure that she could remember more, especially after she'd had a good night's sleep. Harder had no difficulty with her. Roach was a good hypnotic subject.

After describing, again, how she was moved from the house to the ship, Roach said, "They put me on a table and they hooked me up on one leg and one arm. I didn't like their examination."

"Was it like a G-Y-N exam?" asked Harder.

"That's part of it," she said. "I don't like what they do with my head."

"What are they doing?"

"Taking my thoughts . . ." Then angrily, she said, "They don't have the right to take them."

She and Harder discussed exactly what she meant by taking her thoughts. The aliens were making her relive past events as if building a catalog of human emotion. Roach said, once again, that they didn't understand human emotions.

Roach leaped over a span of time and Roach said, "I'm getting dressed. They don't know."

Harder asked, "Don't know what?"

"They don't know how we humans are. I called them stupid." Roach laughed about that.

"What did they say to that?"

"They weren't angry. They just do what they want to. The man was a regular man."

Neither Harder nor Randle was ready for that revelation. Harder said, "What? What was that? You thought there was a regular human being with them?"

"Yes."

"Was he taller? Bigger?"

"Yes. He was bald."

"Was he the one who did the examining?"

"He helped."

Harder questioned her closely about the human being. She was sure there was a human with them. He was different from the aliens. He had regular eyes and human features.

Roach began describing other features of the abduction and finally said, "They need us. . . . I don't know why they need us. They're very intent. They need information quickly. I don't know if it's my imagination but they limit time."

Harder asked for her to say it again, and Roach responded, "They limit time and we don't."

"What gave you the impression they had a limited amount of time?"

"Not amount of time. I don't know what it means. Just that they limit time."

Roach began to talk about her children and started to cry. In seconds she was awake again. She sat for a moment, as if thinking about what she had just seen, and then wanted to talk about the experience. She said that the human was about fifty-five, had a fringe of gray hair, wore glasses, dressed in black, and wore at least, one glove.

When Roach left the room for a few moments, Harder said that he had been worried because Roach had failed to show any emotions during the first session. For Harder, Roach's emotions during the second session had added a dimension of realism to the story. He was convinced that Roach had been abducted by the crew of a flying saucer.

During the afternoon, Harder thought that everyone should get away from the house for a while. He wanted to move to neutral ground where everyone would have an opportunity to relax. Sitting in an ice cream parlor, Harder discussed some of the other abduction cases that had been reported over the last decade, including the Hill case. Harder told Roach about Betty Hill's belief that a needle had been pushed into her stomach and that eggs had been removed. She had said, more than once, that she believed there were lots of little Betty Hills running around in space.

The final session was held on the evening of July 9. Of Roach's children, only Bonnie seemed to slip into a hypnotic state easily. In interviews conducted with the other children, Randle believed that everything they had to say had been uncovered. Further attempts with hypnosis would be of no value.

Using a room at a local hotel, both Roach and her daughter would be put under. While Bonnie was left alone to concentrate on her experiences, Roach was given a pen and paper and asked to draw one of the aliens. She sat for a moment, as if looking at something, and then sketched, quickly, one of the creatures.

With that accomplished, Harder again questioned Roach, asking for more details about what she had seen on the ship. She described the interior of the craft, mentioned a "clock" with lots of hands, and told of the human who worked with the aliens.

Again, after she had been floated back to the house, Roach began to worry about her children. She began to cry, and slipped out of the hypnotic state. Now she remembered the needle and thought that it had been pushed into her stomach.

With her mother awake, Bonnie too, slipped out of the hypnosis. Now she remembered being on the craft. She was standing near a wall and could see her mother on a table that floated, surrounded by alien creatures. She said that she didn't watch too closely because her mother had no clothes and she was frightened.

Then Bonnie said one thing that excited both Harder and Randle. She said, "I can see a human with them." She went on to say, "He was taller and he had an ear like a regular ear."

Bonnie then took the paper and sketched the scene as she remembered it. It agreed with what Pat Roach had said earlier. The numbers of beings and the positions of them were all correct, just shown from a different angle. Harder and Randle questioned her, but there was nothing new to be added.

After the two left, Harder and Randle discussed the case in detail. According to Harder, it matched several other reports, some of which hadn't received any wide circulation. Only someone who had studied the phenomenon would be aware of the reports. There certainly was no way for Roach and her children to be aware of any of those cases.

Harder was impressed by a couple of details. First, because the majority of the story was reported while Roach was in a hypnotic state, Harder believed it contained a note of authenticity. Harder was aware that a subject can confabulate under hypnosis, but he was impressed by her emotions. Her emotions, and her repeated worries about the children, suggested to Harder that the abduction was real.

But now let's analyze the case from a slightly different angle. If we study the whole report, we find a number of very disturbing aspects. First, and foremost, is the way the case reached the hands of researchers. Roach, after having read the story of an abduction in *Saga's UFO Report* wrote to Martin Singer. Although Roach said she had read no books about UFOs and abductions, it is clear from her first letter that she had read magazine articles about them because she mentioned that.

Coincidentally, or maybe because of it, the article Roach mentioned in her letter was written by Randle. It concerned Dionisio Llanca, the South American who claimed to have been abducted during the wave of sightings in 1973 that included the Hickson-Parker abduction. There are a number of parallels between what was reported in that article and what Roach said. For example, both report a domed disc and male and female beings involved, with long hair and elongated eyes. There are other similarities as well.

What we have here is a possible source of contamination. We can't then assume that because there are similar items in both stories they must be true. What we can say is that Roach could have picked up that information through her reading of the Llanca abduction.

The other point that must be made is that the family had discussed this among themselves for two years. The events began on October 17, 1973. That can be checked because of the police report the Roaches filed that night of a prowler. Randle did, during his investigation, visit the police station and review the record. A call of a prowler in that area was logged just after midnight. That certainly verified an event.

But almost from the very beginning, the family was talking about alien intruders. The story of Hickson and Parker was being reported nationally at that time. They claimed an abduction on October 11, and according to various records, news of the case was reported, nationally, the following morning.

According to *The APRO Bulletin*, September-October 1973, it was at 9 A.M. on October 12 that APRO headquarters received the first call about the Hickson-Parker abduction. After learning the details, Coral Lorenzen tried to find a psychologist to go to Pascagoula to interview Hickson and Parker, but none of the consultants could get away fast enough. The job fell to James Harder.

Harder interviewed both men and used hypnosis to attempt to learn more about the sighting. After the sessions, he told APRO headquarters that it would be nearly impossible for the men to simulate the feelings of terror while under hypnosis without some kind of outside stimulus. According to Harder, the terror both men displayed seemed to be quite real.[2]

This was almost the same thing that Harder would tell Randle about the Roach case two years later. In fact, during the first session, there was a lack of any real emotion. Roach related the material and answered the questions, suggesting there were things that she didn't remember. Harder, of course, assumed that she had been ordered not to remember these things.

But more important, throughout the first session, Harder told Roach, "It may be a little bit frightening." Later he asked, "Is there something that you think would be frightening to remember?" Not long after that he said, "It might have been a very frightening experience at the time."

In the first session, there is almost no emotion. Throughout Harder told Roach it was frightening. In later sessions the fright, the fear, is evident. It is clear that Harder, through his technique and questioning, suggested to Roach that she was to be frightened.

---

[2]Ralph and Judy Blum, *Beyond Earth: Man's Contact with UFOs* (New York: Bantam, 1974); Jerome Clark, *High Strangeness: UFOs from 1960 through 1979* (Detroit: Omingrapics, 1996), 389–95; Philip J. Klass, *UFOs Explained* (New York: Random House, 1974); Randle, *Scenario*, 11–16.

He also implanted the idea that Roach's children were frightened and she would have to reassure them. He also said to her, "You were worried about your children." And later, he asked, "And you were worried about them?"

These two points, which were apparent in the later sessions, were almost nonexistent in the first. Roach was relating the tale as if it were something she had read elsewhere. There was no emotion in it for her. In fact, there was so little that Harder commented on it to Randle. And later still, he told Randle that the emotion finally displayed by Roach suggested that the case was real.

Harder also provided other information to Roach. For example, Roach mentioned there were machines and buttons. Harder then asked, "What kind of machines? Did they look like typewriters, computers?"

When she responded, "They looked like computers?" Harder asked, "What made you think they looked like computers?" Although Roach said, "Because they had wavy lines going through them," a better answer might have been, "Because you just mentioned them."

That's a little point, however. Implanting the idea that there were computers on an alien spacecraft isn't of much importance. During the interview, Roach said, "I don't remember being examined but I know I was."

This contamination can be traced directly to the Llanca article published by *Saga's UFO Report*. Llanca mentioned some of the things that Roach had described during her session. The examination by the aliens is an obvious one. The elongated eyes that Roach mentioned several times were also mentioned by Llanca. He mentioned the eyes several times.

There is one other interesting parallel between the Llanca story and Roach's report. Llanca said, "There are many viewing devices, many . . . two viewing screens. In one, stars can be seen."

Roach, in her first session said, "It's very bright [in the room]. . . . Door's on my right-hand side and I look out, you can see out at the stars, not the top but through the side, toward the top."

Harder asked, "You can see stars? Is it clear?"

"No, I can see stars. It's as if you could see the stars. It looked like a lot of technology."

Later, as Harder, Randle, and Roach discussed what she was talking about, she said that she could see the stars on a screen. She wasn't looking outside the ship but at a screen near the top of the room in which she stood. In other words, she is describing a scene straight out of the article about Llanca.

In addition, Harder would ask a somewhat leading question. After Roach mentioned that she knew that she had been examined, Harder said, "You think you might have been physically examined?" Roach had said nothing about a physical exam and to that point had been talking about a mental examination.

Later, he asked, "Did you get the impression that you were up on a table?" He also told her, "They probably were just taking a little skin sample or something superficial, cells or something?" There had been nothing in the interview, to that point, to suggest that the aliens were collecting any kind of tissue samples, but Harder implanted the idea.

Worse still, during the interview, Harder asked, "Did they put a needle into your stomach or anything like that?" Roach said that she remembered nothing like that during the first session. Eventually, she did say, "They put needles in me in places." But she said nothing about needles until Harder asked his question.

Later, as mentioned, Harder told Roach about Betty Hill's experience with needles into the stomach. After she awoke from the final hypnotic session, Roach told Harder that a needle had been pressed into her stomach. Clearly this was an implanted detail.

In a review of the transcript of the session, it appears that Harder was looking for something specific. He seemed to be working for a detail that Roach had a needle pressed into her. He was trying to draw a parallel between the Hill abduction and the Roach case.

The one area that Harder and Randle believed to be an important area of corroboration probably demonstrates the suggestibility of abductees. When Bonnie Roach mentioned a human with the aliens, both Harder and Randle thought it important. However, looking at the transcripts and notes carefully, it is clear that Bonnie was present during one session when her mother described the event.

Remember, both Roach and her daughter were in the room for the final session. Harder put both under, telling Bonnie to concentrate on what she could see. He then interviewed her mother, who provided a description of lying on the table, of the human with the aliens, and the scene as she remembered it. Later Bonnie told the same story with the same details. It's no mystery how she learned of it if she hadn't witnessed it during an abduction. She had just heard her mother tell Harder all about it.

We can take this even further if we want. Harder, throughout the sessions, was suggesting to Roach exactly what he wanted to hear. At one point, he said to her, "That's a very intelligent thing for you to recognize." Later, he told her, "It would be very helpful for me to know, as a scientist, what kinds of things that they are looking for. . . ." He also told her that he found some things interesting or very interesting.

As we look at the other sessions, we can see the influence that Harder exerted. He mentioned something, either in the first session or in private conversations held between the sessions, and those things appear later. Studying the transcripts now, it is very easy to see what ideas were unconsciously implanted by Harder and what ideas were contamination by the Llanca article she read before writing to Martin Singer at *Saga*.

Even if not recognized by Harder, the Roach abduction is a clear case of contamination. The event that precipitated it was the prowler in October 1973. But with the country talking about UFO abduction, and headlines from various newspapers telling readers that two scientists (Harder and J. Allen Hynek) believed the tale, it is not a stretch for Roach to leap from a prowler to alien intrusion.

Later, when she watched *The UFO Incident,* she wrote Randle, saying, "We watched the Hills' story on TV last night and were amazed how closely it resembled our experience. Some very insignificant details were alike."

Of course, the problem is the Hill story had been published in books and articles for years before Roach wrote her letter, and finally, most damagingly, Harder had been pressing her for details that he remembered from the Hill case. Discussing Hill with Roach before he finished his work was a major mistake.

This case reveals how easily a subject who wishes to learn about a "forgotten" experience can be subtly led. It demonstrates just how carefully the questioning must be done, and how easily, in today's world, contamination can happen. The Roach case is one of the most important to the understanding of alien abduction.

13.

BUDD HOPKINS

If there is one individual who is responsible for bringing the tales of alien abduction to light, it is Budd Hopkins. He has been described by some, with justification as a world-class artist who has given up his art to investigate claims of alien abduction. Publication of his 1980 book, *Missing Time,* and then the publication of *Intruders,* alerted UFO researchers, the media, and the public to the idea of alien abduction. *Intruders* was made into a CBS miniseries that was seen by millions.

Hopkins didn't start out to become an expert on UFO abduction. In fact, according to him, he had little or no interest in the topic even after he saw a saucer himself. Hopkins, writing in *Missing Time,* explained that even after his 1964 experience, his interest in UFOs was casual. When he read the 1966 *Look* magazine account of the Barney and Betty Hill abduction, he didn't believe it. He could accept the idea of alien spacecraft, even that there might be a living crew in the ships, but not that they were kidnapping humans for bizarre medical procedures as described by Betty Hill.

But Hopkins was beginning to believe. He bought John Fuller's *The Interrupted Journey,* and, according to him, was impressed by the fact the tale emerged from two sources while each was under independent hypnosis. And when Charles Hickson and Calvin Parker claimed to have been abducted in 1973, Hopkins found that he didn't reject the idea out of hand.

Even with his reading, he still wasn't actively investigating sightings. He mentioned that he would bring up the topic at dinner parties and would hear some interesting stories. He gradually moved from the passive to the active. Hopkins began to work with Ted Bloecher, who, at the time, was well acquainted with UFOs. With Bloecher, Hopkins began to hold "rather sprawling, unstructured gatherings." At one of these meetings, a man Hopkins identifies as Steve Kilburn was introduced to the use of hypnotic regression in relation to UFO sightings, and recovering memories about them.

What became interesting, looking back on it, is a statement Hopkins makes in *Missing Time*. "The next day, when I talked to Ted by phone, I mentioned the conviction in which Steve had described the uneasiness he felt whenever he was on that particular length of highway. At this point in our understanding

In folkore, a female demon that has sexual intercourse with sleeping men was called a *succubus*.

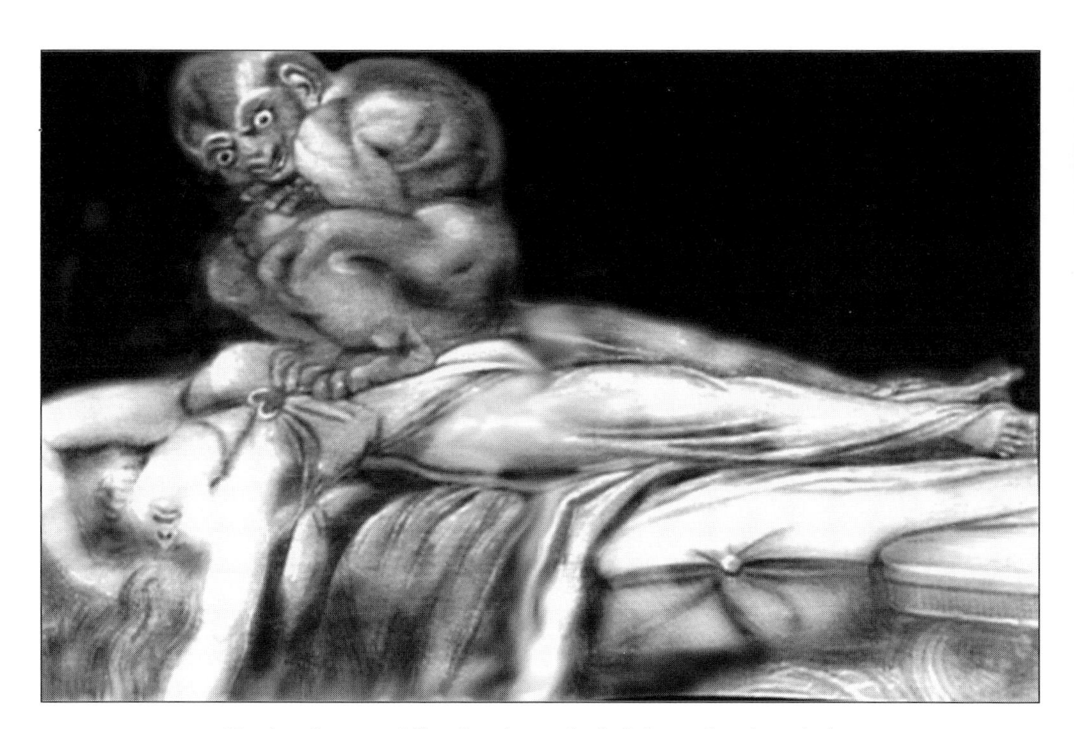

The incubus straddles the chest of a helpless, sleeping victim.
The incubus is the opposite of the succubus and preyed on female victims.

The mythological origins of alien abduction can be seen in this early painting of a satyr.

This human abduction by an alien creature is depicted
in this eighteenth-century painting.

An eighteenth-century pen-and-ink drawing depicting Dr. Mesmer hypnotizing a female subject. It is interesting to note that most eighteenth-, nineteenth-, and twentieth-century drawings that dealt with hypnosis showed the hypnotist as male and the subject as female.

An eighteenth-century representation of false and screened memories induced by hypnosis.

Dr. Franz Anton Mesmer, an Austrian physician, is credited with developing Mesmerism, which is the forerunner of modern hypnotism.

A classic example of sleep paralysis is portrayed in this eighteenth-century painting.

Professor of psychology Robert Baker is well-known for his research on the pitfalls of hypnosis.

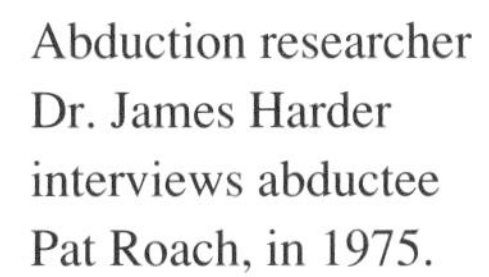

Abduction researcher Dr. James Harder interviews abductee Pat Roach, in 1975.

Professor Stephen Cece, an expert on the implantation of false memories.

Philip J. Klass

Budd Hopkins, artist and author, is considered by some to be the abduction guru because of his books *Missing Time*, *Intruders*, and *Witnessed.*

Dr. John Mack, M.D., the Harvard professor of psychiatry and Pulitzer prize–winning author of *A Prince of Our Disorder: The Life of T.E. Lawrence*, began his study of alien abduction under the tutelage of Budd Hopkins. Mack is a strong proponent of the positive effect abduction.

Professor Elizabeth Loftus is the foremost authority on implanted and false memories.

Professor Richard Ofshe, author of *Making Monsters*, is an authority on the implementation of false memory.

Philip J. Klass

Abduction researcher Budd Hopkins takes the podium with "Linda," a very controversial abductee and the focus of Hopkins's latest book.

Contactee George Adamski's unbelievable story of alien contact is unacceptable to researchers who accept abductees' unbelievable stories of alien contact.

of abduction cases, we accepted the Hill case and others like it as a general model."

When it was arranged that Kilburn would undergo hypnotic regression, all he knew consciously was that a particular stretch of highway bothered him. Each time he was forced to drive it, he became anxious. As they began the session, Kilburn apologized to each of them, Hopkins, Bloecher, and Dr. Girard Franklin, who would perform the hypnosis, for dragging them out.

Under hypnosis, Kilburn related a tale of abduction that was very similar to that told by the Hills. And Hopkins is quick to point that other cases emerging at the time reflected what Kilburn had said. Travis Walton mentioned the same kind of short, pasty white, balding creatures. Charles L. Moody said much the same thing. The patterns, according to Hopkins, were beginning to emerge.

In fact, Hopkins, writing in *Missing Time,* made much of these patterns. They are what makes the abduction phenomenon so real to so many. One tale of alien abduction could be ignored as an aberration. But when dozens of witnesses in widely separated parts of the country tell the same story, a weight of evidence is created. There are so many people that it is impossible to believe that all would be confabulating their testimony. Hopkins is quick to point out that a good quarter of the abductees have conscious memories of the abduction *before* the researchers interview them, which to him is another proof of the reality of the phenomenon.

Hopkins even made that point in an interview conducted by Jerry Clark, editor of the CUFOS *International UFO Reporter*. Hopkins told Clark, "The first point is, of course, that we have 20 to 30 percent of the abduction accounts recalled without resort to hypnosis."

Clark was pursuing the idea that hypnosis could lead to confabulation, suggesting that abduction reports were similar to past life regressions in that they both emerged under hypnosis. The subject, for example, entered into the session believing in past life and under hypnosis told of a past life that was rich in detail. Hopkins told Clark, "A person enters a past-life regression out of a desire to believe in reincarnation. He's entered a situation—allowed himself to be hypnotized—for that reason alone."

Yet in reading what Hopkins wrote about Kilburn, we find the same thing is true in that case. The whole investigation of Kilburn's experience came out of UFO research. Bloecher brought him around because of his fear of the stretch of highway, but Hopkins had noted the "pattern" of victims being taken from deserted highways. During the discussion that preceded the hypnosis, Kilburn said, "It was a very dark road, that was for sure, and it did occur to me as I was driving that this is the kind of place that something like that (a UFO encounter) should happen."

What that says, simply, is that Kilburn had entered the situation expecting some sort of a UFO encounter. They had discussed it before the hypnosis began. And they found a UFO event. This was a point that seemed to be lost on Hopkins, though he had already demonstrated that he understood it when he was describing the past life regression therapy.

Hopkins in that same interview with Clark also said that stories of reincarnation are pleasant, but tales of alien abduction are "at least unpleasant and sometimes horrifying, something where the person has every reason to feel terrified, humiliated and upset about the whole experience. Who would want something like that? Yet the stories emerge nonetheless."

Hopkins is suggesting that the tales must be true because the victims of alien abduction would not make up horrifying tales. They must be true, because, as John Mack has said repeatedly, "This is not a club that anyone would want to join."

But these arguments are based on the logic of those making them. There is no evidence that such is the case. When the victims appear at the therapist's office, or at Hopkins's group meetings, they expect to find an alien abduction and they do. The relative pleasantness of the experiences is irrelevant to those telling them.

And the parallel to reincarnation can be drawn even further. Although Hopkins suggests the tales of reincarnation are pleasant, often they are not. In one case, a woman afraid of the water learned, during a hypnotic probing of a past life, that she had been killed by an alligator as she jumped into a river. Hardly a pleasant experience.

Hopkins has spent the years since the Kilburn case interviewing others who claim abduction. During the Clark interview, he pointed out that he had a number of psychiatrists, psychologists, and psychotherapists who had come to him to explore segments of missing time with hypnosis. If these people, and their clients, aren't already "programmed" to believe in an abduction, then why search out Hopkins for assistance? Wouldn't anyone versed in hypnosis be able to help them learn what happened during this supposed missing time? Aren't each of them "primed" for an abduction, just as those who go to a reincarnation specialist are "primed" for learning about a past life.

And as others must admit, there really is no solid, physical evidence that the abductions are real. There is a great deal of testimony, much of it gathered under hypnosis, that suggests abductions are based in reality. It is this preponderance of evidence that seems to prove, at least to the believers, that alien abductions are real.

Even that testimony is somewhat contradictory. In Hopkins's own work we find evidence that the subjects do not easily accept the notion of abduction. Hopkins reports the tale of a man he called Philip Osborne. Hopkins wrote, "I noticed his interest in the subject had a particular edge to it. It was almost as if he accepted too much, too easily." Hopkins believed "that someone with a hidden traumatic UFO experience might later on be unconsciously drawn to the subject."

Osborne called Hopkins after an NBC UFO documentary and said that he had been struck by Kilburn's remark that anyone could be the victim of abduction. Osborne had been searching his memory for anything in his past that would indicate some sort of strange experience. Then, one night after the NBC program, Osborne awoke in the middle the night, paralyzed. He could not move, turn his head, or call for help. The experience was over quickly, but it reminded him of another, similar event that happened while he was in college. That earlier event had one other, important addition. He felt a presence in the room with him.

Hopkins, along with others, met Osborne a few days later to explore these events using hypnosis. During the initial hypnotic regression, Osborne gave only a few answers that seemed to direct them toward an abduction experience. According to Hopkins, Osborne told them that he "had more or less refused to describe imagery or events that seemed 'too pat,' too close to what he and we might have expected in a UFO encounter."

During the discussion after the hypnosis, Osborne told Hopkins that "I would see something and I would say to myself in effect, 'Well, that's what I'm supposed to see.' "

And while under hypnosis in a second hypnotic regression session held a few days later, Osborne said, "I'm not sure I see it. . . . I think it's my imagination. . . . It's gone now."

Osborne, it seems, had recognized one of the problems with abduction research, had communicated it to Hopkins, and then ignored it. Osborne was wondering if the "memories" he was seeing under hypnosis were real. Hopkins believed they were, so appears to take no notice of Osborne's concern. Hopkins believes in the reliability of hypnosis as a method for uncovering the truth. We, however, see those statements by Osborne as extremely important in attempting to understand alien abduction.

Osborne's initial experiences are also classic forms of sleep paralysis. Even the belief that an entity is in the room happens in about 80 percent of the cases of sleep paralysis. This simple explanation is all too often ignored by UFO researchers.

All of this pales against one of Hopkins's newest cases. He believes this is the case that will convince skeptics that there is something real to the abduction phenomenon. At the MUFON symposium in Albuquerque, New Mexico, in July 1992, he told the audience of an abduction that was witnessed by a high-ranking public official and that involved some of the security forces of the United States. It was a case of multiple abduction and multiple witnesses and would provide the final proof of alien abduction.

According to Hopkins, and to Linda Napolitano (identified by Hopkins as

Cortile), she originally learned of his work when she bought a copy of *Intruders* in 1988, believing it to be a mystery rather than a book about UFO abductions. When she read that the aliens sometimes left implants in their victims' brains and noses she began to suspect that she had been implanted. Thirteen years earlier Linda had found a bump on her nose she couldn't explain. A doctor said it was the buildup of scar tissue after surgery, but Linda never had any surgery that would account for the scaring, at least according to her.

A year after buying his book, Linda contacted Hopkins. Under hypnosis, Linda told of seeing Casper the Friendly Ghost at age eight. Hopkins believed this to be a "screened" memory, and learned by his probing that she had seen a top-shaped object flying above another apartment house near her home. Hopkins invited her to join his support group for abductees.

It wasn't long before Linda called Hopkins to tell him she had been abducted from her apartment early in the morning of November 30, 1989. It was about three when she went to bed. Before she fell asleep, she felt a numbness and tried to awaken her husband. He failed to respond. She saw a humanoid creature. When Linda reacted to it, she was paralyzed.

Little of the rest of the night remained with her. She had the impression of white fabric floating in front of her and of being seated at a table while small hands or instruments probed her.

Under hypnosis, three days later, Hopkins learned more about the abduction. Three or four alien beings had moved her through the apartment and into a bluish-white beam of light. She was apparently floated out the window, or through the wall, and then upward, into a large circular craft. She did mention that she knew that she was floating in the air, twelve floors above the street.

Once on board the craft, a fairly standard tale of abduction took place. She was returned to her apartment, where she found her family still unconscious. She reported that she at first believed them to be dead, but after a few minutes they returned to what Linda described as a "normal sleep pattern."

More than a year later, in February 1991, Hopkins received a letter that was signed "Police Officers Dan and Richard." They wrote that they had witnessed the abduction while sitting in their patrol car. They confessed that they were concerned about the welfare of the woman they'd seen abducted. Hopkins was, quite naturally, delighted. Here was independent corroboration of the Linda abduction. Unfortunately, Hopkins did not know their full names and they refused to meet with him, though they did supply additional letters and an audiotape recounting their beliefs and experiences.

What is interesting about this is the men's reason for contacting Hopkins. They claimed concern for the woman, but apparently had not checked on her during the year that had passed. After all, they had seen the abduction and knew which apartment in her building was involved. It wouldn't have been too difficult to ask a few discreet questions about the fate of the woman. That, apparently, never occurred to either Dan or Richard.

They did, however, visit Linda a few days *after* they wrote Hopkins. Linda said that she hadn't been concerned when they appeared because the police often canvass the building after a crime. Even after meeting with Linda, finding her to have survived her ordeal, both police officers refused to meet Hopkins.

The introduction of the police officers complicated the story. Hopkins learned later that they were not police officers but security men assigned to guard a very important person who was traveling through midtown Manhattan early in that fateful morning. This VIP, later identified by some as the former Secretary General of the United Nations Javier Perez de Cuellar, would eventually be questioned about the abduction.

The role of Dan and Richard began to change. Linda claimed that she was abducted by the two, taken to a beach house and then asked to wear a white nightgown not unlike that she had worn the night of the abduction. She escaped from the house and Dan chased her down the beach, catching her. He forced her into the ocean and pushed her head under water. Linda thought she was going to be drowned. Some kind of force hit Dan, knocking him back to the beach. Linda ran again, but glanced back as she heard a pistol being cocked. It was Dan taking pictures.

Linda kept running and Richard appeared next to her, apparently out of nowhere. After Richard explained that he could control Dan, Linda agreed to return to the beach house. While Richard put Dan in the shower to wash off the sand and mud from the beach, Linda searched the house, discovering stationery with a Central Intelligence Agency letterhead.

More corroboration of Linda's claims came about two years after the events. In November 1991, Hopkins received a letter from a woman he named Janet Kimball who claimed to be on the Brooklyn Bridge when, at about three o'clock in the morning on November 30, 1989, she witnessed Linda's abduction. Kimball wrote that she could hear the screams of horror from people in other cars stalled on the bridge near her. She included a number of drawings of what she had seen that Hopkins believed provided further confirmation of Linda's strange tale.

What is interesting is that we have a host of witnesses, if Kimball is to be believed, who witnessed the Linda abduction. At the time it happened, their cars were stalled on the Brooklyn Bridge. Although there are other examples of vehicles being stalled by flying saucers in the UFO literature, this case seems to be unique. The witnesses panicked, screaming in terror. And once this horrifying experience was over, not a single person reported what they had seen to the police, the Air Force, the newspapers, or any UFO researcher. All returned home and said nothing about the experience until Kimball wrote to Budd Hopkins reporting it to him.

It is also interesting that none of the alleged witnesses, and we use the term "alleged" on purpose here, contacted any of the other UFO investigators in New York or the United States. There are certainly many of them who have appeared on national television shows, who have written books and articles on the topic, and who have achieved a certain amount of fame. Yet none of these witnesses wrote or contacted any of the other researchers. Instead they were all drawn to Hopkins, providing the much needed corroboration for the Linda case.

And now there are more mysterious witnesses who are coming forward, providing additional detail and corroboration. Hopkins asked for patience, wanting to explore the case quietly himself. A fair request in the beginning and one that each of us can understand.

But Hopkins was the one who took this abduction into the public arena. Since the MUFON symposium in Albuquerque, Hopkins has lectured about the case throughout the United States. It has been written up in *OMNI*, the *Wall Street Journal, Paris Match*, and the *New York Times*. Hopkins and Linda have appeared on national television. With that sort of reporting, it is only fair that critics of the case be allowed to voice their opinions, though those in the UFO abduction community have been unanimous in their opinion. Any criticism is bad, unethical, and distorted.

The arguments against the case have grown but have not been fully refuted. Instead, as usual, the rebuttal has degenerated into name-calling. Hopkins wrote in the *International UFO Reporter* about his three main detractors, Joe Stefula, Richard Butler, and George Hansen. In the article, he claimed, "I had no idea if he [Stefula] was or was not a government agent (I still do not know). . . . I was told later that, far from being 007 or an officer in G2, Stefula had been an enlisted man in the Army. . . . He had acquired his habits of distrust and heavy-handed interrogation, I assumed, from too many months of listening to soldiers lying about who stole cigarettes from the PX."

Hopkins took it further, suggesting that Stefula might be a government agent. This is the one allegation that proves someone has made it in the UFO community. You can't listen to him because he is a CIA agent, or Air Force intelligence, or with the government. Everyone who has ever written, investigated, or lectured about UFOs has had this allegation hurled at him or her at one time or another.

But the fact of the matter is that Stefula ended his military career as a warrant officer, not an enlisted man. He was a member of the CID, the Criminal Investigation Division of the Army. He technically wasn't a military policeman, the distinction being somewhat similar to that between a homicide detective and a patrol officer.

Hopkins also told Butler that he suspected Butler of being a government agent as well. He warned that he planned to tell others of his suspicions. And a few weeks after Stefula, Butler, and Hansen met with Hopkins on October 3, 1992, Hopkins told several people that he believed Hansen was a CIA agent.

Then, to question their reliability so that people would not credit what they had to say, Hopkins attempted to link them all to Philip Klass, the well-known UFO debunker. Hopkins wrote about a critical white paper authored by Stefula, Hansen, and Butler, "Hansen's words could be mouthed with equal ease by CSICOPers and Air Force PR men. . . . Together the three set about, in cooperation with Philip Klass, to attempt the poisoning of a sensitive and important ongoing investigation. . . ."

Hopkins, however, in his *International UFO Reporter* article wrote, "Philip Klass is the first known figure they approach with their ideas." They say that they never approached Klass. At the MUFON symposium held in Albuquerque, Klass happened to overhear Stefula commenting negatively on the Linda case. He asked Stefula questions about it, which Stefula answered. The "meeting" between Stefula and Klass was a lucky accident for Klass and not a planned rendezvous by any of the participants. Philip Klass has said that no official contact has been established among the parties.

In fact, in his response to the white paper written by Stefula, Hansen, and Butler, Hopkins, rather than giving the facts as he knew them, attacked the authors. He wrote, "They have talked with none of the witnesses whose accounts corroborate those of the Cortile [Napolitano] family."

The question to be asked of the three is "Why not?" But the answer is quite simple. Hopkins had hidden the witnesses behind a veil of secrecy, preventing anyone from talking to them and learning exactly what they had to say. The witnesses were unavailable. There are too many questions about this case to allow us to ignore this gap.

In one of the few points that is somewhat relevant, Hopkins wrote, ". . . [the Linda case] has been publicly vilified as a probable hoax by Philip Klass, Joe Stefula, Rich Butler, and George Hansen, though they have not presented a shred of convincing evidence. . . ."

Overlooking, for the moment, the fact that they are not required to prove the case a hoax, but that Hopkins must prove it authentic, we find the statement by Hopkins just plain wrong. The three had caught Linda in a couple of lies about the case, including the claim that her husband was born in the United States. That turned out not to be true. She later acknowledged that she had "misled" Hansen.

It was Linda who told Stefula that she had a financial arrangement with Hopkins. She claimed that she would share in the royalties from the book, and the profits from movies or miniseries. During the October 3 meeting, Hopkins denied that any such arrangements existed. Linda then claimed that she had "deliberately planted disinformation." In other, more plain terms, she had been caught in a second lie, while, at the very least, suggesting an arrangement with Hopkins.

Penelope Franklin, the editor of the newsletter, *IF—The Bulletin of the Intruders Foundation,* and, according to Hopkins, a "coinvestigator of the Linda case," thought the lies made no difference. In fact, she told Hansen during a break in the October 3 meeting that she thought Linda was justified in lying to them about the abduction.

Stefula, Hansen, and Butler believed that the whole case was beginning to collapse. Linda had been caught lying about some aspects of the report, and although she had eventually admitted to these lies, the abduction *investigators* were providing her alibis for it. In most investigations, when a witness is caught lying, the investigators don't provide excuses for them. They become suspicious of everything else that witness says.

Stefula, Hansen, and Butler had even found what they believed to be a literary basis for Linda's tale. According to the three opponents, Linda's story seemed to be a composite of two characters in the science fiction novel *Nighteyes*. The similarities are, in fact, startling. Hopkins dismisses them, writing in the *International UFO Reporter*, "Now the truth is that apparently no one except Butler finds any similarity between the novel and Linda's account, and I invite readers to test this assertion by giving the novel a read."

Later Hopkins wrote, "What poor Butler failed to notice about the so-called similarity of *Nighteyes* and the [Linda] Cortile case is that the latter has as its essential feature an observed abduction in New York City, in which a major political figure, two government security agents, and another witness at a separate location all attest to watching a woman and three aliens floating 12 stories up in the midst of an abduction. Nothing remotely like this extraordinary event occurs in the book."

Yet, many of the elements from the book do appear in Linda's abduction. Linda claimed to be abducted by a UFO hovering over her high rise apartment. Sarah, one of the characters in the book, was abducted from her high-rise apartment. Dan and Richard originally claimed to be on a stakeout and in the novel the two government agents were on a stakeout. Linda was eventually kidnaped by Dan and Richard. Wendy, another of the characters in the book, was kidnaped by the two government agents. Linda was taken to a "safe house" on the beach and in the book one of the safe houses used is on the beach.

Greg Sandow, a journalist living in New York, explored some of these allegations as he searched for the truth. Sandow noted that some of the items on the list of similarities were something of a stretch. He points to others suggesting that the stories are dissimilar. For example, in the book, the two men who plot the kidnapping are not two active intelligence agents but are, in fact, a renegade intelligence agent and a newspaper reporter, and kidnap not the abductee but an abduction researcher. But the real point is that the kidnapping does take place by two men.

Sandow, however, wrote in the summer 1997 issue of the *International UFO Reporter*, "But to see how fanciful these supposed similarities are, just look at the first one, which—once you read the book—turns out to be a mighty stretch, verging on a fabrication. There simply is no scene in *Nighteyes* where a UFO hovers over a building in New York."

Sandow's investigation into the Linda case did have one attribute that the Stefula, Hansen, and Butler investigation did not. Hopkins cooperated with him from the beginning, allowing Sandow access to some of the most important of the documentation. Sandow's conclusion is that the case is either a hoax or it is real without any middle ground. In the end, he believed Hopkins, noting, "The hoax is getting cumbersome; its dramatis personae keeps multiplying. And where's the evidence for all of this? There isn't any. It's not enough to say 'de Cuellar couldn't witness an abduction' or 'Richard and Dan don't even exist' or 'this whole abduction thing is nonsense' and think you've proved the whole thing a hoax. The hoax, complex as it is, should leave a trace. Skeptics have a right to ask for evidence that the case is real. So why can't believers—or for that that matter, people on the fence—ask for proof that it isn't? As I've said, the proof doesn't exist. There's nothing conclusive on either side of the 'Linda' case, and even the skeptics will have to live with that."

But Sandow, in his investigation, did provide new evidence to support the case. He didn't talk to Kimball about her involvement but did listen to the tapes that Hopkins had made of his interviews with her. Sandow was impressed with her emotions as she described the abduction from her vantage point on the Brooklyn Bridge. But before Sandow entered the investigation, Kimball had withdrawn, explaining that her family thought she was "crazy" for confiding in anyone. They wanted no part of a flying saucer story, so Kimball reluctantly ended her communication with Hopkins.

Sandow, who said he knew who she was, respected her desire to bow out of the UFO arena. He didn't call in an attempt to corroborate the tale, satisfied with listening to the tapes. While we agree with Sandow's reasoning for avoiding contact with Kimball, we must also point out that as a witness, she is useless. Sandow noted, ". . . her withdrawal from the ranks of active witnesses clearly is a weakness for the case."

And her description of the other drivers on the bridge that night opens up a major question. According to her taped statement, Hopkins asked of those drivers, "Were they screaming?"

Kimball responded, "No, no, they were kind of like, they were blowing horns, you know what happens when there's a traffic jam. Everyone was kind of blaming everyone else for their car stalling."

And that is a major inconsistency. If the cars are stalled, and Kimball also made it clear that she believed the lights on the bridge were out, then how could the car horns be blaring? The implication is that the proximity of the UFO suppressed the electrical systems, stopping the cars, and dimming the lights on the bridge. Yet the horns, part of the electrical system, continued to work. And if the horns worked, then couldn't someone have used a cellular telephone to call for help?

Another of the witnesses that allegedly corroborates the Linda case is "Francesca," who saw a bright light on the night that Linda was abducted. Or believes that she saw a light on the night Linda was abducted. Sandow suggests that the convoluted way they determined the data tends to corroborate it. And he did talk to Francesca's ex-husband, who did confirm that his wife had seen something frightening in the time frame of the Linda abduction.

Of course, it should be noted that seeing a bright light in the courtyard of an apartment is not the same as seeing an object hovering over the building. And it could be argued that the date of Francesca's experience is just a best guess.

As mentioned, former Secretary General of the United Nations Javier Perez de Cuellar has denied his involvement in the case. Linda, in one of her sessions with Hopkins, suggested that de Cuellar was also abducted, as were Richard and Dan. In fact, Richard has a long history of abduction, according to Linda. They have been together on board alien craft since childhood and have been lovers.

De Cuellar sent a fax to the PBS science show *Nova*. He wrote, "I cannot but strongly deny the claim that I have had an abduction experience at any time. On several occasions, when questioned about this matter, I reiterated that these allegations were completely false and I hope that this statement will definitely put an end to these unfounded rumours."

The true believers suggest that de Cuellar would say this even if he had been abducted, so they suggest that the statement isn't the last word on the topic. However, when we begin to look into the history of public statements, we often find that when an allegation is true, the denial is usually made with much less force. During the Nixon administration this was called a nondenial denial. But here, there are no weasel words. De Cuellar is telling the world that he was not involved and that is just not good enough for the believers. They sense some sort of conspiracy to hide the truth.

Finally, one of the witnesses that was featured in Hopkins's book about the case seems to have recanted her testimony. A woman identified by Hopkins as

Marilyn Kilmer, who, according to Hopkins, at one time claimed to have been abducted with Linda and de Cuellar, now won't agree that the abduction took place. Sandow reported, ". . . Kilmer won't agree that this abduction ever happened. Or so I'm told: I called her, and she preferred not to speak with me, saying only 'Budd and Linda know the truth' (and offering no further explanation)."

Hopkins, defending the Linda case, wrote, "To sum up, I believe that the true nature of the New Jersey trio's [that is, Stefula, Butler, and Hansen] house of cards must be looked at this way: I have a mass of evidence—physical samples, audio and videotapes, letters, photographs, affidavits, hundreds of typewritten pages of testimony, and so on, involving 13 key witnesses and covering about 15 separate incidents in this case. This material, *known only to me* [emphasis added], comprises, let us say 5000 facts. It is accurate to say that these 5000 facts form an extraordinarily consistent account, in which everything makes sense. . . ."

This seems to demonstrate the whole problem here. Hopkins claims to have five thousand facts. Yet, all the corroboration seems to have been driven by Linda to Hopkins. In other words, the story comes from a single source.

But this case, and the commentary written by Hopkins, provides a good look at his work in the public arena. Linda came forward *after* reading his book, claiming to have a bump on her nose that might by an alien implant. X rays taken seem to confirm the implant, but when arrangements were made to remove it surgically, it had mysteriously disappeared.

Hopkins has shown slides of the X rays at his lectures. But even this bit of physical evidence is in dispute. The technician who made the X rays was a relative of Linda.

Hopkins wrote in *Missing Time* that he hadn't been interested in UFOs until he saw one. Then that he didn't accept the idea of abduction until he read Fuller's *The Interrupted Journey*. Now the stage was set. All Hopkins needed was a victim of abduction.

The pattern of evolution of the abduction is interesting. Hopkins learned, from Fuller, that abductions took place on lonely highways at night. That is what he found. But then, as he tried to confirm the stories by adding others to them, they began to evolve slowly. A victim of one abduction suddenly became the victim of a second or third and finally it was a lifelong problem. The aliens spotted their victims at an impossibly early age, studying them throughout life just as human scientists studied the life cycles of the animals. And finally, it was a multigenerational problem. The aliens were abducting the members of families back into the nineteenth century.

Now the next step in the evolution has been presented. Hopkins spoke at the Fifth New Hampshire MUFON UFO Conference on September 9, 1995.

He had discovered the aliens now "co-opted" their abductees, forcing them to assist their captors.

Hopkins reported that Linda was "the very first person I ever heard this from . . . from her own personal experience . . . It could almost be called the Linda Cortile Syndrome."

He also reported that a woman from Connecticut "remembered beings inside a UFO. She was wearing a blue, skintight uniform without buttons, without zippers, and she doesn't know how she got into it. And she's with the Aliens and they instruct her telepathically what she has to do. . . . She walks out of the craft, which is on the ground, and there was a car that had stopped. And she walked over to the car and sitting in the car is an absolutely terrified large man, an African-American man, and he is obviously paralyzed. She takes him out of the car, zip . . . And he is looking at her with terror . . . she's a tiny woman. And she leads him up to the entrance of the craft and that was it. That's what she had to do. Now, while she's doing that, she's upset with herself . . . 'Why am I doing this?' "

Hopkins described another woman, also wearing the skintight blue outfit, who walked into a room of the craft where there are four or five naked humans. She gestured at one and then led that person from one room to another. That was all that she had to do.

But what is interesting here is that something new has been added, first by Linda and then by others. Clearly, none of these women knew one another, so they couldn't have shared the details. It was Hopkins who discovered this in his work with Linda and other abductees. And then the others corroborated the story, giving Hopkins some of the same details that Linda had mentioned. The question that Hopkins asks is how these three women could all come up with the same detail if the abduction and the experience wasn't real?

An examination of the sessions printed in Hopkins's two books, *Missing Time* and *Intruders,* reveals that he does, in fact, subtly lead the witness. During the sessions, leading questions are asked. Ones that Hopkins does not view as leading but which direct the course of the session and the information recovered nonetheless. C. D. B. Bryan, who could not be classed as a skeptic, and who was invited to witness a number of sessions, wrote about Hopkins's leading of witnesses in *Alien Abduction, UFOs, and the Conference at M.I.T.* On page 417, he expresses his concern, noting, ". . . of course there were moments when I felt Budd Hopkins might have been leading them. . . ."

Bryan went further, reporting on one of the session with a woman named Alice, who claimed to have multiple abductions. The transcript provided in Bryan's book is a shining example of how easy it is for a hypnotist to inadvertently suggest things to a subject. The transcript is of Budd Hopkins regressing Alice. While Hopkins has often claimed that he is very careful not to lead his subjects, the transcript of this session with Alice is rampant with highly suggestive statements.

When searching for genetic experiment memories he says. "Now being very systematic, you're lying there and we're moving up to your female parts, to your genitals. What do you feel in that area that feels different in any way?"

This statement, of course, prompts Alice to tell a story about pain in her genitals. "It feels tight," she replies, to which Hopkins says, "Is the pain in your abdomen, or is it down lower."

These words are a direct suggestion that it is lower. The stage has now been clearly set to recover a genetic experiment.

"Down lower," Alice replies, and Hopkins once again says, "In your genitals?" reinforcing the suggestion.

This avenue of suggestion leads Alice down the path to recovering memories of artificial insemination by the aliens. Hopkins, searching for the insemination memory, says, "If you were pregnant at some point we don't know whether you were but if you were, your body is going to know that a seed was inserted into it somehow to produce that baby. There are different ways it could have occurred. Through intercourse the normal way a penis. People can be artificially inseminated, all kinds of things can happen." Now he has suggested artificial insemination, and Alice then produces the required memory.

"Tube. It's a long tube like Dr. Fulton uses." she says immediately. According to Bryan, Dr. Fulton was Alice's veterinarian. Alice's statement would suggest that she had seen Dr. Fulton using such a tube on an animal. It is common for people under hypnosis to incorporate what they believe and what they have seen into their recovered memories. Humans, however, are not inseminated with a tube.

Alice describes the tube as long, clear, and flexible, and slightly thinner than a pencil. She then tells Hopkins that the tube is inserted through her vagina into her fallopian tube. At this point, Hopkins suggests that this may not be her first insemination when he says, "Have they done this to you before?" This prompts Alice to tell Hopkins that her first encounter was at twelve years old and fishing with her father. Her father stood by while the aliens inseminated her. She says that she felt great hatred toward him for not helping her.

Hopkins then tells her that her father was paralyzed by the aliens and could not help. Of course, he could not possibly know this because he wasn't there at the time, but his statement clearly indicates that he wishes Alice to accept this as fact.

Alice described the insemination. Earlier she had described the tube that was used when she was an adult as thinner than a pencil, but Bryan reports that she now says the tube used when she was twelve was thicker than a pencil. The entire episode takes place not on a craft but right on the riverbank.

Alice had been in therapy before to deal with an alleged rape by her father when she was twelve. After the session Hopkins told Alice that her therapist had been wrong, her father did not rape her. It was the aliens. "You were being treated for the wrong disease," he said. Of course Hopkins could not possibly know if Alice's father raped her or not.

What this demonstrates for us is that Hopkins, like so many other investigators, leads the witness to the point he wants to reach. He started learning about those abducted from a lonely highway, and finally reached the point where the aliens came into the house. In *Missing Time* he mentioned the Patty Price (actually Pat Roach) case, where she reported the beings inside the house. That is the first case where it was reported the aliens had entered the victim's home.

It wasn't long before everyone was reporting the aliens inside the house. It wasn't long before the aliens had developed the ability to walk through walls and to defeat any attempts to videotape their activities.

In fact, Hopkins wrote, in *UFO* magazine, "In the later 1970s, my own ongoing investigations replicated the Hill case findings, but I discovered several new and startling patterns which earlier investigators apparently overlooked."

Looking at the whole, complex history of alien abduction, as reported by Hopkins, we see he is exploring new ground. Alien abduction tales are old hat now and people with professional credentials in psychology and psychiatry have entered the arena. This does mean that Hopkins is lying or inventing tales.

What we see now is the aliens not only abducting people repeatedly, not only stealing genetic material from their victims and forcing the females to bear children. They are now forcing them to assist in the abductions. They are forcing their victims to perform some kind of limited task while wearing the "standard uniform of a one piece, tight-fitting garment."

The question that must be asked is if this makes any sense whatsoever. Repeated abductions of the same individual makes little or no sense. Overlooking the impossible task of tracking the individual without being detected, what would be the point? Doesn't abduction make more sense on a limited basis? And why not abduct those who will not be missed if they disappeared forever? A hitchhiker on a lonely road would be the perfect victim. Who would suspect an alien race?

So now Hopkins is asking us to believe in repeated abductions and forced servitude by the aliens, all based originally on the Linda case. She provided Hopkins with the first such report and then Hopkins set out to confirm it with other witnesses. Just as his original research confirmed to his satisfaction the reality of the Hill abduction, his current research has confirmed, to his satisfaction, Linda's claim of forced servitude.

What we have here is a man who came into the field slowly, dispassionately, looking for confirmation of his own UFO sighting. While it could be argued that he failed to find that, he did find evidence that UFOs are extraterrestrial, which is sort of the same thing.

But more important, he learned that aliens were abducting humans for a variety of reasons. That research has pushed him to the front of the pack of abduction investigators. The problem is not that he looks for alien abduction, the problem is that he works so hard to find it.

14.

DAVID JACOBS

David Jacobs officially entered the UFO field in 1975 with the publication of *The UFO Controversy* in America. It was, in fact, an expansion of his Ph.D. dissertation and has become one of the classic works in ufology. His academic training provided him with the background and expertise to examine the UFO field from the perspective of a historian. He brought his talents and training to his examination of the history of UFOs.

But his interests changed and he became fascinated with alien abduction. He spent several years working with Budd Hopkins. This work resulted in the publication of *Secret Life* in 1992. It contained, according to Jacobs, firsthand accounts of alien abduction gathered during his personal investigations.

His conclusion in that book was that alien abduction was real. He based this on the hypnotically recovered memories of the victims, many of whom he had hypnotized more than a dozen times. To him, the conclusions were inescapable. Unless there was some form of psychopathology that science had failed to detect, Jacobs knew, based on his own research, that his conclusions about alien abduction were valid.

These conclusions were fairly benign when compared to some of the reports being made by other researchers. Jacobs was suggesting that some form of experimentation was taking place, that a large number of people had been victims of this experimentation, but that it was, in and of itself, while essentially negative, harmless. That is, if the trauma to the victim and the ramifications of the abduction were ignored, it could be considered harmless.

But his latest book, *The Threat*, subtitled *The Secret Agenda: What the Aliens Really Want . . . And How They Plan to Get It,* has taken us far beyond those early ideas. Alien abduction is now a threat to the survival of the human race. The aliens are creating a race of hybrids, that is, a cross between humans and themselves so that they can move into human society to take it over. Never mind that this genetic manipulation comes from the realm of science fiction. He talks of true alien rape in which hybrid men have raped and tormented the human women to whom they have been assigned.

He takes it further, suggesting that the creation of these hybrids is just one of the steps in taking the planet away from us. Alien abduction, according to

Jacobs, now has absolutely nothing positive about it. Jacobs has determined, through his own, personal research, and in direct conflict with many other abduction researchers, that alien abduction is completely negative.

And to make his case even stronger, Jacobs attempts to make the point that he is doing well-reasoned and very carefully controlled scientific research. To prove that he is right, he has spent considerable time countering arguments that have been raised by many critics of his methods. He attacks such psychologists as Michael Yapko and Elizabeth Loftus for suggesting that alien abduction is the result of false memory. Jacobs writes, "Neither they, nor any other critics have ever presented evidence that abduction accounts are the products of false memory syndrome (or, for that matter, of any causative factor other than what the abductees have experienced). The reason they have not presented the evidence is that they do not understand the abduction phenomenon."

Yet he readily admits that his own research has shown that hypnosis can cause false memories. He knows that hypnosis handled improperly will lead to confabulation, invention, and imagination. He writes, "That people can have false memories is beyond doubt. Given certain circumstances, they can, for example, invent complex accounts of sexual and physical abuse. The False Memory Syndrome Foundation in Philadelphia is filled with members who have been unfairly accused of sexual abuse. People can convey false memories with such conviction and sincerity that they have fooled many investigators. Uncovering false memories of sexual abuse can also lead to major emotional upheavals in people's lives. Families are torn apart, siblings are estranged, lawsuits are instituted, innocent people are unjustly accused and even jailed."

Jacobs also recognizes that the victims "relive their experiences with the emotional impact of real events." He is aware that the lack of hard evidence for these stories of abuse lessens their credibility, and knows that this lack of physical evidence suggests many of these stories are not real. Finally, he is was aware that the victims in these cases will go out of their way to manufacture excuses for the lack of that evidence.

To that end, he writes, "When the 'victims' are confronted with facts (investigators have not found dead babies; no babies were reported missing at the time and place of the ritual abuse cases), they angrily provide explanations—such as that the mothers themselves were Satanists who gave up their babies for sacrificial purposes and did not report them missing."

Curiously, he also seems to realize that it is the mind-set of the therapist that causes these false memories to emerge. He wrote in *The Threat* that "uncovering false memories is usually facilitated by a therapist who is convinced that a client has been sexually abused (or whatever abuse the false memory recounts), even though the client has no memory of it. Through insistent persuasion, the therapist inculcates the idea into his client that all his emotional problems stem from the repression of the memory of some earlier trauma."

All this being said, Jacobs still believes that none of these factors apply to recovered abduction memories. Jacobs writes that "[no] critics have ever presented evidence that abduction accounts are the products of false memory syndrome."

To bolster his argument, Jacobs states that abduction accounts differ from false memory syndrome in five significant areas.

1. "Abductees do not recount only childhood experiences. They do, of course, recall abduction events during childhood, because the abduction phenomenon begins in childhood, but they also recall abduction events as adults. In fact, many abduction accounts, unlike false memory accounts, are of very recent events. . . . The immediate reporting of this event does not fit the description of false memory syndrome."

Jacobs seems to be arguing that abductions are real because some have only recently occurred. It indicates an ongoing problem for the victim, unlike with reports of satanic ritual abuse. Jacobs seems to believe, for example, satanic ritual abuse comes from false memory because the instances of abuse are not recent, at least from the victims' perspective. Unfortunately this is absolutely false. The case of Paul Ingram, examined at length later, is one proof of this. His daughters, who had accused him of sexual abuse, rape, and satanic worship, when asked when the last abuse had taken place, answered, "Last week." The abuse, according to them, was an ongoing thing, and had the same immediacy as the abduction reported by others.

Cone has worked with many victims of Satanic abuse who claimed to be tortured every week. They often feared for their lives because the abuse was continuing unchecked. Some claimed that they had nowhere to turn, because they were periodically victimized by the police and the clergy in their towns, who were, they believed, members of the cult. In other words the ritualistic abuse was an ongoing thing just as are the alien abductions.

2. "In contrast to victims of false memory syndrome, abductees have indirect corroboration of events. For example, I [Jacobs] was on the phone with Kay Summers, whose abduction experiences began while we were talking. She described a roaring noise sometimes associated with the beginning of an abduction, and I could hear this noise over the phone. Hypnosis later revealed that soon after she hung up the phone, she was abducted. False memories do not take shape simultaneously with the occurrence of actual events during which a researcher is an indirect corroborator."

He accepted this roaring noise as evidence of an abduction just because she said it was. He accepted the abduction as real because he was able to recover it under hypnosis. But in reality, this is not indirect evidence of anything, other than a noise over the telephone. Dr. Alvin Lawson reported much the same sort of indirect corroboration during his investigation of what he called the Garden Grove abduction, which is examined in detail later. Lawson, initially impressed with the indirect evidence, eventually concluded the evidence had been manufactured by the victim.

And even if the victim was not engaged in an elaborate hoax, the noise that Jacobs reported could have been anything. That is the point that he seems to have missed in his rush to accept, as real, the abduction phenomenon.

3. In contrast to victims of false memory syndrome, abductees often remember events without the aid of a therapist. They can remember events that happened to them at specific times in their lives. They have always known that the event happened, and they do not need a therapist to reinforce their memories.

He is wrong again. In fact, the majority of satanic abuse victims that Cone has worked with claimed to always have been conscious of their abuse. Some claim to have been raised since infancy in a cult. Several patients have even told Cone that they could take him to the cult meetings, if he wanted proof. So this refutes, not only this point, but also the claim that the ritualistic abuse took place in the victims' distant past.

Many seek therapy because of memories that suggest, to them, that they were abused as children. There are those who have such memories, have believed they have been personally abused, and do not need the therapist to reinforce the memories. The situation, as described by Jacobs concerning false memory, is inaccurate.

4. In contrast to victims of false memory syndrome, abductees are physically missing during the event. The abductee is not where he is supposed to be; people who search for him cannot find him. The abductee is usually aware that there is a gap of two or three hours that neither he nor anyone else can account for. Such physical corroboration does not exist in false memory.

This is the same argument made to support the validity of the Travis Walton abduction. Walton was physically missing for five days while family, friends, and police authorities were searching for him. That does not prove that he was abducted by aliens, only that he was missing for a period of time. In the Garden

Grove case, the victim disappeared for twenty-four hours, but again, this was not proof that an abduction took place.

Jacobs adds his own similar event to the list. He reported that Janet Morgan's sister, Beth, had been asked to baby-sit six-year-old Kim. While Kim slept on the couch, Beth took a bath. She experienced a strange sensation when a "mental haze" came over her. She suddenly came out of it and thought of the child. She dressed and rushed downstairs, but couldn't find Kim.

She searched the house. She searched outside the house. She ran up and down the street shouting for Kim. A neighbor confronted her, learned of the trouble, and went into the house to search it himself. Kim was asleep on the couch, right where she was supposed to be.

To Jacobs, this meant that Beth had been "switched off" by the aliens. They had rendered her unconscious while they abducted the child for their experimentation. But Beth had broken out of her "trance" too soon. She had searched for the child, failed to find her, and panicked. Before the neighbor could search the house, the child had been returned to the couch.

This, to Jacobs, is more indirect evidence. The child was physically missing from the house for a short period of time. But it is actually proof of nothing of the sort. It is only evidence that Beth had somehow overlooked the child asleep on the couch and drawn the conclusion that Kim was missing.

And just as important is the fact that Jacobs accepts these claims with little foundation. He was told that the abductee, during the abduction, was missing, but there was no documentation to support it directly. There are alternative explanations for this that do not require alien abduction.

5. In contrast to victims of false memory syndrome, abductees can provide independent confirmation of the abduction. Approximately 20 percent of abductions include two or more people who see each other during the abduction event. They sometimes independently report this to the investigator.

In the Paul Ingram case, the two daughters tended to corroborate one another. How could, and why would, these two young women make up these horrific tales about their father if they weren't true? What this demonstrates is that Jacobs's argument here is not accurate. There are cases of "multiple witness" sexual and satanic abuse that were later found to be invention or false memory. It means nothing without corroborating physical evidence, which is so frequently missing, especially in alien abduction.

The McMartin Preschool case in Southern California is another example of multiple witnesses to sexual and satanic abuse. After seven years of prosecution, after years of investigation, after all the interviews conducted, in the most expensive case ever tried in the state of California, no one was ever convicted of abuse of the children. There were, literally, dozens of "victims," which means that the corroborative testimony, while interesting, must be coupled to some kind of physical evidence. In alien abduction there is no direct physical evidence. There is never the direct physical evidence.

Continuing the arguments in that same vein, Jacobs believed that, "unlike victims of false memory syndrome, abductees do not usually experience disintegration of their personal lives after they become aware of their situation."

Sadly, and very obviously, this is not true. We have seen dozens of cases where the person's life absolutely fell apart as a result of their abductee status. They were divorced, they lost jobs, their schooling ended, and they became estranged from friends and family. The sad tale of Leah Haley is just one such example.

Jacobs believes that before false memory syndrome came to prominence most therapists assumed that abductions were due to repressed memories of sexual abuse in childhood. Therapists postulated that because the sexual abuse was so traumatic, the victim may have unconsciously transformed the abuse into an abduction account. But he believed the opposite was actually true; that is, people remembered sexual abuse when in reality they were victimized by alien abduction.

One of Jacobs's claims is that he is able to ferret out real abductions from imaginary material, or false or implanted memory, during a hypnosis session. Among his techniques is his requirement that anything he has never heard before be considered highly suspect. For example, Jacobs recounted a story of an abductee being asked to touch an alien's head. This bewildered him, because, "I never received, nor did I ever hear of, another report of an abductee who had been required to touch an alien's head and receive loving emotions. Also, Melissa [the abductee who made this statement] would never, in our more than thirty abduction sessions, recall a similar event. All this suggested that she might have unconsciously absorbed a memory fragment from her abductee friend and been confused about other details."

He continued, "The first time Melissa had told me about the small alien with great conviction and emotion. Now when I asked her about the encounter, she was not sure that it had ever happened. I then questioned her about the council meeting with the aliens. Melissa thought for a second and said that perhaps *this had happened to another abductee with whom she had been friends* [emphasis added]. She was pretty sure that it had not happened to her. This experience taught me an invaluable lesson because I realized that, in all sincerity and honesty, abductees might sometimes remember things that were not true."

THE MCMARTIN PRESCHOOL

On August 12, 1983, Judy Johnson, a single mother, complained to police that her son had been molested by Ray Buckey, a teacher, at the McMartin Preschool in Manhattan Beach, California. Although there was no corroborative physical evidence, Buckey was arrested on September 7. The lack of evidence forced the district attorney not to prosecute.

But that certainly wasn't the end of it. The chief of the Manhattan Beach police circulated a "confidential" letter to the parents of students and former students at the McMartin Preschool. He alleged that Buckey might have forced the children to engage in very specific sexual acts, and that they should question the children in an attempt to find confirmation.

By the spring of 1984, hundreds of children had been interviewed and 360 children were believed to have been abused at some point at the school. Medical examinations failed to find physical evidence, but the doctors thought that as many as 120 children had been sexually assaulted. A television station reported a possible link to child pornography and sex industries in Los Angeles.

Under the pressure of questions by parents, child abuse experts, and police, the children eventually began to tell tales of abuse. As time passed, the stories became unbelievable. They claimed that they had seen mutilations and murders, dead and burned babies, they saw movie stars, flying witches, and local politicians. Children said that they were forced to engage in satanic rituals including the drinking of blood. They were taken to airports to be flown to other cities. Sometimes they used airplanes and sometimes hot air balloons. Children were flushed down toilets, traveled by sewer, were molested in a small market and a car wash. And they were always at the school when parents returned to pick them up without a hint that anything had happened.

Finally, in March 1984, 208 counts of child abuse were handed down against seven adults including the owners of the school, four female teachers, and Ray Buckey. After twenty months of preliminary hearings, the prosecution offered deals to some of the defendants if they would testify against the others. None of them took the offer.

In January 1986, Ira Reiner, who had recently been elected district attorney, dropped all the charges against five of the defendants. Fifty-two of the counts were left against Ray Buckey, twenty against Peggy Buckey, and a single count of conspiracy was placed against both.

On January 18, 1990, after nearly three years of trial testimony and nine weeks of deliberation, the jury acquitted Peggy Buckey of all counts remaining against her. They acquitted Ray of thirty-nine of the fifty-two accounts but were deadlocked on the remaining thirteen. Rather than give up, the prosecution retried Buckey, but the second jury also deadlocked.

Peggy McMartin filed suit against the city, the county, a child protective agency, and an ABC-TV affiliate for a variety of improper behaviors after the failed prosecution. Virginia McMartin and two of the teachers who were never charged also filed suit. California law prohibited the suits because of statutes designed to protect social workers and those involved in the prosecution of child abuse cases.

Hundreds of children, because of the leading questions asked, the desire for testimony, and the improper methods of questioning, now believe that they were abused during satanic rituals at the school. But no corroborative physical evidence was ever found. The children's wildest tales were believed because of a notion that children don't lie. And though it can be argued that children do frequently lie, child abuse experts and ritual abuse experts seem to believe that none lied about this. The children were led to their beliefs by investigators who used improper questioning, a system of rewards and punishment, and a knowledge of what the "right" answers were.

This was a case that many believe was created out of whole cloth in which none of the allegations was ever corroborated by physical evidence. It was a case that showed just how easily children could be manipulated into believing they had been abused, and how the system failed to understand the psychological evidence about it. This was a clear case of false memory that was careful detailed on the videotapes made of the sessions with the children. Techniques designed to be used only after abuse had been confirmed were applied to children who had made no allegations of abuse until the questioning began.

The mistakes made here, in a legitimate and real effort to protect children, provided a new framework for the conduct of such investigations. This sort of wild arena should not be allowed to recur unchecked.

Jacobs at this point has come amazingly close to the truth—events that never happened can be told with complete conviction. But sadly, here he missed an opportunity to do some actual research. If, in fact, Melissa did hear this story from another abductee, Jacobs should have found out who told this story, originally and interviewed her to learn if it had evolved with the retelling.

Jacobs was grateful to Melissa for helping him with his discovery. He wrote, "She had taught me the dangers of hypnotically recalled testimony. It was a lesson I was grateful to learn, and one that all abduction hypnotists and researchers have to learn."

Unfortunately, his solution to this dilemma was not to question his assumptions but to find a way to sort these false memories from real ones. Somehow he found a method. He wrote, "[When] an abductee reported something I had never heard before, I would wait for confirmation by another abductee unaware of the testimony. I carefully questioned every inconsistency, gap, or logical leap. I worked for a complete chronology and tried to obtain a second-by-second recounting of each abduction event, with no skips, no gaps, and no omissions." Jacobs reported that if he did not hear of the event again, he concluded that it must be false. Conversely, if many people reported an event, it must be true.

This, of course, means that any recovered memory that does not fit Jacobs's theory is rejected as a false memory, while those that do fit his theory are accepted. Jacobs has discovered that aliens are creatures of habit and apparently never do anything just once.

So Jacobs was arguing that he could identify the false memories by methods he had discovered during his work. Science certainly operates in this fashion. Those doing the fieldwork make the discoveries and report them to their colleagues so that the method can be replicated and verified. Of course, we are not aware of any evidence that such verification of Jacobs's methods has been done. He appears to be at odds with researchers who have studied human memory, false memory, and hypnosis.

First, we have found no scientific evidence in the psychological journals that "screen memories," such as Jacobs is suggesting here, exist at all. This is, in fact, an obsolete Freudian term based on some of Freud's loose theoretical concepts that are seldom used by mental health professionals. The early Freudian theory was that childhood trauma was the root of our problems and that these traumatized children repressed the memories, creating "screens" around them. By drawing these alleged repressed memories out, they could be confronted by the victim and their power to traumatize would be dissipated.

Not one study has ever shown the existence of screen memories, or that the mind can produce them. There is no evidence that people use sexual abuse as a screen memory to protect themselves from the trauma of being abducted. The only evidence of this has come from "abduction researchers" repeatedly using hypnosis. A well-trained therapist knows that traumatic memories are not repressed, and that screen memories do not exist, according to current psychological thought.

Later in *The Threat,* Jacobs, after dispensing with the problems of false memory, screen memory, and poorly trained hypnotherapists, goes on to describe a new problem—instilled memories. He wrote, "What is not addressed here is that if in fact aliens can instill memories, how can one be sure that the entire abduction event consists of these memories? Maybe people aren't actually undergoing any of the procedures they recall. The whole experience could be instilled."

But Jacobs has an answer. He can tell the difference between a real recovered memory and an instilled one. How does he do this? He wrote, "Indeed, the aliens have created, perhaps unwittingly, a unique obstacle to learning the truth about abduction events. It is the problem of 'instilled memories'—images aliens purposely place in the abductees' mind. During visualization procedures, the aliens might show an abductee a multitude of images: atomic explosions, meteorites striking Earth, the world cracking in half, environmental degradation, ecological disaster, dead people bathed in blood strewn about the landscape, and survivors begging the abductee for help. Or the aliens might show abductees images of Jesus, Mary, or other religious figures. These images have the effect of being so vivid that abductees think the events 'really happened' or they 'really saw' the religious figure. [We should note here that these images are scenes from the film *A Clockwork Orange*. It is the procedure used for brainwashing.] This can be a problem, especially when the investigator is not familiar with visualization procedures and fails to identify instilled memories."

What is not addressed here is that if, in fact, aliens can instill memories, how can one be sure that the entire abduction event doesn't consist of these sorts of memories? Maybe people aren't actually undergoing any of the procedures they recall. The whole experience could be instilled, and not necessarily by alien abductors.

But Jacobs has an answer for this criticism. He can also tell the difference between a real, recovered memory and an instilled one. According to him, "I would probe further to determine whether this event happened. I would look for contradictions or inconsistencies by going over the incident from different temporal perspectives, asking questions that move the abductee forward in time and then back again. I would ask the abductee to describe the sequence of events on a second-by-second basis, searching for slight disjunctures in the account. I would ask whether the aliens were standing or sitting, precisely where they were looking, and exactly what they were looking at. In other words, I would search for the alien visitation procedures that might have instilled this image in the abductee's mind. . . . If the abductee were inconsistent in his answers, I would regard the incident with skepticism. If he held to his story, at the very least, I would put it in the 'pending' file, waiting for another abductee to confirm the same experience independently." In other words, if the abductee changes his story at all, if he is even slightly inconsistent, or if he remembers that the alien was staring at him, it must be because of instilled memories. Real memories, according to Jacobs, are consistent, unwavering, and finely detailed.

Jacobs used an example from psychologist Edith Fiore in her 1989 book, *Encounters*. In the book she described the hypnotic regression of a man named Dan, who "remembered" being a member of an alien military attack force destroying enemies on other planets and visiting the planets "Deneb" and "Markel." Fiore believed that each of Dan's experiences "actually happened very much as they were remembered."

But Jacobs objected that these memories cannot be real. "Clearly, this scenario in no way fits the abduction scenario as we know it." In other words the fantasy doesn't fit his particular model of the abduction phenomenon, and because it doesn't fit his model, he rejected it out of hand.

Here we gain more insight into Jacobs's theories about being a good hypnotist. Fiore faltered in her investigation because, according to Jacobs, "Rather than focusing on one incident and gathering data carefully and critically, Fiore skips to nine different encounters in her first hypnotic regression with Dan, which, in the hands of an inexperienced abduction hypnotist, can lead to a confused and superficial accounting." Apparently Jacobs believes that you should only examine one encounter per session.

Jacobs also objected to the accuracy of Dan's memory. "Furthermore," according to Jacobs, "Dan knows the answer to virtually every factual question that Fiore asks about life on board the military vessel. This omniscient factual assurance is usually a strong indicator of confabulation."

In other words, a good abductee should have poor recall. This is interesting in light of the fact that it is the recounting of exquisite detail that seems to convince Jacobs that some abduction stories must be real. Remember, to verify the accounts as real he would ask the subject for a second-by-second account of the abduction. If it was inconsistent, then he would question the validity of that account. But when Dan gave precise answers to all questions, this detail suggested to Jacobs that the story must be a confabulation.

So, having explained how to verify that the memories are real, having explained why false memory and hypnotically induced hallucination is not part of the problems that he faced, Jacobs moved to another criticism often leveled by skeptics and debunkers. Though they have suggested that the media has influenced the abduction phenomenon as it is now reported, Jacobs is firmly convinced that media contamination does not play a part in abduction research. He is convinced that abductees are aware of the dangers of cultural influences and work to avoid them.

He wrote, "When they examine their memories with me, they are acutely conscious of the possibility that they might have 'picked up' an incident and incorporated it into their own account. In the first few sessions of hypnosis, self-censorship is so heavy that it becomes a problem. People do not want to say things that make them seem crazy, and they do not want to parrot something back to the researcher that they picked up in society. They will tell me during hypnosis when they think they might have mixed in something from the culture. They are so worried about this contamination that very often I have to tell them to verbalize their memories and not censor themselves."

This, however, doesn't mean that they are not victims of media contamination, it simply means that they don't want to be. The way to tell media contamination from a real story, according to Jacobs is "When abductees tell me what they remember, their accounts usually have a richness of detail that could not have come from media contamination." Yet he is quick to reject Dan's military tale because of the richness of detail in it.

It is Jacobs's belief that the mass media has disseminated very "little solid information about abductions." Abductees must be telling the truth because they remember and describe so many specific aspects of procedures—details that have never been published. This is an astounding statement in light of the fact that there are dozens of books, movies, and radio programs about abductions that have gone into the tiniest of detail. Some of these books have been written by people like Betty Hill, Travis Walton, Kathy Davis, Leah Haley, and Richard Boylan who are abductees themselves. Others, put together by investigators, contain every tiny detail, including those reported by people like Jacobs, John Mack, Budd Hopkins, Jim and Coral Lorenzen, and Ray Fowler.

And even if these sources were not available to a specific abductee, every talk show has devoted at least one episode to abductions. Abductees are allowed to tell their tales to national, and often international, audiences with little in the way of skeptical discussion. The information that Jacobs believes to be obscure is, in fact, everywhere in the full richness of detail that he doesn't believe exists.

In fact, as pointed out earlier, the 1954 science fiction movie *Killers From Space* contains all the elements of the abduction scenario, including underground bases, big-eyed aliens, missing time, and a mysterious scare. And is it necessary to point out that the aliens had established themselves on Earth because their planet was wearing out? They were going to take our planet to make it into theirs. This is the new and frightening scenario that Jacobs has recovered from his abductees.

To eliminate all these problems, Jacobs has concluded that once the first few hypnotic sessions are completed the abductee feels more comfortable and therefore his or her memories become more accurate. Jacobs seems to believe the more often you hypnotize people the less they confabulate, but he offers no scientific evidence that this is true. In fact, research on hypnosis suggests just the opposite.

Critics say that abductees should be able to consciously remember their experiences and to provide investigators with accurate information. Jacobs reveals that abductees do consciously remember abductions. Often these accounts are accurate and detailed and closely match those recovered under hypnosis. He used Marian Maguire as an "excellent example" of conscious recall.

He wrote, "*She woke up one morning* [emphasis added] and consciously recalled an instance in which she was with her daughter during an abduction years before. She remembered holding hands with her daughter and, along with other people, being 'plugged into' a special apparatus on a wall. This is all remembered but she was certain that this event really happened."

This suggests that the memory was not actually consciously remembered, but it was, in fact, a fragment of a dream. This is not an example of a conscious remembrance but sounds more like a "forgotten" event that popped back into consciousness through dreaming. Jacobs continued, "I had not heard about abductees being plugged into a wall before. A few weeks later Marian and I explored this event with hypnosis. During the hypnotic regression, Marian found it difficult to remember walking up to the wall, being plugged into it, and becoming unplugged. The more I probed, the less sure she became about what had happened. She realized that the wall contained small black squares. And as she looked at them, I asked her to tell me what she saw beneath them. I expected her to say the wall or the floor. Instead, she said, 'Funny hands.' The hands were attached to wrists, the wrists to arms, and so on. She then realized that she was staring into an alien's black eyes. She had not been plugged into a wall. She was standing in a room with her daughters and a being came up to her and stared into her eyes. Over time, the black eyes in her mind had transmuted into an 'encasing' on a 'wall,' and her inability to avoid them transformed into being 'attached' to them. During hypnosis, the encasing transmuted to 'squares.' Although there was a real basis for Marian's memory, the details that *she consciously recalled had not happened.* [emphasis added]"

So this is not only not a totally consciously recalled event, it is also incorrect, and more closely resembles a "screen memory." More important is the fact that this consciously recalled event only surfaced after she had begun to explore, hypnotically, her experiences as an abductee. It is simply not the same as the recalled event of someone unfamiliar with alien abduction who has not been exposed to the memory enhancement techniques of a researcher or a therapist.

Jacobs provided additional information about consciously recalled abduction. "Another example is that of Janet Morgan, a single mother with two children, who consciously remembered a bizarre abduction experience. As she was lying on a table, she saw small beings struggling to bring a live alligator into the room. They put the animal on the floor next to her table, turned the reptile on its back, and then took a knife and slit its underside from top to bottom. The unfortunate alligator groaned and looked at Janet in shock. In hypnosis, Janet's memory turned out to be part of a complex abduction event in which aliens performed many different procedures upon her . . . she began to realize that the animal did not actually look like an alligator; she did not see an alligator's head or legs. In fact, it was a man in a green sleeping bag. When the aliens unzipped the sleeping bag from top to bottom, the man looked up at Janet and groaned. There had been no alligator. The aliens had not slit its belly."

Here again, the memory of the event was wrong. Jacobs appears to have trouble with the definition of conscious memory. He wrote, "Some of the most common consciously recalled memories are of the first or last few seconds of an abduction when the person is still in a normal environment. Abductees often remember waking up and seeing figures standing by their beds. But instead of remembering aliens, they recall deceased relatives and friends or religious figures."

But what Jacobs is describing here could also be sleep paralysis. In fact, Jacobs in *Secret Life* provides a textbook example of sleep paralysis, apparently without realizing what he has done. This suggests that the conscious memories of the victims, in some cases, are not actual abduction experiences but are a well-understood psychological phenomenon, which we will examine at length later. It is the use of hypnosis that changes the normal and mundane sleep paralysis into a terrifying tale of alien abduction.

And he provides a new example to verify this. He writes that Lily Martinson, a real estate agent, recalled being asleep in a hotel room and waking up to see her deceased brother standing at the foot of her bed; she clearly remembered what he looked like and found this memory comforting and reassuring. But when Jacobs examined this memory under hypnosis Lily's description was of a small, thin person without clothes, having no hair and large eyes. This was in all likelihood a hypnogogic hallucination, something often associated with sleep paralysis, not a conscious memory of a real event.

Jacobs also takes issue with the fact that skeptics have accused researchers who use hypnosis of "leading" people into believing that they have been abducted. "When inexperienced or naive hypnotists listen to an abductee's story, they often do not recognize dissociative fantasies, confabulation and false memories, or alien-instilled memories. The result is that the subject leads the naive hypnotist into believing an abduction scenario that did not, in fact, occur. But there is no way to define, delineate or differentiate between these categories."

The problem for researchers, according to Jacobs, is that Jacobs is the preeminent authority on being able to tell a real memory from an alien induced memory.

These memories are discovered by the researchers and therapists that Jacobs has termed the "Positives." Researchers such as Dr. John Mack, Dr. Leo Sprinkle, Richard Boylan, and John Hunter Gray, among others, suggest that the abduction experience is essentially positive. The trauma of abduction, when set aside, is replaced by the vast positive aspects of it. Jacobs rejects the positive outlook, suggesting that it is the result of incompetent hypnosis.

At this point, Jacobs dances close to the real explanation for alien abduction. He suggests that the orientation of the hypnotist, of the researcher or therapist, influences the type of experience related by the abductee. Jacobs writes, "When unskilled hypnotists regress an abductee, they fail to situate him in the event's minute-by-minute chronology. Without links to a temporal sequence, the abductee can interpret the events without the facts necessary to guide his thoughts, which leads to confabulation and other memory problems."

In other words, Jacobs has discovered that those conducting abduction research, if not as versed in hypnosis as he claims to be, created the false memories and encourage confabulation. The experiences being related by the abductee cannot be trusted. According to Jacobs, "The inadequate hypnotist and the abductee engage in a mutual confirmation fantasy: the abductee reports the fantasy; and the hypnotist assumes that the abductee's narrative is objective reality."

Jacobs is admitting that confabulation, false memory, and fantasy do influence the abduction field. Of those doing research, those with a positive orientation are most guilty of these errors. Their research is tainted by their own beliefs in the benevolent space brothers. These researchers are leading the witnesses to the positive orientation because that it what they seek.

He goes even further when he writes that those with the positive orientation refuse to accept information that is counter to their point of view. They see only the data that confirms their own belief structures. They lead the witness to it. According to Jacobs, for example, "While Mack does not lead the witness in the classic meaning of the phrase, he embraces the 'positive' therapeutic technique that leads to mutual confirmational fantasies and easily steers the abductee into dissociative channeled pathways . . . it represents the antithesis of the goal of scientific research—to uncover facts. . . . John Mack accepts 'recollections' at face value."

So Jacobs has demonstrated that he understands the problems of confabulation and fantasy while under hypnosis, but fails to see how it could affect his work. He searches for it in others, finds them leading the abductee to the reality of positive abduction, and points out how they lead the witness to that point.

But Jacobs believes in negative abduction. He believes there are no positive aspects to it, and to find them is the result of the researchers' imprinting of their own points of view on the abductee. But isn't Jacobs guilty of just the same sort of thing?

According to him, Pam Martin (a name he invented for the witness) reported a lifetime of experiences. Jacobs wrote, "As a result of her UFO experiences, Pam had come to believe over the years that she was leading a 'charmed' life with 'guardian angels' helping her overcome life's difficulties. . . . After one particularly vivid abduction experience, she decided that aliens were wonderful beings visiting her from the Pleiades constellation. She felt certain that she had been given 'powers' that enabled her to manipulate time and reality to her benefit."

Jacobs had just told us that Pam Martin came to him with a set of beliefs about her abduction experiences. These were positive events. Jacobs had long ago realized that abductions were negative. Her beliefs were in conflict with his beliefs.

Jacobs wrote, "I have had over thirty sessions with Pam, and during that time she has come to have a less romantic idea about what has happened to her. She was initially disappointed that what she remembered under hypnosis [conducted by Jacobs] were not the pleasant experiences she had imagined, but she now accepts the [Jacobs] reality of what has been happening to her."

To spell it out carefully, what Jacobs writes is that Martin's positive experiences were reversed after she began to work with him. He led her from the positive to the negative.

He goes on to create many other problems. As noted, he believes that some false memories are actually "screen memories" of real events. The child was not sexually abused but abducted by aliens, and the sexual abuse tale is a screen for the real events.

Psychologists have long suggested that hypnosis can lead to "wild confabulations." Freud stopped using hypnosis because he realized this about a century ago. Abduction researchers who practice hypnosis should be aware of Freud's early work, in which he developed his theories of screened memories, repression, and other facts, but they have failed to follow through, reading about Freud's later rejection of these theories as flawed. What we end up with are flawed and oversimplified theories that are not understood by the researchers, hypnotists, and investigators of alien abduction.

Psychologists recognize that a hypnotized abductee, or anyone else for that matter, when confronted with these pseudomemories, is convinced that they are real. These created, confabulated, and false memories are impossible to tell from the real thing. If the abductee is given a polygraph examination he or she will pass because of that belief. And it is something that Jacobs understands because he addresses the issue, but only as it applies to the "naive" hypnotist and researcher. He doesn't see it as expanding into the entire field as a problem for all those doing abduction research.

Jacobs is angry that others are skeptical of his findings. He writes, "I believe that anyone who puts his or her reputation on the line and ventures into this treacherous area deserves the plaudits of all who value the search for the truth."

Jacobs then admonishes his colleagues for being too lax in their method, writing ". . . even the most prominent researchers sometimes fall into investigatory traps such as mutual conformational fantasies." In other words, he has once again identified the problem but failed to see exactly how it affects abduction research.

Instead, Jacobs writes that in Mack's 1994 book, *Abduction,* Mack related a hypnosis session he conducted with a woman, "Catherine," in which aliens allegedly showed her images on a screen of a deer, moss, deserts, and other "nature things." Then she reported that she saw Egyptian tomb paintings and felt certain that she was watching herself in a former life.

They showed her a picture of tomb paintings with paint flaking off. "But then," according to Mack, "it switched to me painting it." But in that incarnation she was a man and as she watched this scene she said, "This makes sense to me . . . this is not a trick. This is useful information. This is not them, pulling a bunch of shit like everything else." Catherine now felt that her insistence upon a more reciprocal exchange of information had been affirmed.

Writes Mack: "I then asked Catherine to tell me more about this image of herself as a painter in the tomb of an Egyptian pyramid. In response to my question she provided a great deal of information . . . about the man and his methods and his environment. What was striking was the fact that . . . she was not having a fantasy about the painter. Instead, she was [him] and could see things from totally his point of view instead of from one watching it."

Catherine went on to "remember" many details of Egyptian painting and life. And later in the session she told Mack that an alien had asked her if she understood the meaning of the Egyptian scene. She then realized that " 'everything's connected,' canyons, deserts, and forests. 'One cannot exist without the other and they were showing me in a former life to show that I was connected with that, and I was connected to all these other things.' " Catherine also appreciated that she was connected to the aliens. Resisting them only meant that she was struggling against herself, and therefore there was no reason to fight.

Jacobs complained, "Mack not only accepts the validity of this 'dialogue' but embraces Catherine's interpretations of it as well. Rather than treating the entire episode with extreme caution and skepticism, he does not question her acceptance of a previous life, her sense of connectedness, her sense that a previous request for reciprocal information was answered affirmatively, and her decision not to resist."

Jacobs was upset because Catherine's narrative contained a past life, "dialogue," alien attempts to help the abductee, an environmental message, and personal growth. He wrote, "For the skilled abduction hypnotist, every aspect of this narrative should be suspect. Catherine could have easily slipped into a dissociative state in which she regarded internal fantasies as external events happening to her." However, there is no research that delineates the difference between hypnosis and a "dissociative state," or how anyone can tell the difference.

But Jacobs didn't stop there. He claimed that the aliens made Catherine believe she had a past life. Jacobs wrote, "If the Egyptian past life imagery happened at all, it might have taken place during an imaging sequence, and that automatically means that an instilled mental procedure was in process."

In the next sentence Jacobs makes another important statement, "Sometimes abductees combine imaging procedures, dreams, and fantasies for memories of external reality. Their interpretation of these 'memories' is often more dependent upon *their personal belief system than on the actual occurrences* [emphasis added]." We've known this for some time now—abductees cannot tell reality from fantasy.

He continued, "Unless properly versed in the problems that these mental procedures can create, the hypnotist can easily fall into the trap of accepting fantasies and confused thinking as reality."

He then admonished Mack for being duped by the story. "Mack displays no skepticism about this story. He admires her 'straightforward articulation' of the narrative."

Jacobs, a historian, explained why his work is superior to that of Mack and Fiore. "Fiore and Mack were trained as therapists and not as investigators. Their approach to abduction accounts is very different from that of researchers who are more empirically oriented."

But the very training that Jacobs suggests inhibits their abilities as researchers also provides them with an insight into the human psyche. Those trained in the mental health profession should be aware of various psychopathologies that could mimic the tales of alien abduction. Both Mack and Fiore have been trained to search the various psychological databases for information and research conducted by others.

Author David Hackett Fischer wrote a book called *Historians' Fallacies: Toward a Logic of Historical Thought* that outlines the mistakes historians can, and often do make as they attempt to reconstruct the past through the documentation, letters, diaries, and other artifacts left by our predecessors. It cautions the historian to avoid fallacies of "question-framing," "factual verification," "composition," and "false analogy," among many others. Unfortunately, because Jacobs is dealing with a current event rather than a historical event, he has ignored all of these warnings against error in his rush to prove that alien abduction is real.

But in other aspects he brings his training as a historian to bear. How else to explain his discussions, at length, about the development of a "hybrid" society.

Jacobs seems to believe, based on his hypnotically recovered information, that a secret breeding program between the alien creatures and the people of Earth is in effect. There are several stages to it, beginning with a cross between species that looks more alien than human, which he terms "hybrid.1" At the next stage is a cross between hybrid.1 and a human, or hybrid.2. These individuals too are quite alien in appearance. The cross between the human and hybrid.2 is another hybrid (hybrid.3) that is more human in appearance and according to Jacobs could "pass" as human for a short period of time. Finally is a cross that is amazingly human and whom Jacobs has labeled as the "Nordic."

These hybrids interact with the abductees, providing clues about their function. The earlier-stage hybrids, the ones and twos, "often help aliens with the abduction procedures and are an integral part of the alien workforce." They have been reported taking care of the hybrid babies, as well as completing other important tasks.

The early hybrids are relegated to work on the alien ships because they are so obviously alien. The later hybrids, the threes and fours, are human enough to "pass." The threes have been described with slightly larger than normal pupils, the only clue that they are not human. They are allowed out of "gray" control for periods of up to twelve hours. The threes and the fours are those that engage in sexual activity with human females, and it is at this point the logic of Jacobs's construct begins to disintegrate.

Jacobs has pointed out, repeatedly, that this program of abduction and breeding must be carried out in secrecy. It is the reason that the memories abductives have are "wiped" from the conscious mind. It is the reason that the aliens come in the night, "switch-off" those not to be abducted, defeat the various surveillance techniques attempted by the abduction victims, and why there is, in reality, no physical or corroborative evidence that abduction is taking place. The breeding program is "classified."

With all this in mind, Jacobs then asks that we accept one more theory to top all the others he has suggested. He introduces what he calls Independent Hybrid Activity (IHA). Because these late-stage hybrids can survive for about twelve hours in human society without being recognized for what they are, Jacobs has found that most sexual activity between hybrids and human abductees is between male hybrids and female humans. The male hybrids are used to impregnate the human females.

But if the aliens are conducting a program designed to reproduce their population, and they want to maintain secrecy, this plan is worked backward. A male abductee could impregnate a female hybrid, or a number of hybrids, and be returned to his environment with no real clues that an abduction has taken place. The aliens would remain in control of the experiment. This is the logic that was suggested by Coral Lorenzen in describing the Villas-Boas abduction. He was introduced to a human-like alien woman because she would remain with them as the baby developed.

But if they have a male hybrid impregnate a female human, they have lost control of their experiment. They keep the male with them, but the female is returned to her environment, where doctors and investigators can begin to document the case. The female is returned to an environment where she is exposed to the dangers of life on Earth. There are drunk drivers, murders, disease, and the ever-present abduction investigator searching for the proof that alien abduction is real.

Even with dozens attempting to gain that evidence and even while the alien creatures must realize how ineffective their attempts to conceal their presence is, they continue to abduct women and impregnate them. They continue to conduct the same sorts of experiments over and over, never learning from them.

Jacobs attacks the work of Mack, Fiore, and other mental health professionals, never questioning why they, as well as he, can break through the hypnotic blocks erected by the aliens. Instead, their grand scheme is laid out for him to see. Using the tools he developed as a historian of reconstructing histories by the evidence he can find, he now creates the whole of the alien abduction scenario.

And then, having disposed of the work of Mack and Fiore because they are therapists rather than researchers, Jacobs tackles those with a "New Age" orientation. He believes that their "New Age agendas" create problems because people uncritically accept such things as past lives, future lives, astral travel, spirit appearances, and religious visitations. He feels researchers who believe in these events are flawed because "all assume legitimacy even before the believing hypnotist begins abduction research. When the abductee relates stories with false memories, the believing hypnotist is unable to recognize them and is therefore more than willing to take them seriously." However, Jacobs never makes it clear why a past life is not as legitimate as an alien abduction. If one used all of Jacobs's techniques in recovering a past life, wouldn't that verify that it must be real, just as those techniques have convinced Jacobs that abduction is real?

Jacobs believes that you must be a good hypnotist, and if you are then you can get accurate information. He has again stumbled on a truth about the whole of the abduction phenomenon but he doesn't understand the overall ramifications of it. He writes, "The accuracy of abduction accounts depends, to a large degree, upon the skill and competence of the hypnotist. Memory is fallible and there are many influences that prevent its precision. Hypnosis, properly conducted and cautiously used, can be a useful and accurate tool for uncovering abduction memories. Competent hypnosis can illuminate the origin of false memories and can untangle the web of confusing memories. What emerges are accurate, consistent, richly detailed, corroborated accounts of abductions that unlock their secrets and add to our knowledge of them."

Sadly, not one study has been done to correlate hypnotic skill and accuracy of memory. Furthermore, Jacobs does not offer an empirical definition of hypnotic skill or any objective way to define the accuracy of memory. And finally, the medical, psychiatric, psychological, and legal professions all vehemently disagree with his statement.

And sadder still is the fact that Jacobs, like so many of the other abduction researchers, has identified the problem once again. He saw it, reported on it, but failed to understand it.

15.

DR. ALVIN LAWSON

In a search for the truth about alien abductions there have been dozens of theories offered to explain the phenomenon. Dr. Alvin Lawson, a professor of English at California State University, has been investigating UFO abduction as long as Budd Hopkins and has come to a conclusion that is in conflict with what Hopkins and many of the others believe. His theory is unique, and his research methods supply us with some interesting facts about alien abduction.

Lawson himself claims to have participated in the hypnotic regression sessions with more than one hundred people who claim alien abduction. By the sheer number of subjects he has treated he is one of the leaders in research in the field. The difference is that Lawson's theory has not received much popular press and he explains abduction without invoking an extraterrestrial presence.

Lawson established a hotline for the reporting of UFO reports about twenty years ago. Although the sightings reported ranged from nocturnal lights to abductions, the case that Lawson would label as the Garden Grove abduction caught his attention. It was, according to Lawson, one of the most spectacular abductions ever reported.[1]

The call came into Lawson's hotline in the fall of 1975, prior to the airing on NBC of *The UFO Incident* and before Travis Walton's abduction claim hit the national media. Alien abduction was a topic being discussed in UFO literature, and the national UFO magazines were carrying stories about it, but, for the most part, people were not aware of the phenomenon. For those interested in UFOs, the information about alien abduction was readily and easily available. It wouldn't be until after the airing of the NBC film, and the publication of Budd Hopkins's *Missing Time* and *Intruders* that the idea of alien abduction would become widespread in the consciousness of the general public.

Lawson interviewed the sole witness of the Garden Grove abduction and identified him only as B. S. The man claimed that during a camping trip with a friend to Arizona in 1971, they had seen a UFO appear overhead. A beam of

---

[1]Alvin Lawson, "UFO Abductions and Birth Memories: Scientific Proof That Close Encounters Are Fantasies," manuscript, 1991.

light flashed down on them, paralyzing them. In moments both B. S. and his friend were being levitated toward the saucer-shaped craft. Here, as Lawson is quick to point out, was a rarity in abductions report. There were multiple participants, at least according to the the man Lawson was interviewing.

B. S., under hypnosis conducted by Lawson's partner, Dr. W. C. McCall, told of being met by seven-foot-tall, crocodile-scaled creatures with elephantine feet and hands with three fingers and a recessed thumb. While on board the alien craft, the men were quickly stripped of their clothes, then separated and taken in opposite directions.

Like so many others, B. S. told of a physical examination. He was placed against some kind of X-ray screen. It was described as box with blinking lights on the surface from which two beams of light were played across B. S. Although not painful, the experience was uncomfortable and he felt that he was bleeding, urinating, and that his stomach was being drained, almost all at the same time. He also believed that his chest was opened and his heart was pulled out briefly. B. S. even believed that his brain was temporarily removed. It should be noted that many other abductees have reported similar experiences and that some of this has been played out in various science fiction movies.

When the examination was over, B. S. saw another being about nine feet tall enter the room. The creature smelled and had bad breath. This being communicated with B. S. through telepathy, giving him a message about the aliens' purpose, and telling him that the aliens would return to visit Earth again.

Finished with this, B. S. was given a tour of the craft that included a second floor, where he observed an advanced computer system. He visited a huge birth laboratory that held row upon row of transparent cylinders that contained alien-looking embryos. He later saw chemical formulas and the atomic weights of various elements carried out to four decimal places. There was even a message written in pre-Homeric Greek. And, again, many of these same elements are suggested by dozens of other abductees.

Undergoing repeated hypnosis, B. S. reported that he had once been visited by a ball of light that materialized as a humanoid. After seven weeks, B. S. vanished, only to return twenty-four hours later. He claimed that he had been abducted once again, this time to the Plains of Nasca in Peru. He also reported that he had experienced his first abduction at the age of sixteen.

Lawson reported that the Garden Grove case would eventually involve "at least six abductions, two with separate second witnesses, and loads of hypnotically retrieved data, including blueprints of the UFO . . . chemical formulae for the UFO's propulsion system . . . a mysterious computer-like alien language, and a brief message from the aliens. . . ."

This abduction became more important not for the details reported under hypnotic regression but because further investigation by Lawson and McCall revealed it all to be a hoax. They were originally impressed with the report because, as Lawson has said, they wanted to believe it. B. S. even provided physical evidence in the form of a "spider figure" drawn on the ground at the site of a UFO event in the Anaheim Hills. And he had provided the indications that there were other witnesses to aspects of the case that would offer the necessary corroboration.

Lawson and McCall, so impressed with the story told by B. S., and the promised corroborating detail, pressed further, hoping to prove alien abductions were real. But then the whole abduction scenario began to fall apart. They believed they had, for example, found a witness who had seen B. S. scratching the marks in the ground that he had used to suggest that his Peruvian abduction was real. The physical evidence of the landing of the craft, or the markings on the ground, so important to the story, now seemed to suggest some kind of a hoax.

Lawson and McCall also learned that the two witnesses who were supposed to confirm part of the abduction could not do so. Other physical evidence, which included scarring that was alleged to have happened during the abduction, had been done by the witness himself. And recordings of strange beeps and voices that seemed to suggest strange monitoring of B. S. by the alien creatures were reproduced, according to Lawson, by placing a tape recorder in the clothes drier. B. S. had been caught attempting to create a mystery rather than being a victim of it.

Although Lawson noted that many of the events described by B. S. during the abduction were well known in the UFO literature, Lawson reported that some of these events had not published and were known only to UFO and abduction investigators. He said, "His [B. S.] narrative contained dozens of images and events that I knew from other investigators' then unpublished abduction cases, material that B. S. could not possibly have known about." The case fit the mold, according to Lawson, and had no "antecedents in UFO lore, science-fiction, or any other source."

Those points, raised by Lawson, would become important in his later investigations. He wondered how these widely separated witnesses, who had not read the UFO literature, who were not privy to the private communications of the researchers, could come up with the same stories. Did this confirm some of the reality of B. S.'s story or was there another mechanism working here?

To test the idea, Lawson, along with Dr. McCall, designed an experiment. They would use a number of people who had no interest in UFOs, who had not had an abduction experience, put them under hypnosis, and see if they could find an abduction. Lawson expected the results to be wildly different. He expected highly divergent accounts with all sorts of alien descriptions.

While the tests didn't prove that anyone was inventing a tale of alien abduction, and most of those who claim to have been abducted seem to be sincere, it did raise some alarming questions. Why would the abduction scenario remain so consistent in the accounts of those who had not been abducted? Lawson had already dismissed science fiction and pop culture as the culprits. Still, he believed there had to be some connection that involved both honest abductees and those in his test. What event would they have all shared?

His answer was the trauma of birth. Everyone is born, and if these tales of abduction were reliably reported memories, then it was possible they were of birth trauma. Even the exceptions, he noted, reinforced his views. Those without a definitive birth trauma seemed to have been born through cesarean section.

In an article, *A Testable Theory for UFO Abduction Reports: The Birth Memories Hypotheses,* published in *Archaeus* in 1989, Lawson outlined his theory. In that article, Lawson writes, "In an attempt to make an objective evaluation of abductee claims, we induced imaginary abductions in 16 volunteers who had demonstrated no significant knowledge of UFOs. . . . Although we had expected major dissimilarities, an averaged comparison of data from four imaginary and four putatively real abduction narratives show no substantive differences.

"In the continuing absence of unambiguous physical and dual-witness evidence, our study concluded tentatively that abduction experiences were poorly understood hallucination." Lawson, by using the qualifier "unambiguous," was saying that all evidence for the reality of the abductions must be questioned and the source of the experiences would be found somewhere other than alien intervention.

The theory then grew from a search for a common factor that would explain why both abductees and non-abductees would recount similar experiences under the influence of hypnosis. Lawson credited Stanislav Grof, whose experiments conducted with hallucinogenic drugs produced revivified birth memories that were similar to the imagery of UFO abduction.

According to Lawson, "In time, we developed our Birth Memories Hypothesis (BMH), which asserts that an abduction experience is an involuntary, fantasized sequence of images and events unconsciously based on the prenatal or birth memories of the witness."

To prove his point, Lawson suggested that his unconventional theory was based on DeMause's work looking at the "long-term effects of birth experiences in obstetrical literature, anthropology, psychology, history, and sociology. He has carefully linked the perinatal connections in years of media headlines and editorial cartoons with the careers of several U.S. presidents. . . ."

The main thrust of DeMause's work seemed to be that code words, for example, those used during the Second World War, contained birth images. The Japanese, wondering about the bombing of Pearl Harbor had asked, "Does it seem as if a child might be born?" Or that the discussion by Americans about the atomic research in New Mexico, and the news that the weapon had been detonated, was, "The baby is born." Or that those atomic bombs were named "Little Boy" and "Fat Man."

Writes Lawson in support of his theory, "In these passages the explosive and destructive upheaval of war is repeatedly equated with birth. The consistent perinatal language cannot be accidental and it certainly is not irrelevant."

Other researchers, looking at the theory could ask, "Why isn't it irrelevant?" Merely stating that it isn't doesn't make it so. And it can be argued that the terms were selected simply because they conveyed not only the important message but masked in the language of the most common of events. There is nothing in asking if the child might be born that would tip the secret to those on the outside.

What this means, simply, is that the language surrounding these events is irrelevant to the whole idea that there is a birth trauma or that some people remember it. Too often, while searching for evidence that a theory is accurate, everything that even remotely suggests that theory is correct is lifted, out of context, and plugged into the formula. When it moves too far afield, it is time to reject, at the very least, that portion of the theory.

Lawson's theory of birth trauma is interesting because, as Lawson points out, it is falsifiable. That means that it can be disproved if the theory is not solidly grounded. To show this, Lawson presented three tests and drew conclusions on them.

The first test, according to Lawson, was to see if there were any UFO abduction cases "in which absolutely no prenatal data appears." He wrote, "Unfortunately for true believers, abduction cases that lack perinatal data are about as scarce as aliens. As a result, the evidential birth/abduction score currently favors probable birth hallucinations over verified abductions by 300 to zip—more or less."

But Lawson has drawn the category so broadly that almost anything could fit into his theory. The mere presence of a doorway, especially one that "irises" open, is a birth memory. This overlooks, completely, the science fiction connection to that sort of door played out in dozens of films from the 1950s.

Or the presence of a hallway is a reflection of the vaginal canal. Of course, the question to be asked was if the presence of a hallway in any structure is an unconscious desire to return to the womb, or is the hallway the most functional of the architectural designs for creation of a number of rooms off a given area of floor. In other words, the whole concept is irrelevant because there aren't many viable ways of creating the structure without incorporating a hallway.

Lawson's conclusion, then, that all abductions contain memories that resonate with the birth trauma is flawed. Yes, there is what Lawson would consider perinatal imagery present in abduction narratives, but it is the interpretation of the researcher (Lawson) that must be questioned.

Lawson frames his second test in the following terms: "If abductees whose narratives describe late-stage perinatal associations (such as extreme head and/or body pressure, fetal rotation, breech or other delivery positions, compression followed by sudden relief, etc.) prove to have been born by cesarean section and so have not undergone such natal experiences, the BMH will be weakened, if not proved false. If there are no such cesarean-born abductees, it will be supported."

According to Lawson, there is no good research that has been done here. He believes, and rightly so, that medical records would be necessary to prove the point. Although the abductee could be asked the question, and most should know their own birth history, it is important to document the facts. Lawson suggests that recovery of those records, especially going back to before 1950, might be difficult. However, he offers no evidence that his statement is accurate, only reasons why the research hasn't been conducted.

Lawson's final test suggests investigators can "search through a witness's abduction narrative for specific echoes of his/her birth history, and (depending on whether or not they find them) the Birth Memories Hypothesis will either be strengthened or weakened."

He then writes, "For instance, suppose an abductee describes (as in fact one of them did) how a UFO's huge metal clamp held him, twisting his back; and say we discover that his birth was forceps-aided. Based on that evidence, we will be entitled to doubt the abductee's claim and to assume that the BMH may be valid. But suppose again that later we find other forceps-delivered abductees whose narratives give no hint of their birth history. In such cases, their Close Encounters will gain credibility, and the Birth Memories Hypothesis will become uncertain."

Lawson, while discussing his theories on "The Randle Report" on KTSM-AM radio in El Paso, Texas, suggested that one abductee who spoke of metal clamps to the head was a forceps-aided birth. However, in his 1989 *Archaeus* article, the man who claimed that was part of Lawson's test group who had been "*given an imaginary abduction*" (emphasis added). In other words, Lawson had described the event as if it had come from someone who claimed abduction, not as part of the test Lawson was conducting. It rendered his conclusion irrelevant to the study.

That same article also produced a number of other "abductees" whose births were not easy or smooth. Using this sample, each having been given their abduction scenario by Lawson and his investigators, showed that unusual aspects of birth, such as the use of forceps or breech births, all related, during the abduction, events that reflected those birth traumas.

While the study is interesting, it is completely irrelevant. Lawson and his team provided the abduction scenario to the subjects. They were not dealing with people who actually claimed to have been abducted, but those who had the abduction implanted by hypnosis. Lawson also wrote that they used ten volunteers who were selected because they claimed birth-related problems. Because of these flaws, and because he had tested the theory with those who were actually claiming abduction, his conclusions here were also flawed.

The real value of Lawson's work is not in his theory of birth trauma but in his "implanting" of abduction scenarios. He wrote (and to repeat here), "In an attempt to make an objective evaluation of abductee claims, we induced imaginary abductions in 16 volunteers who had demonstrated no significant knowledge of UFOs. Eight situational questions comprising the major components of a typical abduction account were asked of each subject. Although we had expected major dissimilarities, an averaged comparison of data from four imaginary and four putatively real abduction narratives showed no substantive differences."

This led to the development of Lawson's theory. Searching for a common event to explain this finding, he developed the flawed birth trauma theory. His own methods, however, included another mistake, one that James Harder was quick to point out. Lawson, according to Harder, had engaged in leading the witnesses. Lawson provided the information, unconsciously, to his subjects, to prove his point. The experiment was flawed and the results skewed.

There is, however, another major problem with Lawson's research. Psychologists suggest that the human brain is not completely formed at birth and that it continues to develop long after birth. Language isn't possible until many months after birth. The baby cannot remember the trauma of birth. Theories about birth memories and birth trauma are based on poorly drawn data. Because of this, there are no birth trauma memories on which an abductee can base an experience, at least according to psychologists and researchers who study human memory.

Carrying on with his theory, Lawson postulates that the creatures in alien abduction accounts are based on the human fetus. He even provides a diagram of both the "human embryo/fetus and a typical UFO humanoid entity. . . ." There is a remarkable similarity between the two, if we accept Lawson's comparison.

But there is a problem that Lawson seems to ignore, one of size. The human fetus is a tiny thing, but the alien is, according to Lawson, two to five feet tall. Even ignoring the fact that most abductees do not describe aliens as small as two feet, it is still much larger than the human fetus.

Another problem with which Lawson fails to deal is the cultural heritage of the abductees. When asked, specifically, about the influence of science fiction, Lawson was quick to suggest that *Close Encounters of the Third Kind* and Whitley Strieber's *Communion* did a great deal to fix the description of an alien in the public consciousness, but those two works didn't appear until after 1977 and a great deal of the abduction material preceded them.

Ironically, when it was suggested that science fiction films of the 1950s contained such items as an alien implant and the abduction scenario, Lawson ignored it, calling them "rarely seen." If these films or the science fiction literature that preceded them provide the basis for some abduction reports, then an alternative theory has been produced. This explains, in simple terms, how some of the witnesses can report similar events though they have never spoken to one another. It can also explain how the subjects of Lawson's studies could provide the same imagery. Rather than seriously look at this point, Lawson dismissed it.

Of course, the most damaging problem for his theory is that the fetus cannot see itself. There is nothing inside the womb that is reflective and there is no source of light. Granting that a bright light played across the belly of a pregnant woman *might* provide illumination in the womb, wouldn't the perception then be of a creature that was reddish and not gray or white? And that supposes that the eyes would be functioning in the womb and that the brain would be able to record anything at all.

When asked about this on "The Randle Report," Lawson glossed over the question by saying, "The fetus knows what it looks like." He told of a picture of a newborn baby, maybe two hours old, that was shown sticking its tongue out at a doctor who had stuck his tongue out at the baby. Somehow this was to confirm that a fetus knows what it looks like.

Ignoring for the moment that a full-term baby doesn't look like a fetus, and that the doctor certainly didn't look like a fetus, the action on the baby's part has no relevance. It is doubtful that it was reacting to the doctor's gesture. And even if the baby was, it proves nothing about the baby knowing what it looked like, let alone what a fetus looks like. If we follow the logic of the situation, the baby must think it looks like a full-grown doctor and not a tiny human with soft bones, little hair, and chubby appendages.

So the basic flaw in the BMH is that a fetus is "required" to know what it looks like, but there is no mechanism provided that would support that. There are no reflective surfaces in the womb, there is no light in the womb, and the organs for discerning the images are not fully developed. If the fetus can't see itself, is it enough to suggest, as Lawson does, that it knows what it looks like? Or is this the basic flaw that renders the theory untenable and impossible?

On "The Randle Report" Lawson complained that he could not get his theory published. He believed that everyone was interested in the mystery, but no one wanted to hear a solution that didn't involve extraterrestrial life. But even CSICOP, which seems to delight in deflating the ideas of those who believe in the paranormal, had rejected his work. Lawson believed it was because they did not want to publish anything that took UFOs and alien abductions seriously.

It may be that his book was poorly written and or that those looking at it believed that the theory was flawed. CSICOP might have ignored the work because they didn't find it convincing. Certainly it would seem that CSICOP would support any theory that didn't involve extraterrestrial life, but they might not support one that has no good scientific basis.

But in the process, Lawson, though he doesn't recognize it, touches on an explanation for the corroborative statements of abductees who do not know one another. He wrote, "In short, Hopkins's 'witnesses' have active imaginations, a wild tale to tell, and a sympathetic and eagerly ETH-oriented hypnotic guide to help them tell it. Under active investigation—with or without hypnosis—such a situation can trigger an *imaginary* UFO encounter in subjects; and they may tell wilder stories, but not necessarily truer ones."

In condemning Hopkins's work, Lawson also pointed out, "A more objective view might see Hopkins's subjects as resembling our imaginary abduction subjects and our few hoaxing claimants in too many ways for his work to be taken seriously. Hopkins admits that his subjects were often recruited because they had a skin blemish or a vague memory of a years-old UFO sighting."

Having pinpointed a primary problem in abduction research, Lawson then allowed his vision to be clouded by his own theory. He reported that Hopkins [as of 1989] had regressed over 160 supposed abductees whose narratives were loaded with reproduction imagery. "He believes that UFO aliens are conducting a longitudinal genetic study of human beings."

So Hopkins's theories of alien reproductive and genetic experimentation become one more "proof" of Lawson's own theory. Although it seems that everything about an abduction can be bent into the Birth Memories Hypothesis scenario, Lawson doesn't explain how genetic experimentation fits into a birth memory.

Lawson did some good work. His research into the Garden Grove hoax pointed out just how believable a hoax can be. That man told his story with the appropriate emotion, added the little details that researchers look for, and under hypnosis, gave a more detailed account.

16.

DR. JOHN MACK

Of all the people studying the alien abduction phenomenon, none carries more impressive credentials that Dr. John Mack. He is a tenured Harvard professor, a medical doctor, and a psychiatrist. He is the winner of the Pulitzer Prize for his autobiography of T. E. Lawrence. He has made it his life's work to understand the human mind and the reasons for the actions taken by people. He understands psychopathology, brain trauma, disease and its effects on people, and all the related problems generated by the human condition. Surely, here is man who should be well acquainted with the workings of the human psyche.

So how did a man who had been trained in the sciences, who is a professor at Harvard, end up studying alien abductions? According to Mack himself, "It came out of an interest in what might be called human transformation which goes back to 1979, 1980, when I began to question a lot of assumptions about our minds, our psyches, whether we're able to take in as much reality as our minds, our bodies, our souls are capable of. Five or six years ago [that is, 1993] I began to work with Stanislav Grof, a Czech psychoanalyst who's exploring, remapping the mind, first using LSD and then without any drugs, exploring the deep rapid breathing, a method he calls holotropic breath work. Opening up the psyche to realms of the unconscious, that Jung talked about as the collective conscious or you could call the mythic conscious, which goes way beyond our own biographies [and] extends us to our capacity to identify ourselves with spirits, with animals, with gods, with mythic beings. So I was kind of open when somebody asked me in the fall of 1989 would I like to meet Budd Hopkins."

In his book *Abduction* Mack wrote that when he first learned of the work being done by Budd Hopkins, and that Hopkins was working with people who claimed they had been taken into spaceships, he thought that both researcher and subject were crazy.[1]

---

[1]C. D. B. Bryan, *Close Encounters of the Fourth Kind: Alien Abuction, UFOs, and the Conference at M.I.T.* (New York: Alfred A. Knopf, 1995), 5–6, 11–13, 131–3, 156–62, 254–78, 420–2, 445–6; John Mack, *Abduction* (New York: Charles Scribner's Sons,

In January 1990, while in New York on business, he met Budd Hopkins for the first time. According to Mack, "It was . . . one of those dates you remember that mark a time when everything in your life changes. . . ." What happened when he met Hopkins apparently changed his way of thinking about UFOs and alien abductions.

Hopkins also introduced Mack to some of the abductees he had been interviewing. After meeting with four such people, three women and a man, Mack noted that there was nothing about them to suggest they were delusional, experiencing a misinterpretation of dreams, or that their tales were the product of fantasy. When Hopkins asked if Mack wanted cases referred to him, he agreed and in the spring of 1990 he began to see abductees himself.

In his book Mack reported that in a three-and-one-half-year period that followed his meeting with Hopkins he had seen more than one hundred individuals. He wrote, "Of these, seventy-six fulfill my quite strict criterion for an abduction case: conscious recall or recall with the help of hypnosis, of being taken by alien beings into a strange craft, reported with emotions appropriate to the experience being described and no apparent mental condition that could account for the story."

Mack wrote that his own investigations into the abduction phenomenon had led him away from his orientation toward Western science, although his apparent disdain for the Western world was already well known. His work led him, according to his book, ". . . to challenge the prevailing worldview or consensus reality which I had grown up believing. . . . It has become clear to me also that our restricted worldview or paradigm lies behind most of the major destructive patterns that threaten human future—mindless corporate acquisitiveness that perpetuates vast differences between rich and poor and contributes to hunger and disease; ethnonational violence resulting in mass killing . . . ecological destruction on a scale that threatens the survival of the earth's living systems."

Mack also made this point: "The type of hypnosis or nonordinary state I employ has been modified by my training and experience in the holotropic breath work method developed by Stanislav and Christina Grof. Grof breath work utilizes deep, rapid breathing, evocative music, a form of body work, and mandala drawing, for the investigation of the unconscious and for therapeutic growth."

While this sounds interesting, and certainly scientific, it is, in reality, mind

---

1994); John Mack, "Studying Intrusions from the Subtle Realm: How Can We Deepen Our Knowledge?" in *MUFON 1996 International UFO Symposium Proceedings* (Seguin, TX: MUFON, 1996).

expansion techniques that were developed in the 1960s. It is an outgrowth of an Eastern philosophy that embraces the mystic and the paranormal and certainly doesn't require the evidence and objective proof demanded by a Western orientation. Mack is quick to caution us against requiring that sort of objective proof.

In studying the abduction phenomenon, having studied the history of human abduction, that is, "the phenomenon of humans being transported into other dimensions," Mack realized that UFO abduction is nothing new. Unfortunately he doesn't seem to understand the significance of this revelation, ignoring it or using it to build the theory that UFO abduction has been going on for centuries.

His studies and his examination of abductees suggested to him that he "was dealing with a phenomenon that I felt could not be explained psychiatrically, yet was simply *not possible* [emphasis in original] within the framework of Western scientific worldview."

In his work with the abductees, Mack began to develop what for him became the "proofs" that abduction was a real event, though not necessarily a physical event, as suggested by Western science. This is a theme to which he would return on a number of occasions, blaming science for its failure to explore and investigate alien abduction. He would suggest that a "scientific paradigm" might be necessary to understand the phenomenon.

To prove his point, he has created a number of questions or made a number of statements about alien abduction that, to him, suggest the reality of the situation. He presents these questions and statements as if they somehow underscore the validity of the research he has done into the phenomenon of alien abduction. Yet they do nothing other than obscure the facts.

He has asked those who are skeptical of abduction, or those who would simply question the reality of abduction, "Why is it so difficult for this culture to accept the notion that there could be another intelligence, however strange, that has found a way to enter our world?"

In other words, Western science has developed a methodology for the examination of questions and the observation of the world around us. Rather than embrace that objective thought, Mack is telling us to reject it because it does not fit with his view of his investigation. When science demands proof for extraordinary claims, Mack is quick to reject that as a paradigm of the Western mind. We should be more open to thought even if that thought is outside the realm of what we see as the world around us.

He continues his theme by suggesting, "One must consider that this experience may be beyond the bounds of normal science. It may be a paranormal experience."

But this is actually trying to have it both ways. He suggests that there is solid evidence of a physical reality such as the scoop marks reported by some abductees, the UFO sightings that sometimes coincide with an abduction, and other physical traces of UFO reality. But when these points are challenged on a scientific basis, Mack suggests that many people believe that the abductions are paranormal experiences that produce no physical evidence. The objective researcher, or the skeptic, can't win under these rather arbitrary conditions.

When you begin to enter the realm of the paranormal, you leave the land of normal science. This is not an unreasonable realm in which to dwell, but it is one where faith is more important than scrutiny, which is what Mack is suggesting. Once you have accepted the paranormal explanation for abductions, scientific method and scientific thought not only become relatively useless, they become a hindrance to the investigation. Several centuries of research have shown this to be the case.

Having dispensed, at least in his mind, with the problems with the lack of any sort of corroborative evidence, he switches to the logical. He asks the skeptics, the debunkers, and the scientific community, "Why would anyone make up a story like this?"

There are, or course, several possible and some very good answers to this question. First, there is the sense of self, or rather the lack of it exhibited by abductees. Those with a poor sense of self attempt to find anything that will bring them the attention they seek, even if that attention is negative. The class clown is disruptive because his or her antics brings the attention of the teacher and the other students. So one answer for Mack's question would be that many people with a poor sense of self would make up the story simply because of the attention it brings to them regardless of the spin of that attention.

Second, it seems, based on the objective evidence available, that some people are simply making it up. The lure of personal gain in the form of media attention certainly plays a part in this. There is also the possibility of book contracts that can help line the pockets of those who are telling these tales of alien abduction. Sometimes the more lurid the tale, the more attention is showered upon the abductee. Anyone who has spent any time in the field of ufology knows that there is no shortage of charlatans, quacks, hoaxers, and frauds who are there simply for the attention that is brought to them.

In fact, one man visited William Cone in 1996 to explore his abduction experiences so that he could finish his book. Another woman offered a piece of the movie she planned to make if we would help her undergo chemical regression. When her offer was refused, she became quite angry, and screaming that only fools would pass up the millions that could be made off the project. After all, how many movies have been made recently with alien abduction as the theme?

Third, the lure of the spotlight, is a powerful force, and should never be underestimated. Many of the people we have interviewed came to us saying that they absolutely wouldn't want anyone to know that they have had this experience, only to show up on a talk show, featured in a magazine article, or doing books of their own shortly thereafter. These are people who would never appear on a television program, or receive any sort of national, and sometimes international, recognition if it weren't for the abduction story they now can tell.

Maybe we should take a moment here to reinforce this. All the abduction researchers tell us repeatedly that the abductees seek no publicity. Some of them are camera shy, but once again it should be noted there is no shortage of abductees to appear on television or to be the subjects of UFO research.

Mack also points out that this is not a club that anyone would want to join. The experiences are so horrific, and the emotion displayed during the sessions to recall the abduction so terrifying, that it is impossible to believe that anyone would consciously invent the tales.

But the statement is demonstrably false. Hundreds, if not thousands, have told tales of violence and horror and believe, without a doubt, that they are the victims of satanic ritual abuse. During their psychological treatment, they describe horrible tales of murder, mutilation, and human sacrifice. Surely, this is a club to which no one would want to belong, but apparently thousands join voluntarily as evidenced by the stories that are circulating.

We have now seen that these same people, with their horrifying ritualistic abuse tales, sometimes recant their experiences as false memories. In some of these cases, we have been able to prove that the events that they remember could not possibly have happened. Even with this, even when the physical reality of these memories is properly challenged and the lack of corroborating evidence is demonstrated, many of these people cling to the horrible memories. In light of properly obtained evidence, they reject the evidence in favor of their newly retrieved memories.

The desire to believe these things is not always a conscious choice. It's actually the combination of a weak sense of self and the desperate need to belong somewhere. Some of these people come to wear their abuse, whether ritualistic or abduction, like a badge of honor. It is the strongest sense of identity they have ever had. In reality, it has become a club that people will fight to join. Since this is clearly true of the satanic abuse and multiple personality population, there is no reason to believe that it wouldn't apply as strongly to the abduction population. Please note here that we have indicated that it is true of many abductees, but by no means all of them.

In fact, as we have noted, there are support groups all around the country that focus on alien abduction. It is a weekly or monthly meeting that becomes the primary focus of life. There are often potluck suppers, group field trips, and an opportunity to meet with others who claim to have had similar experiences.

It is not a time of therapy or a time to deal with reality but a time for socializing. Here, too, is a reason for wanting to join this club. It provides, not only a sense of self but a new identity.

Mack as well as many of the other abduction researchers note that these people are telling the stories with appropriate effect for the content. The emotion displayed by the abductees, according to Mack, is appropriate for the experiences they believe they have had. Mack, in fact, has said that before they can join the club, or at least join the club that he runs, "They have to persuade me with the intensity and quality of the emotions with which they describe the experience that it is authentic."

Anytime a person remembers something frightening or horrible, they react with appropriate emotion. People having nightmares wake up terrified. Being afraid of a memory says nothing about the veracity of that memory. It's important to remember that one's emotional investment in a belief is not necessarily correlated with the validity of the belief.

In fact, we know of one researcher, studying post-traumatic stress syndrome who had a client who told horrific tales of heavy combat while serving in Vietnam. The tales were told with the appropriate emotional investment. These were detailed, descriptive tales that would have frightened the most hardened combat veteran. And none of them were true. The man had served in the Army during the Vietnam War but he had never been in Vietnam. The emotional investment by the subject is simply no indicator of the reality of the memories being discussed or retrieved.

Mack then suggests that he has exhausted all of the diagnostic possibilities to explain why some people tell tales of alien abduction. These people, the abductees, are not suffering from any mental disorder that he can find.

The Diagnostic and Statistical Manual-IV is a book used by all mental health practitioners to categorize and diagnose mental disorders. Many people with abduction experiences do not fit the criteria for any specific mental disorder. But just because you can't find these people in the *DSM-IV* doesn't mean that they don't have a mental health problem. There are several diagnostic possibilities.

For example, many of the people appear to have a boundary deficit problem. They have a weak sense of identity and have a difficult time telling reality from fantasy. Many of them admit that they can't tell the difference between dreams and actual memories. Most of these people also have sleep disorders. But this is not a mental disorder. It is an artifact of character structure.

Once again we have returned to a poor sense of self. These are people who desire attention regardless of how they get that attention. Suddenly they are the center of attention, telling their stories to the abduction researchers, other abductees, and even television audiences. Their importance in the world is increased because people are seeking them out to hear their incredible stories.

Several of the people fit the criteria for identity disorder. They are unable to hold a job or make a career choice. They are unable to establish or keep a relationship. They often have sexual identity problems, or are unable to have sex at all.

Think about that for just a moment. They have sexual identity problems. This is a condition that has been often overlooked or ignored by abduction researchers. Mack, when asked specifically about sexuality, said, "Nor [have] I found any link with sexual abuse or with peculiar gender issues." He also said, "Reproductive might be a better [term]."

But one abductee, in describing his sex life, said that he had never had sex before. When asked if his previous therapist, not Mack, knew this, he said, "He never asked." This is also quite common. The researchers get so caught up in the content of the abduction story that they fail to take a thorough history and fail to recognize the sexual dysfunction when it is a critical part of the problem.

Mack, in fact, when asked about this, rejected sexual dysfunction as unimportant. He was aware of it but felt it had nothing to do with the research he was conducting. He suggested that the percentage of people exhibiting some form of sexual dysfunction was no larger among abductees than in the population as a whole. Other researchers, including us, have found that sexual dysfunction in the abductee population runs as high as 90 percent. This is certainly outside the limits of the general population.

A few of the people who claim to be abductees do, in fact, have a serious mental disorder. Some exhibit symptoms of histrionic personality disorder, schizotypal personality disorder, or delusional disorder. Several people show all the signs of paranoid personality disorder. Often, their family members agree with this. Every abduction researcher should become familiar with these disorders before he or she begins interviewing abductees, or suggesting that no psychological pathologies exist.

When this has been suggested to abduction researchers, that is, when it is suggested that it is a fact that some abductees do seem to show signs of a mental disorder, the response has been classic. We are told, "Just because one is mentally ill doesn't meant that they haven't been abducted."

Which begs the question, as a researcher or practitioner who is heavily invested in a belief system, isn't it very easy to induce that belief system in someone who has come for help? Mack seems to recognize this, but fails to grasp the importance of it. Interviewed by C. D. B. Bryan in *Close Encounters of the Fourth Kind,* Mack said, "And there's another interesting dimension to this which Budd Hopkins and Dave Jacobs and I argue about all the time, which is that I'm struck by the fact that there seems to be a kind of matching of the investigator with the experiencer. So what may be the archetypal structure of an abduction to Dave Jacobs may not be the uniform experience of, say Joe

Nyman or John Mack or someone else. And the experiencers seem to pick out the investigator who will fill their experience."

And when interviewed by Estes, Mack said the same thing once again. "One of the interesting aspects of the phenomenon is that the quality of the experience of the abductee will vary according to who does the regression."

Mack is suggesting that abduction researchers are finding exactly what they want to find as they conduct their research. Hasn't Mack just explained the whole of the abduction phenomenon in a way that is consistent with Western thought and science with no need to create a whole new paradigm?

Mack and the other researchers are quick to point out that abductees are particularly unsuggestible. Researchers have tried repeatedly to trick abductees, only to be met with direct contradiction of their efforts. Again Bryan reported that Mack said, "Furthermore, abductees are particularly unsuggestible. To meet the above criticisms I and other investigators have tried repeatedly to trick abductees by suggesting specific elements—hair on the aliens, corners in the rooms on the ships, for example. Proponents of the controversial 'false memories syndrome' as an explanation for abduction memories need to account for this."

But we have seen just how subtle this coaching can be. A ham-handed question about hair on the alien, from a researcher who *knows* there is no hair on the alien, is just not sufficient evidence that the researcher isn't leading the witness. Instead, Mack tells us that the quality of the abduction experience comes from the researcher who conducts the interview. Is it likely that the "experiencers seem to pick out the investigator who will fit their experience," as Mack suggested, or is it more likely that the belief structure of the researcher influences the abductee. In other words, the researcher "leads" the abductee into the type of experience he or she reports and that is, in fact, the implantation of that memory by the researcher. In other words, Mack's suggestion appears refuted by statements he has made himself.

Mack made it even clearer for us when he said, "It seems to me that Jacobs, Hopkins, and Nyman may pull out of their experiencers what they want to see."

Hasn't Mack just answered his own point, accounting for this idea that researchers have tried to trick the abductees, telling us that experience is based on the researcher? The key here is the researcher and not the abductee, a point that Mack notes and then apparently fails to understand.

We also know, however, that people with an intact belief system tend not to be suggestible. While working with ritual abuse victims Cone, on several occasions, tried to make suggestions about the rituals they witnessed and abuse that they experienced. They are very clear about what did and did not happen, and cannot be swayed by these suggestions. The point, however, in some cases, is that it is later brought to light that none of it actually happened. The whole thing was a false memory. Scientific research has established this beyond reasonable doubt in both the ritual abuse and multiple personality populations. Since this is true for these two populations, why would it not hold for abductees?

The final point for Mack is that these stories are remarkably similar. They often contain the same elements, with those elements appearing in the tale in the same order. Descriptions of the procedures and the aliens themselves are remarkably similar. And all of that is, in essence, irrelevant to understanding abductions.

But once again research and personal experience have shown that ritual abuse victims also tell similar stories. Some of these stories are intricately detailed and remarkably complex. Nevertheless, they later turn out to be pure confabulation. This shows clearly that the fact that people tell similar stories, in similar ways, is not conclusive evidence that the stories are true.

One of the reasons for this might be the media exposure. For research being conducted in today's world, it cannot be denied that tales of alien abduction have been told repeatedly in the public arena. The elements for the tale are out there and the argument that a specific abductee hasn't watched the television shows, seen the movies, or read the books simply no longer works.

In a recent *Simpsons* episode, Homer was abducted by alien creatures. He dropped his pants and bent over only to be told that they didn't do anal probing anymore. They had learned about everything they could from those sorts of examinations.

The point is not that *The Simpsons* mentioned this but that the joke could work. For it to work, a large portion of the population who are not interested in or who have not studied alien abduction had to understand what was happening to Homer. This just proves that our culture is filled with references to alien abduction and the sorts of things, sometimes in incredible detail, that happen during an abduction. And for someone to suggest they have not been exposed to this information in any meaningful way is to miss the real point here.

But there are other explanations. Social pressure might explain it as well. People in support groups routinely take on each other's symptoms. Those involved with the abduction phenomenon often discuss their specific cases with each other, eventually incorporating points from one story into another. And although researchers are often careful to suggest that the various members of the group do not discuss the details with one another, they often do. Psychology has understood this for at least a century.

And finally we come back to the researcher. He or she knows what is expected during an abduction and searches for that information. The similarity of the tales can be explained by the researcher who is looking just for that similarity. While the abductee might not have access to the data, all the abduction researchers share it. Is it surprising, then, that what one researcher finds is often corroborated by another? The contamination is not always through the media but sometimes through the networking of the abduction researchers.

So Mack's theories and questions about the abduction phenomenon can be answered. The criticisms he levels at the skeptics and debunkers are, for the most part, without merit. And as we begin to examine his investigations and writings, we find that he seems to know the answers. He has danced around them, failing to see them in the proper light. He identifies the sources of the contamination but won't acknowledge them. He should know why people would invent these abduction tales but refuses to admit it. He must know exactly why people would want to join this abduction club but he won't see it.

Mack should know that emotion in telling a story is not an indication of the truth, or a validation of the tale, but seems to have forgotten that as well. He told Bryan a number of years ago, "If you were God and you were trying to reach the Western mind, you could not reach it with anything other than that which shows up in conventional physical reality, because we don't have the senses to know anything else. . . . And it's ironic that this rather crude, harsh intrusion of the abduction phenomenon into our senses is forcing certain people, the abductees, to open up additional realities with great terror. But if they have such terror, imagine the nonexperiencers, who, when confronted by this phenomenon, respond 'It's nonsense!' because they have no means of relating this to their notions of reality. They don't because it doesn't fit."

Or perhaps they understand that terror does not confer truth.

17.

THE ALIEN HUNTER—DERREL SIMS

Derrel Sims bills himself as an "Alien Hunter." He is more than just an abduction researcher. He is more than just a ufologist attempting to prove the reality of abduction through the recovery of implants left by the alien beings. He is a man who challenges the aliens, fights them, and hunts them as his life's work. He believes that someday the artifacts recovered through surgeries performed by his colleague, podiatrist Dr. Roger Leir, will change ufology as we now know it. Sims will be the pioneer who leads us to the answers about all aspects of the UFO phenomenon, charting the way for us, whether we like it or not.

Sims comes from a background that is fairly mundane. He says he was born and raised in Alamogordo, New Mexico, and attended New Mexico State University for about a year and a half. He dropped out, according to him, to fight in the Vietnam War, but apparently was sidetracked. While in training to become a military policeman at Fort Gordon, Georgia, he was subjected to a special screening process and was, again according to him, trained by the CIA. Sims won't say much about this training, only that after two years he left the CIA and completed his tour as a military policeman in Korea. He apparently never made it to Vietnam.

Sims also claims that his family tree originated in Europe and that he is descended from an English king. He fails to provide evidence for this and does not mention which king it is. If it is self-promotion, it is harmless.

He is also the force behind Saber Enterprises, which he founded. He teaches hypnotherapy to those interested, and he conducts hypnotic regression sessions on those who believe they have been abducted. These activities are conducted under the banner of Saber.

Sims' interest in UFOs may stem from his own experiences as an abductee. As regarding his CIA period, he won't say much about this, only to suggest that the whole story will be told in his autobiography, which has yet to find a publisher. He has told some that he was first abducted at three but the abductions ended when he was seventeen. His family has apparently been the targets of alien interest for about a century.

His fame in the UFO field, however, seems to be an outgrowth of his often expressed search for physical evidence. Sims has suggested, quite correctly, that lack of hard physical evidence is one of the major problems with any type of UFO research. According to him, around 1991 he confronted the leadership of the UFO community, whom he failed to identify, and asked them "Where's the beef?" meaning, quite simply, where is the physical evidence that UFOs are extraterrestrial? He told them that they had been in charge for forty years and had failed to prove the point. It was not an enviable track record.

According to him, they all turned red and then one of the panel members admitted that there was no proof. Rather than convince Sims that his time might be better spent in another endeavor, it just meant that he should take over the search for the physical evidence and the truth. He believed he would be able to do what they had failed to do in their forty-year reign.

So Sims turned his talents to becoming an alien hunter. In fact, and according to a background sheet that he hands out, "Derrel Sims' pioneering work in the alien and abduction phenomenon and his discovery of alleged artifacts may someday be the foundation for major discoveries in the fields of science and technology."

The Sims philosophy is one of "aggressive" research. Before Sims, the UFO investigators were passive. They would wait for something to happen and then go out to interview the witnesses, analyze the photographs or any physical evidence, and file their written reports with one of the many UFO organizations. This was the approach that had not provided the physical evidence that there was something to the UFO field. Sims believed that research could be taken to a new level. He believed that investigators should attempt to make things happen. He had a plan for doing that and he implemented it as quickly as he possibly could.

Part of that plan was a confrontation with an alien Sims named Mondoz. According to Sims, Mondoz had thought that his experimentation on his human subjects had gone unnoticed. He thought that he had fooled the human race because no one knew what was happening. Sims had, of course, discovered Mondoz's plan through his use of hypnotic regression and aggressive UFO investigation.

Another part of the Sims plan was to attempt to communicate with the abductors. Sims began to implant hypnotic suggestions in the subconscious minds of the abductees, one of whom, according to him, is abducted every eight weeks. He believed that when these people with a history of multiple abductions were taken again, the hypnotic suggestions would allow the victim to gather more information, to observe the surroundings with a more critical eye, and would allow the establishment of some sort of communication with the abductors.[1]

---

[1]Robert Terrell Leach, "Abduction Pioneer," *UFO*, May/June 1996, 28–32.

And according to Sims the plan worked. In November 1992, a subject identified by Sims and his followers as DS92007PH was abducted from her home in Florida. Dale Musser, a close colleague of Sims, reported that the woman, fighting to maintain control, fighting to communicate with the aliens, felt she was losing the battle. She screamed "We know what you're doing . . . We know about . . ."

In another version of the same story, reported by Randall Patterson of the *Houston Post,* Sims said the woman told the aliens, "Derrel Sims knows what you're doing."[2]

The important point, however, seems to be that Sims's aggressive investigation was beginning to pay off. The aliens were now aware that Sims was working to answer the questions and that he was having some luck in gathering data. The aliens, according to Sims, indicated they would terminate him if they were allowed to do it. But according to what he told Patterson, God was on his side. The aliens could not stop Derrel Sims.

But they were interested in his work. They wanted to know exactly what he knew, how he had learned it, and how they could counter it. This led, apparently, to the largest mass abduction in the history of the UFO phenomena. It was an event that would provide solid information because it came from several different sources, all unknown to one another. Only Sims and his senior investigator, Dale Musser, knew all the facts as the investigation began.

On December 8, 1992, prior to the Houston UFO Network (HUFON) meeting in which various abductees would participate in a panel discussion, "Alien Abductions: Working with Abductees," several of them reported dreams of possible abductions. Some of them reported nosebleeds or sinus problems, all believed to be symptoms of abduction. The reason for these new abductions, according to those various sources, was so that each member could be implanted with some type of recording device. The aliens wanted to have a record of the meeting. At the time it seemed that none of the abductees were in communication with one another and some were suffering from what Musser termed post-abduction syndrome (PAS).

On December 10, the meeting was held. The panel members seemed to be reluctant to talk, according to Musser. They looked uncomfortable and said they really shouldn't be discussing their experiences. Other than that, Musser felt the meeting went "quite well."

The next day, one of the panelists called Sims to make arrangements for a hypnosis session, believing he had been abducted yet again. Under hypnotic regression, the subject told of being abducted on the night of the meeting,

---

[2]Randall Patterson, "Space Case," *Houston Press*, May 30–June 5, 1996, 15–20.

though, in this case, "abduction" might not be the right word. The man was awakened in his bedroom by aliens who, in a very quick procedure, removed a nasal implant.

Sims then regressed the man to the previous abduction, which had happened on December 8, just days before the meeting. The aliens took the man from his room, told him to stand in a bright light, and then placed him in a large circular room. He was told to strip and then taken to a small room where he was given a quick examination.

A human appeared then, according to what the abductee told Sims under hypnosis, questioning the man about how he had known that he had been abducted previously, how he had learned there were other abductees, and why they were meeting together in Houston. In a sort of reverse of what Sims and other researchers have been doing, that is, having a panel discussion, the aliens questioned a number of the people they had abducted that night.

He was taken from that room, moved to a conference room, and questioned some more. At some point he was asked about the human subconscious, shown a model of the human brain, and asked to point out the location of the subconscious.

He was also questioned about other abductees, and as he was, the man could see mental images of them flashing into his mind. They were all nude and seemed to be unaware that others were present. But it also seems that each, under the hypnotic regression performed by Sims, was able to identify the others who were there that night. It allowed Sims to identify the other abductees, who when hypnotized by Sims, revealed their abductions and confirmed the mass abduction.[3]

It was on that same day that another of Sims's subjects, identified only as DS92009LT, claimed that she awoke with an eye irritation. While rubbing her eye, a small object, consistently described as the size of a mustard seed, fell out. Naturally she called Sims and then gave him the object.

Sims, following what appeared to be strict scientific protocols, had the object analyzed by independent authorities. Saber Enterprises had a metallurgical analysis performed on the implant at the University of Houston. But Lisa Meffert, a geneticist at the university, told Randall Patterson that the analysis revealed that the object was probably nothing more spectacular than the tip of a bobby pin.

Patterson reported that Sims had entered her office telling her that the object had fallen out of someone's eye. She thought Sims was an attorney working on a lawsuit. She put the object under the microscope and reported that she thought

---

[3]Dale Musser, "Anatomy of a Mass Abduction," *UFO Universe*, Fall 1993, 24–6.

it was the plastic tip of a hairpin piece. She simply wasn't impressed with the "implant."

Unsatisfied with the conclusion, Sims took the implant to the Texas Center for Superconductivity at the University of Houston. According to those test results, there were certain rare metals in the object. They had found a trace of titanium on the tip. That suggested it was something more than just the end of a bobby pin.

The laboratory analysis, written on Saber Enterprises letterhead, was reprinted in *UFO* magazine in 1993. Under the comments section of that letter, it was noted that "It was impossible to tell precisely what this object is with any certainty. We are unable to identify . . . any natural or man made object it could be."

An implant recovered by a known abductee who claimed that she found it shortly after an abduction, as reported under hypnosis, and the physical evidence that science has demanded appeared to have been found.

Unfortunately, the University of Houston was not excited about their involvement in the analysis of the object. Again according to Randall Patterson, after his original story was published, the university's public affairs office issued a statement about the "implant work." It was suggested that no more work would be done for, or with, Derrel Sims. They felt that they had been misled, though it seems that the conclusions they drew about Sims were their own.

There was still the analysis by Saber Enterprises. Nowhere in that *UFO* magazine story was it revealed that Saber Enterprises was founded by Derrel Sims. It must be noted, however, that the analysts listed on the report were Dale Musser and Derrel Sims, and the connection to Sims was mentioned in another article in that same issue.

Undaunted, Sims continued, telling various researchers that some top-notch scientists had joined his team for the analysis of the alien implants. But the names of the scientists have not been revealed. Nor have the names of their organizations been revealed. Nor has any documentation about their work been provided for independent scrutiny.

The other laboratory analyses, conducted on various other implants, some recovered through surgery conducted by Sims's friend and partner, Dr. Roger Leir, have been released. The problem is that the names of the laboratories have been blacked from the reports so that no followup investigations can be made. The reports must be taken at face value because there is no way to verify the accuracy of the documents. It all looks impressive. It seems as if strict scientific protocols are being followed, yet there is no way to verify any of the information.

There has been some "peer" review of Sims's work. After Sims lectured in Israel, a doctor reviewed the lecture for a small group of his friends and colleagues. Somehow, and to the doctor's regret, that message was posted on the

Internet. The doctor had attended a Sims lecture in Israel but was unimpressed with what he had been shown. He noted, "Mr. Sims's stories have medical inaccuracies regarding details of the operations performed for the removals of what he and his organization . . . claims to be implants. . . ."

It could be suggested that the non–medically trained Sims, in listening to the analyses presented by Leir and other physicians, had misunderstood the precise nature of the methods used. The technical jargon and the exact nature of the medical procedures might have confused him. The layman Sims, in attempting to report on surgical procedures, had made some simple mistakes.

The doctor, however, pointed out, "I specifically asked him to provide me with color slides of the actual pathological specimens that he extracted from the subjects studied. He declined under various motives. No other evidence or specimen was available to my specific request. I asked him for such because I have the possibility to have them examined by some of my friends . . . in the pathology Dpt of my Medical Center . . . Israel. No specimen, pathological slide or other material was available from Mr. Sims for examination by an independent specialist or laboratory—as I have asked."

We contacted the doctor to verify the accuracy of the statements that had been posted. We learned that the letter had been posted to a closed discussion list and had been circulated generally through the Internet without the doctor's permission. He had been trying to point out one of the problems with UFO research without moving it into a public arena. He was distressed that his private communication had been circulated without his permission but he did note that he stood by what he had written. In other words, he believed that it accurately reflected the reality of the situation.

He noted that he had been in communication with Leir, and that he had again offered to have implants analyzed by his laboratory at his own expense. Here was an opportunity to add another laboratory to the list that Sims claimed had examined some of the implants. Here was an opportunity for independent corroboration of the data that Sims had retrieved from other sources, but he refused to cooperate. Leir, however, promised to send some samples to the doctor, but, to date, has failed to do so.

Apparently Sims was satisfied with the work done at the University of Houston, and with Warren Laboratories and the independent work they were doing. Sims told one magazine reporter that Warren Laboratories was a very large corporation near Houston. Patterson, of the *Houston Post,* talked to George Warren, the proprietor of the company, and learned that it is, in fact, a small herbal-extract business. Warren apparently met Sims on a plane en route to a UFO conference a few years earlier and has been analyzing implants ever since. Warren told Patterson that he believes in alien abduction and that he does the analyses for free.

When all the data are reviewed, there is little in the way of scientific verification for the work that Sims has done on alien implants. The only laboratories that have been named by Sims turn out to be his own Saber Enterprises and Warren Laboratories, owned by a friend. The question that must be asked is why he refused to provide the Israeli doctor with the evidence of abductions and implants that he had requested. It certainly would have boosted his credibility if the doctor and his colleagues had corroborated Sims's findings.

Sims told *UFO* magazine that he didn't believe in secrets, as he had told various audiences in the past, but he doesn't talk about his own abductions. He doesn't name the names of the laboratories who have performed the analyses of the implants for him. Nor does he talk about the breakup of the original HUFON group. Several of the board members resigned after a controversial meeting held in 1993.

One of the ex–board members reported that the controversy erupted "due to my proposal to institute a protocol for investigations of both sightings and abductions. The biggest point of contention was my suggestion that a mental health professional be appointed to treat abduction trauma." Sims apparently rejected the idea.

Robert Leach of *UFO* reported that "Sims answered this by admitting the suggestion of a mental health professional being involved in the HUFON support group 'irritated' him. 'That sounds harmless enough,' Sims said about the matter, 'but what you're suggesting is there's something wrong with these people because they saw, heard, or experienced something that you didn't and they therefore need a psychiatrist. . . .' "

Max Washburn apparently also thought the idea of the mass abduction was dangerous. Implied in this was the idea that Sims might be operating out of his depth. Implied was the idea that someone with specific training in a mental health field could be valuable in the research being conducted, not to mention the fact that it might help the people who had been traumatized.

According to Patterson, Vince Johnson and Rebecca Schatte, former HUFON members, were also concerned about the mental health of those in the abductions group. Schatte believed that some of the people needed real psychological help and both she and Johnson questioned Sims's qualifications to provide it.

The meeting degenerated at that point, but Sims was apparently content to sit back, letting the dispute swirl around him. When Rod Lewis suggested that one of them could kiss his ass, Sims lunged across the table. It was time for the meeting to end.

Lewis, who appeared to inspire the confrontation, left the group. He now seems to believe that the government is somehow involved in the abductions, or government secrecy is covering it up, or that the government is trying to control everyone's mind through the use of the implants that have been recovered. To some, the idea of alien abduction has evolved into the idea of government abduction. Alien creatures are not responsible. Government agents or military officials are. They implant, with hypnosis, the "screen" memories that it is alien creatures that are responsible, to hide the truth from abductees and researchers.

Lewis then mentioned that Sims might be a government agent. But once again it is a charge without foundation. Everyone who has ever done any sort of UFO research has been accused of being a government agent at one time or another. Each of us has had that charge hurled at us. It means absolutely nothing to anyone.

So Sims continued his research, enlisting the aide of Leir, who said that he was, at first, unimpressed with Sims. Leir thought him strange with his "box of treasures," that is, the implants removed from patients, that he displayed at every opportunity. But then Sims showed Leir the X rays, and the report from the University of Houston that suggested rare metals in one of the implants. That interested Leir.

It probably also should be noted that Randle has seen the box of implants. Sims told him that they were not all alien, but some of them were from the CIA. Apparently the government is also interested in keeping track of those who have been abducted by alien creatures.

In fact, at one of the sessions to remove implants, one of the patients identified only as a man in his mid-thirties claimed no alien contact. He thought the object was implanted during routine dental work. At the time the man worked for a Department of Defense subcontractor and had used a dentist recommended by his employer. After the dental work, he began to hear two distinct voices in his head, but it was only after an MRI and X rays that the implant was discovered in his neck.

The implication was that the man's implant was not alien in origin but was, in fact, from some governmental agency. It is not clear why the government would place an implant in the neck of a man who only worked for a subcontractor. Nor has it been established that the government has the technology to make and implant these devices. Test results have yet to be revealed.

Leir has also suggested that the implants are discovered without the inflammation that would be expected around foreign objects found in the human body. The biopsies showed the presence of peripheral nerve and pressure receptors, which suggested to Leir that he was dealing with advanced medical techniques. Leir, in various lectures, has suggested this is indicative of an alien presence. He found all of this quite persuasive.

Others in the medical community weren't so impressed. At least one molecular biologist suggested that the evidence cited by Leir, and by Sims, was not all that unusual. The man pointed out that a small object, to be implanted into the body by earthbound doctors, could be coated with a nerve-growth material that would mimic the evidence found by Leir. It was not evidence of extraterrestrial science as Leir had suggested, but evidence of advanced earth-based medical techniques, if Leir's conclusions about the implants were correct.

Leir, at the 1997 Omega Communications conference held in North Haven, Connecticut, reported that he had removed implants from five separate abductees. Again, he claimed the chain of custody had been established, and that photographic as well as video evidence of the removal had been created, but

THE CIA AND IMPLANTS

Like so much of the abduction phenomenon, there is another phenomenon that parallels it. UFO researchers believe that alien beings are implanting humans with some sort of tiny monitoring or controlling device. Outside the UFO field there are those who believe that it is the CIA that is responsible for implanting humans with these devices.

The information from these other sources matches that produced by UFO researchers. Those suggesting CIA responsibility note that the devices, implanted during routine surgical procedures, are sometimes discovered on X rays of the bodies of the unsuspecting victim. Doctors who have examined the X rays can offer no explanation for the devices.

In a case from Sweden, Robert Naeslund claimed that he had been used in some kind of medical experimentation that had caused him pain and suffering. He claimed that the implant had been inserted through his right nasal passage and lodged in his brain. Doctors who examined him after the procedure and who X-rayed him denied that anything had been implanted. They suggested that psychiatric care might be warrented.

Naeslund traveled to the United States, where a single doctor confirmed that a "brain transmitter" had been implanted. From that point, Naeslund said that he had statements from ten other doctors who confirmed an implant, but his own doctors in Sweden refused to look at the evidence, insisting there was nothing there. He has not been able to find a doctor to remove it.

Naeslund was not one of those who believed in alien abduction. He was convinced that the source of the implant was something other than alien creatures. In fact, he suspected either the Swedish government or the American government.

there still has been no proper peer review of the material and no information provided that could be corroborated by independent investigations.

Leir suggested there that he could not understand how the implants had been introduced because he found no evidence of scarring on the surface of the skin, and that he saw no evidence of the rejection of these foreign objects by the body. To him, it was impossible to understand, and that, to him, suggested an extraterrestrial source.

But once again, other scientists, concerned with the rejection of foreign material by the body, suggested that Leir's opinions might be formed on erro-

In another case, a man claimed to have undergone an operation to remove some sort of polyp from his right nostril. Although the operation was supposed to last about ninety minutes, according to him, it last nearly seven hours.

The man also related that his son, who had undergone minor surgery to remove his tonsils, was also implanted with some sort of device that was part of a "Mind Control Experiment." He believed that it was part of the CIA's attempt to fight its battles.

None of this, however ties in directly to the UFO abduction phenomenon, other than the similarities of the cases. UFO abductees report nosebleeds, and a belief that something has been shoved up their nasal passages. X rays sometimes reveal the objects, but most often they do not. Or when a second set of X rays is taken, the object is missing.

What does tie this to UFO abductions, however, is the belief and reports by some that after the abduction by alien creatures the CIA or the Air Force abducts them again. Or, in a few cases, military personnel have been seen working with the alien beings. Air Force intelligence and the CIA are apparently working in concert with the aliens not only to abduct people but to help monitor their activities.

Derrel Sims, the Houston-based abduction researcher, carries a small box to lectures with him. It contains odd-looking bits of debris that he claims are implants removed from various abductees. He points to some suggesting they are alien, and to others claiming they are from the CIA.

However, each time implants recovered by any researcher have been analyzed by reputable independent laboratories and scientists, they have been found to be organic material or slivers of glass or other completely terrestrial debris. Nothing has been discovered to suggest they were any kind of mind control device or that they had any detectable function.

neous assumptions. The Veterans Administration, for example, has conducted surveys of soldiers who have had shrapnel in the body for decades and found tissue and nerve growth as well as a lack of inflammation. In other, more precise words, it could be that Leir was overreacting to what he thought of as extremely unusual or as extraterrestrial. Medical doctors have not been quite as impressed.

With Leir's help, and surgical expertise, Sims has seen additional implants removed. Leir still claims, as he did during the Omega Communications conference, that they have been sent on to reputable independent labs for analysis. But when the reports are shown to those of us outside of Sims's camp, as they were at the conference, the names of those alleged prestigious laboratories and universities have been blacked out.

It is all reminiscent of the stir Sims, Leir, and Paul Davids, a man interested in UFOs, created in Roswell, New Mexico, during the Fiftieth Anniversary celebration during the summer of 1997. Sims claimed that an unnamed source, who had been in Roswell when the UFO crashed in 1947, had recovered some of the strange metal from that alleged alien craft. He had given it to Sims, who had sent it off to be analyzed by a number of high-powered universities. Sims and Davids promised that the chain of custody of the debris had been preserved, that the analysis proved that the metallic object had been manufactured on another planet, and that all this would be revealed, along with the proper documentation, at a special lecture scheduled during the UFO celebration in Roswell.

Davids introduced a number of the players, displayed some of the documentation about the debris, and then introduced Dr. Russell VernonClark, a chemist at the University of California in San Diego. During his presentation, Dr. VernonClark mentioned that a review of the isotopic makeup of the sample suggested that it was a manufactured object and that the manufacturing had taken place on another world. VernonClark was saying that here, at last, was good evidence that extraterrestrials had visited our planet.

But during the question-and-answer period held after the lecture, with VernonClark conspicuous by his absence, Davids couldn't provide the chain of custody as promised, could not identify the names of the other high-caliber universities where the results would be confirmed, and could offer no proof that the metal analyzed could be tied to the events in Roswell fifty years earlier.

VernonClark, interviewed by us several months after the Roswell announcement, admitted that the results had been only preliminary. If he had it to do all over again, he would not have allowed himself to be pressured into making the announcements. He also suggested that the while the isotopes revealed in his analysis were not naturally occurring on Earth, they certainly could be manufactured on Earth. The existence of the isotopes were not, in and of themselves, proof of extraterrestrial visitation. The chain of custody, and the name of the man who had originally supplied the sample, was necessary. That was not forthcoming.

A year after the Roswell event, we are no closer to the scientific analysis of the material than we were in 1997. The material is still in the hands of Sims and there are still promises that the final analysis will be completed soon. At that time, it will all be released to the public and to the media.

Dale Musser, when asked why it was taking so long for the analysis of the implants and other material, and told that the public was becoming impatient, said that he didn't care about the public impatience. They, meaning Sims and his group, were going to wait to release the data. They were going to make sure that all the proper scientific protocols were followed so that no criticism about their failure to follow those guidelines could be leveled by their critics, the scientific community, or the media.

PART 4

PSYCHOLOGICAL PARALLELS

18.

SATANIC RITUAL ABUSE?

There is another psychological phenomenon closely linked to that of alien abduction. It was originally identified by researchers and therapists and revealed to a public that had no clue that such a thing could exist. And just like alien abductions, the existence of it is predicated on, almost exclusively, testimony that has been recovered through hypnotic regression, visualization, and other memory enhancement techniques. And like alien abduction, there is virtually no physical or corroborative evidence that the phenomenon exists at all.

Almost no one had heard of satanic ritual abuse (SRA) until 1980, when the book *Michelle Remembers* first appeared in print. Although there had been earlier reports of satanic worship, indications that some cults were satanic in nature, and many books about satanic rituals had been published, this was the first time that the idea of ritualistic abuse had entered into the mainstream.

Since the publication of that book, hundreds if not thousands of victims who claim ritualistic abuse have been located by researchers, therapists, law enforcement officials, and mental health care professionals. Under the guidance of psychologists, hypnotherapists, police officers, and social workers, these people, who manifest a variety of psychological ailments, begin to tell elaborate and horrifying tales of abuse at the hands of the members of a worldwide conspiracy of Satan worshipers.

These stories, from people who have never met, who live thousands of miles apart, agree in great detail. These people are telling, basically, the same story over and over from their own personal perspective. This agreement in detail, which is cited by nearly every researcher into satanic abuse as quite persuasive, proves to many that something very real and horrifying is happening. The evidence, according to them, is overwhelming.

Over the last few years there have been many articles, books, and television specials about the growing phenomenon of satanic ritual abuse. Dozens of lawsuits have been filed, and hundreds of people have been accused of engaging in these bizarre ritualistic practices, ripping apart families and the social structure of numerous communities. Many of those accused are now serving prison sentences because of the accusations made by those who have been abused, including members of their own families.

Paul R. Ingram, a former member of the Thurston County, Washington, Sheriff's Office, is a good example of the terrors of satanic ritual abuse. On November 28, 1988, he was arrested after allegations that he had raped his own daughters on a number of occasions. The list of his crimes eventually grew long and shocking. It included the rape and torture of his daughters and eventual claims that he had murdered infants during some of the more sadistic of the satanic rituals in which he engaged. The ironic aspect of this case is that Ingram had no conscious memory of any of the activities until his two daughters accused him of molesting them.

Even without a single conscious memory of the events or shred of corroborative evidence, within hours of being arrested, during his official statement to sheriff's investigators, Ingram said, "I really believe that the allegations did occur and I did violate them [his daughters Erika and Julie] and abuse them and probably for a long period of time. I've repressed it."

What is incredible here is that a man who, for his entire life, seemed to be a kind and upstanding member of the community, who went to church regularly, who was a police officer for most of his adult life, was now confessing to some of the worst crimes that could be committed. He confessed because he didn't believe his daughters would invent, or could invent, such terrible tales and therefore those allegations must be true. However, when police investigators asked him for specifics, he told them, "I don't remember anything."

When Ingram failed to provide any details of the alleged rapes, although admitting he had abused his daughters, investigators told him what the girls had said. They also provided Ingram with some particulars from his daughters' statements and asked that he think about the particulars for a while. Unwittingly they had just given him the details that he, himself, could not remember.

After twenty or thirty minutes of meditation in his cell, Ingram, with the tape recorder running in the interrogation room, began to talk. He said that he, wearing a bathrobe, had entered his daughter's room and stripped. In a halting voice, as if he was carefully thinking about each statement as he made it, he said, "I would've removed her clothing, uh, her underpants or bottoms to the nightgown."

The investigators immediately picked up on the word, "would've," which any good defense attorney would use in a trial to prove Ingram innocent. They asked, "Now do you mean would've, or did you?"

"I did," he responded to their pointed question.

He was then asked, "After you pulled down her bottom, where did you touch her?"

"I touched her breasts and I touched her on her vagina."

When asked what he had told her when she woke up, Ingram responded, "I would've told her to be quiet and, uh, not say anything to anybody and threatened her . . . that I would kill her if she told anybody about this."

During the interrogation session, Ingram confessed to raping both his daughters, and eventually impregnating one of them. He confessed to taking his pregnant daughter to have an abortion when she was fifteen. There was no discussion of satanic ritualistic abuse at that point but the confession, even without those graphic details, was disgusting to anyone who heard it. To make it worse, if that's possible, investigators noticed that Ingram seemed to be emotionless as he confessed. He seemed to have no remorse for the crimes that he had now admitted he had committed.

For the investigators, Ingram's admissions wrapped up the case. A crime had been reported, it had been investigated, and the perpetrator was in jail. The sexual abuse allegations have been substantiated when Ingram confessed. No one seemed to think that a search for corroboration should be initiated. If he had gotten an abortion for his daughter, there should have been medical evidence of it in the form of medical records. No one tried to locate those records.

These sexual abuse allegations against their father weren't the first the girls had lodged against an adult male. Both daughters had told similar tales of sexual abuse at the hands of family friends and neighbors in the past. In 1983, Erika told fellow students at a religious retreat that she had been raped by a man she knew. Although sheriff's deputies investigated the accusation, no charges were filed. It seemed, according to the facts discovered by the deputies, a married man had given her a ride and put a hand on her knee.

Two years later, at another religious retreat, Julie said that she had been abused by a man who had once lived on the Ingram ranch. Again a police investigation was initiated, and Erika soon joined her sister in making accusations. But the investigation ended when the story fell apart. Deputies noticed a number of inconsistencies in the testimony of the girls and let the investigation quietly drop.

Then, in 1988, at the religious retreat once again, this time with Erika as a counselor, both listened to Karla Franko tell her stories of spiritual inspiration. During her talk, Franko said that she had the feeling that someone in the audience had been molested by a relative. Several of the girls reluctantly admitted to such molestation.

On the last day of the camp, as everyone was preparing to leave, Erika broke down, crying hysterically. She would tell no one what was wrong. Her friends and the other counselors gathered around her to show support. Finally she blurted, "I was abused sexually by my father."

There is, according to an article written by Lawrence Wright for *The New Yorker,* another version of this traumatic event. Franko, learning that one of the counselors needed some help, approached Erika. Franko told Erika that she had been sexually abused as a child and suggested they seek help. Notice here that it was Franko who made the initial allegation and not Erika.

In the months that followed that meeting, Erika accused her father of abusing her repeatedly. She also accused her two older brothers, and then linked her father's frequent poker parties to the abuse. Eventually it would all be linked to satanic rituals in which Erika and her sister were unwilling participants.

And it didn't end there. A counselor for a rape crisis center, after conversations with Julie, took her to meet police investigators. Julie claimed that the abuse at the hands of her father began when she was in the fifth grade and she told stories of her father raping her or sister on several occasions.

For corroboration of Julie's allegations, the detectives interviewed Erika, who said the abuse began when she was just five years old. She took it further, accusing her father of giving her a disease that was treated by two separate doctors, one local and one in California. Again, no one thought to look for medical records or other documentation.

That was the end for Paul Ingram. Although there was nothing to corroborate the stories, there was no physical evidence or documentation, and even though there were inconsistencies in the many allegations, including the shifting date of the last incident of abuse, the police believed the girls. Although Ingram was one of them, a fellow officer, they just didn't believe him. After all, if it wasn't true, why would the girls make up the tales? And where would they get the incredible detail of incident after incident if it wasn't all true? No one could invent such elaborate stories. Besides that, it was clear that both girls had suffered some trauma. If they were inventing the tales, they should have moved to Hollywood. Their acting was of Academy Award caliber.

Then, even as he was originally denying the events as outlined by his daughters, Paul Ingram surrendered to the sheriff on November 28. Within hours under standard police interrogation techniques, he confessed that he had, in fact, raped them repeatedly. He said he had no conscious memory of any of the events but that the charges must be true. His daughters wouldn't make up those sorts of lies about him.

A psychologist from nearby Tacoma, Richard Peterson, told investigators that it was not uncommon for child abusers and other sexual offenders to suppress the memories of those crimes. It wasn't impossible for Ingram to have no conscious memory of the events, even though, according to his daughters, some of the abuse had taken place only weeks earlier. And child abusers are frequently the victims of similar abuse and therefore it was probable that Ingram had also been a victim as a child.

Then other, seemingly corroborative, evidence began to surface. One of Julie's teachers gave police letters she, Julie, had written. This new evidence, though it was from the same source as the accusations, seemed to strengthen the allegations. The letters helped expand the case outward until some of Ingram's friends, his regular poker buddies, for example, were also accused of molesting the girls.

To this point, the investigation had centered only on the rape allegations made by Ingram's daughters. They claimed to have been raped by a number of men, including Ingram's friends and poker buddies, but that was as far as the allegations had gone. Now Peterson, as well as the original investigators, began a new series of interviews with Ingram searching for more evidence. It was Peterson who finally asked Ingram if the rapes had anything to do with black magic. When Ingram said he didn't know what Peterson meant, one of the investigators said, "The satanic cult . . ." Satanic ritual abuse was suddenly on the table, put there not by Ingram or his accusers but by the police investigators.

Ingram prayed and asked for guidance. Peterson told Ingram that it would be helpful if he stopped asking for God's assistance and if he would just relax. Peterson told Ingram that everyone wanted to help him out of this crisis but first he needed to relax. Just let go.

From that point, it became apparent what was happening to Ingram during the interrogations. He slipped, during his prayers, into a self-induced hypnotic trance. Listening to the questions and the suggestions of those who, only that morning, had been his friends and who were telling him they were there to help him, he began to spin a larger tale that grew from the poker parties. Now the men who attended the poker parties were tying his daughters' hands and feet and subjecting them to various sexual tortures. According to Ingram, one of the men even took several pictures of the events so they could relive them later.

The thought of photographs excited the investigators. If those pictures could be found, then independent corroboration of the satanic cult, and the abuse of the two young women, would be available. It wouldn't be the word of one former police officer against that of two traumatized young women. It would suddenly be the tales corrobroated with hard evidence. But repeated searches of Ingram's house, police locker, and everything he owned failed to produce even a hint of the photographs.

As the investigations continued, now with a more focused nature, specific questions about the cult's activities were asked. As Ingram provided new information, it became apparent to the investigators that he was heavily involved in the satanic cult and his daughters had suffered for years at the hands of that very cult.

With the accusations of the girls, now corroborated by the confessions of their father, two other men were eventually arrested and charged. After an extensive police investigation that failed to find a shred of independent corroboration, those two men were acquitted. No evidence was ever presented that they had engaged in satanic ritual abuse, had raped or abused either Erika or Julie Ingram, or that they had killed anyone as part of their participation in a satanic cult. Ingram, however, to spare his daughters the horror of having to testify in open court against him, and finally convinced of his own guilt, pled guilty.

The Ingram case, which resulted in a single conviction, convinced many that this was only the tip of the satanic ritual abuse iceberg. The stories of satanic ritual abuse were coming from all over the world and from many different witnesses. Stories about SRA were appearing in various magazines and in some books. Psychologists and therapists were finding more people who, under the influence of hypnosis and other memory enhancement techniques, were beginning to remember the horrors inflicted on them by these cults. Satanic ritual abuse was suddenly considered to be an epidemic.[1]

In Pennsylvania, as the Ingram case was winding down, a third grade teacher, Renee Althaus, was called from class and told by her principal and the school psychologist, that her daughter was now in the hands of Allegheny County's Children and Youth Services (CYS). Her husband was being accused of inappropriate touching of his daughter, Nicole, who eventually was placed in a foster home.

The abuse came to light when Priscilla Zappa, Nicole's tenth grade teacher, convinced that something was bothering the girl, talked to her of abuse and gave her books on the topic. Through Zappa, Nicole learned of a worldwide underground of Satan worshipers who abducted, tortured, and sometimes killed children.

With the introduction to SRA made by Zappa, Nicole began to tell similar stories of ritualistic abuse. She told of having been sold to strangers for sex, that she'd had two abortions, and that a fetus had been taken from her by cesarean section. She had been tortured with whips and forced to walk on hot coals or on broken glass. Even her grandmother had participated in the tortures.

Nicole was believed by those interviewing her no matter how wild her stories became. Even when physical examinations failed to produce any evidence of the tortures she claimed or that she'd had a cesarean section, the district attorney continued the investigation. No one could invent such horrifying tales if they weren't true. No one could be so emotionally distraught by such tales if they weren't true.[2]

Ritualistic abuse, like that claimed by Nicole, as well as Ingram's daughters, was now being thoroughly investigated by various authorities, including a number of doctors and therapists who had worked with dozens of others who told similar tales of satanic ritual abuse. Bennett G. Braun, M.D., head of the Dis-

---

[1]Elizabeth Loftus and K. Ketcham, *The Myth of Repressed Memory* (New York: St. Martin's Griffin, 1994): Richard Ofshe and Ethan Watters, *Making Monsters* (Berkley, CA: University of California Press, 1996); Lawrence Wright, "Remembering Satan, Part I," *The New Yorker*, May 17, 1993, 60–81; Lawrence Wright, "Remembering Satan, Part II," *The New Yorker*, May 24, 1993, 54–74.

[2]A. S. Ross, "Blame It On the Devil," *Redbook*, June 1994, 86–90, 114–6.

sociative Disorders Program at Rush-Presbyterian-St. Luke's Medical Center, in Chicago, told police investigators that there is an international organization responsible for the tortures of innocent children. In lectures around the country, Bennett was quick to alert law enforcement officers and mental health care professionals to the danger of satanic abuse. Bennett is one of the leading experts in the field, frequently consulted by those who find themselves overwhelmed by the stories told by their troubled patients.

Because of the credibility of those like Dr. Braun who were spreading the word, law enforcement officials were convinced that satanic ritual abuse was real. When Nicole began reporting her tales, they were corroborated by stories reported around the world. Those who believed asked the skeptics, "If there is nothing to this, how could so many people who are so widely separated tell stories that are so strikingly similar?" None of those telling the tales were in communication with one another, so they couldn't be comparing stories to make them consistent. It was a very good question.

Nicole's story, however, began to unravel when she mentioned that babies had been killed. Erika and Julie in Washington had told investigators that twenty-five babies had been killed in rituals there that involved their father. But when Nicole made the same claim, local homicide detectives entered the case.

Now, suddenly, the cynical detectives didn't believe everything Nicole said. They wanted proof of the crimes and they interrogated Nicole carefully. They looked for evidence that a crime had been committed and they found nothing. There was no corroboration for what she had told police, social workers, therapists, doctors, and Priscilla Zappa.

The final blow to the investigation came when a psychiatrist, selected by the prosecutors, who were pressing to indite those who had abused Nicole, found no evidence of any physical or psychological abuse. In April 1992, after fourteen months of hell, it was all over. The judge dismissed the case against Nicole's parents with prejudice so that it could never be refiled. With the family reunited, Nicole recanted all the allegations of child abuse, including the tales of barbaric satanic ritual abuse, the murder of innocent children, and the rapes by her father.

But other, similar cases were pushed to the front. Many of them came from women suffering from clinical depression. One woman, suffering from just such a depression as well as an eating disorder, entered psychotherapy, where she began to remember sexual abuse by her father. As the therapist probed, the woman began to recall satanic ceremonies in which people drank blood, humans and animals were sacrificed, and people were eaten.

Another woman, in Long Beach, California, checked into a psychiatric hospital because of depression. Within weeks she was recalling abuse by her father, including incidents while she was a toddler.

And still another woman, Janet, an executive for a large corporation who was being harassed by her boss, began to complain of insomnia and feelings of helplessness at work. She was thirty-five, divorced, and was the mother of three children. She had left her abusive husband several years before, and was now the sole breadwinner for the family. She was attractive, intelligent, but about fifty pounds overweight. She was a hard worker, but was submissive, dependent, and nonconfrontational. At the advice of friends, she entered therapy.

For the first few weeks, she improved steadily. Then, in her second month of therapy, she lost her job. The night it happened, her daughter called Dr. William Cone in the middle of the night to tell him that Janet had ingested all of her antidepressants in an attempt to kill herself. Janet was taken immediately to the emergency room for treatment.

Out of danger from the drugs, she was placed in the locked unit of a local hospital. During the first few days, Janet again began to improve. As part of her treatment she attended group therapy sessions every day. Her condition began to deteriorate while in group.

Janet's roommate in the hospital was a young woman who had been diagnosed as having multiple personality disorder (MPD). She believed she had three distinct personalities. That would be irrelevant here except that all of the members of Janet's group had been asked to write about abuses they had suffered in the past. Janet, who had never believed she had been a victim of child abuse, and who had no history of MPD, began writing abuse stories in three distinct types of handwriting, just as her roommate had been doing. Within two weeks of joining the group, Janet began to exhibit the symptoms of multiple personality disorder, post-traumatic stress disorder, and child abuse.

It became clear to Cone, after spending some time on the unit, that a patient was not considered part of the group unless she could produce recovered memories of child abuse. A patient was not considered to belong unless she could show evidence that she had multiple personality disorder. According to Cone, because the patients had a strong desire to fit in with the group, every patient soon manifested new personalities.

Part of the treatment program was a mandatory chemical regression, using sodium amytal. Patients were told that through this procedure they would remember the abuse they had been repressing for years. The staff and patients called this "getting their sodium."

"I can't wait to do my sodium," Janet told Cone, because no one was considered an official member of the group until they had undergone this important rite of passage.

Cone had never witnessed the procedure, and knew nothing about it's efficacy, but was willing to learn. He arrived at the appointed hour to see Janet lying on an examination table, covered with a blanket. The lights were dim. In the room with Cone were an anesthesiologist, the staff psychiatrist, Raul, and the group leader.

As the anesthesiologist began to administer the sodium amytal, Janet began to get drowsy. Soon she was on the brink of unconsciousness and Raul asked, "Can you hear me?"

Janet mumbled a yes.

"Tell me how you feel about your mother."

"She always hit me. She always yelled."

"There were many times when she hurt you," said Raul. "What do you want to do to her?

"I want to kill her."

"There are many times she did bad things to you. As you think about it, several of these incidents will come to mind," said Raul. "Tell us about them."

"She always picked on me."

"What bad things did she do?"

"She forgot my birthday."

"What else?"

"She broke my doll."

"What did you do?"

"She took it away. I wanted it back. I tried to grab it but she threw it on the ground."

"I want you to picture your mother hitting you. Yelling at you. She is calling you names. What kind of names does she call you?"

"Whore," replied Janet.

"Your mother was angry at you. She was going to harm you. What did you think she was going to do?"

This disturbed Cone because it was obviously leading the patient into a specific set of beliefs. To that point in the session, there had been no suggestion of real harm. Punishment administered by an angry parent is not the same as suggesting some kind of harm.

"She was going to kill me," said Janet.

This prompted a curious question from Raul. "Up until this point in your life, had you ever seen anybody dead?"

"No."

"Well, if you'd never seen anybody dead, how did you know you were going to be killed? How did you know what a dead person looked like?"

The session had subtly moved from an angry child and a yelling parent into a discussion of dead bodies. To Cone, this was leading the subject. Raul, it seemed, was implanting ideas into the mind of the patient as he asked his specific questions, almost as if he had an agenda in mind before the session began. There was a direction he wanted to go, and was gently, subtly leading the patient in that direction. But this time, before Janet could respond, the session ended because she had fallen asleep.

After the session Cone asked Raul about his strange line of questioning.

Raul told Cone, "The abuse is deeply repressed in the unconscious. If you don't dig for it, you won't find it."

Raul was telling Cone that he knew there was evidence of abuse even though nothing had indicated its existence to Raul. He was proceeding on his assumptions, forming his questions in such a way as to lead the patient in that direction. His rationalization for this was that it was necessary to dig out the abuse. That it was deeply buried and repressed. Without the proper questioning, it wouldn't be found.

Cone saw this session as revealing nothing about repressed abuse. It indicated to him that Janet remembered an unhappy childhood, which they already knew. It suggested an anger at her mother, but they already knew that too. Nothing suggestive of abuse had been learned, but certainly, the seeds for those beliefs have been planted by the questions of the therapist.

After returning from her chemical regression, Janet slept for an entire day. The next day, she was the star of her group. All sat on the edge of their seats, waiting to hear what Janet had learned about her childhood. As she told her story, the other group members nodded in approval. Cone said that this behavior by the rest of the group validated Janet's experience as being remarkably similar to their own. Janet had finally been fully accepted by the group, one of her goals at that time.

Cone took a tape of the regression home so that he could review the procedure carefully. After studying the tape several times, he could find nothing to suggest that any severe abuse had occurred or that any other personalities existed. Janet, however, was convinced that the tape was proof positive that she was had a multiple personality, and that she had been sexually abused by her father and a baby-sitter.

Since that time, Cone has seen many patients turn benign information on a tape into malignant memories. He has seen therapists with inappropriate questions implant information into the minds of willing patients. A whole industry has grown up around the idea of repressed memories, satanic ritual abuse, and multiple personality disorder, not to mention alien abduction.

One of things that Cone noticed, even in that first experience, was that group sessions no longer marked the beginning of the end of therapy, but the beginning of a social setting that becomes quite important in the life of the patient. It is not seen as a stepping stone to healing, but an event that becomes part of life.

Cone's patients have told him the same thing time and again. When a new patient came to him because her old therapist had died, Cone asked her how long she had been attending the support group the therapist had led.

"Fourteen years."

"Why so long?"

She thought for a moment and then said, "Well, you see, I'm a survivor of sexual abuse. I'm recovering. You know, once a survivor, always a survivor." She spoke with pride.

Cone asked, "How long ago did the abuse happen?"

"Well, it happened when I was two, but I didn't remember it until I began attending the group. They made me realize that the things I was feeling had to stem from sexual abuse."

This wasn't the first time that Cone had heard something like this about support groups. Originally they were meant to help people recover from emotional trauma and solve personal problems. According to Cone, "Something awful has since happened. Today's groups no longer serve this function. The introduction of the idea of survivor has changed the purpose of the support groups. Being a group member used to be part of a healing process. It has now become a badge of identity."

Cone continued, saying, "Group members are no longer people with problems, they are abductees, MPDs, survivors of incest, and codependents. Even worse, recovery never occurs. It has become a lifelong process. The work of a group member is never done. If a member thinks about leaving the group, they are not seen as recovering, but as defecting. There is something dreadfully wrong here."

During his next session with Janet, Cone asked about some of her abuse memories. It was just a way of opening the conversation so that they could begin to talk about her problems. But what happened next shocked him.

Janet fell to the floor, crawled behind the couch, and began rocking back and forth. She screamed, "He's raping me! He's hurting me. He's sticking it in my mouth! It's awful! Stop! Stop!"

Appalled by this behavior, all Cone could do was ask, "Who did this?"

"My father! My brother! My uncle! They all did it!" she screamed.

These abuse memories seemed to spring out of nowhere. At no time during any of the previous sessions had she suggested such horrible experiences. Yet as she writhed on the floor she began talking in the voice of a five-year-old child.

Cone, leaning forward, trying to gain control of the session, asked, "Janet, what's wrong?"

"I'm not Janet," she said. "I'm Jean."

Cone did not, for a moment, believe that Janet was multiple personality or that an alternative personality was suddenly in control of Janet. But Cone did become concerned about the quality of the treatment provided by the hospital. He had already seen how one psychologist had lead the patient into memories of sexual abuse. He had seen the support group make her the star of the session as she began to relate her horrible tales of sexual abuse. Because of those observations, and because of his concern for her mental stability, as soon as she was stable, Cone had her discharged from the unit. It was doing nothing to help her recover but rather dragging her down into depths where she didn't belong.

But despite what Cone had seen and believed and what he tried to tell her, she continued to insist that she had several personalities, violently arguing the point with him. She would claim that these personalities had no knowledge of each other, but when one of them would call for an appointment, the other would show up. The contradiction, when pointed out, was ignored by Janet.

It became apparent that her identity as a multiple personality was the strongest sense of belonging that she had ever experienced. It made her feel special and important. By discovering these multiple personalities, she was getting more attention from those around her than anything she had ever done in her life. Her role as patient was the polar opposite of the experience she had been having at work, where all the attention was negative. As a patient, all attention was focused on her and all the reinforcement was positive.

Cone told her that despite her feelings, he did not see any convincing evidence that she was a multiple personality. Although he was sympathetic with her plight, he refused to reinforce her belief system. She became angry when he refused to acknowledge her other personalities.

While she had been in the hospital, she had heard about several support groups for those with multiple personalities. She told Cone she was going to begin attending one of the groups. That was, after all, where she belonged.

The group that she selected contained several people who claimed to have been victims of satanic ritual abuse. Before she began this group, Cone had told her of his concerns for her mental health. He told her that, based on his experience with her, he was worried that she would take on the symptoms of the other patients. In other words, he believed that she would soon begin to recover satanic abuse memories. She jokingly assured Cone that she had never been ritually abused, and that there was no reason to worry. She was interested in the group because of other aspects of it, not because members believed in satanic ritual abuse.

But Cone had been right. Three weeks into the group meetings, Janet began to recover the ritual abuse memories. Cone, disturbed by this, confronted her. He reminded her that he had expressed his concern, and that if she attended this group, he feared she would begin to recover memories about satanic ritual abuse. He mentioned the conversation in which he warned her, but she completely denied that it ever happened. Cone, in her mind, had never issued any such warnings. Instead, she told Cone, "I've been a member of this cult since I was a baby."

During this session with Cone, she drew some pictures of her experience in the cult. The pictures showed her lying on a ritual sacrifice table, surrounded by torches, knives, and robed figures. She described the murder of several babies during ceremonies she had witnessed. "Now I know why I was depressed," she said. "I've been repressing all of these memories for years. Now I'll deal with them."

Cone said, "I've known you for some time now Janet, and I have never seen any indication from you or your family members that any of this happened."

"You just don't know enough about these things," she said. Instead of listening to him provide a logical framework for the discussion, Janet decided that he was incapable of understanding her memories or her problems. The session ended with her feeling angry and misunderstood.

At the next session, Janet arrived with arms full of books and magazines. She handed Cone copies of *Michelle Remembers, Satan's Underground,* and several magazine articles about satanic cults. "Read these," she ordered, "and then you'll understand." Suddenly her memories had been authenticated by the so-called authorities. What she believed was now validated by those who had studied the phenomenon of satanic ritual abuse. After all, no one could invent such tales. They must be true.

Again Cone told her that she had previously denied any sexual abuse and had laughed at the idea of satanic ritualistic abuse. Instead of understanding what she was being told, she reacted violently. Cone, the clinician to whom she had come to understand her problems, was now trampling on her new belief structure. Rather than listen to what he said, she screamed at him, "You're such an amateur." She stormed from the room.

Cone never worked with Janet again. She refused to see him because he refused to be drawn into her fantasy of satanic abuse. He was more concerned with her stability and mental health than he was in validating a belief structure that had no foundation in fact. Janet, however, had begun seeing a "ritual abuse specialist."

Cone is among a growing number of professionals who don't accept the idea of satanic ritual abuse. Instead, he sees it as a malady that is the invention of others, reinforced by law enforcement officers, mental health care professionals, and the tabloid media. Each of these agencies or organizations validates a belief in SRA by suggesting it is real. Yet there has been no convincing, independent evidence that a worldwide network of Satanists are conducting ritualistic abuse.

However, because of the growing number of cases, with the "eyewitness" testimony of hundreds, if not thousands, of victims claiming to have seen thousands of murders, the FBI, as well as other law enforcement agencies, searched for corroborative evidence. After a seven-year study, the FBI concluded that there was *little or no evidence* of organized satanic activity and conspiracies in the United States or elsewhere.

Kenneth Lanning, an FBI agent, did much of the research into satanic cults. He told all that when he began his investigation he was inclined to believe the victims. Who could make up such horrible stories if there was not some truth to them? How could the victims, in widely separated parts of the world, tell stories that were so similar in detail if the cults and rituals didn't exist? It defied logic to believe that there was some kind of psychological malady infecting the minds of so many people from so many parts of the country that would explain satanic ritual abuse. Something had to be happening.

But Lanning realized that the same logic seemed to argue against SRA. The numbers had continued to grow until conspiracy would have collapsed under its own weight. There were thousands of victims of the murders, but no mass grave sites were ever found. Paul Ingram's daughters had talked of the death of babies, the bodies buried on the family property, but searches by law enforcement officials, with the help of archaeologists, failed to produce a single infant bone. Lanning finally was forced to conclude that there was nothing physical to the tales.

The FBI supplied another agent for the other side of the controversy. Ted Gunderson, who claimed to be a satanic cult investigator, said in Washington State, "I'd like to tell you right now, the next burial ground that we will learn about will be in Mason County, Washington." Not content with stirring the pot once, he added, "We've located a number of burial grounds in Mason County, and they can't possibly go out and dig them all up because there are so many of them."

Gunderson's predictions of mass burial grounds never came to pass. Although the county was searched with some of the most sophisticated equipment available, including helicopters equipped with infrared gear, not a single body was ever located to corroborate the existence of a satanic cult in the area.

In fact, after years of searching, no evidence has been found anywhere to suggest a large-scale satanic operation. Some graveyards and churches have been desecrated by teenagers dabbling in satanic worship, and a few houses have been painted with satanic graffiti, but nothing to prove that hundreds, or thousands have been sacrificed, or that hundreds of thousands, if not millions, have been abused.

The tragedy of the situation comes not from the satanic ritual but from the therapists themselves. They provide the information to their patients and convince them with hypnosis and "visualization" that they will remember long-repressed memories.

Cone said that Janet remained in her abuse group for several years, and became increasingly dysfunctional. Her memories of ritualistic abuse became stronger and more vivid, driving her further into her depression. She eventually was placed on permanent disability and lost her house. She moved into a halfway house with other ritual abuse survivors where she could spend every evening sharing horror stories with them. She had moved from being a functioning member of society who experienced clinical depression to a nonfunctioning individual whose life revolved around the imagined tales of satanic worship and ritualistic abuse.

Cone lost track of her for several years. When he had the opportunity to speak with another psychiatrist who had worked with Janet, he asked about her. "Haven't you heard?" the psychiatrist asked. "Three months ago she killed herself with a shotgun."

Cone is convinced that the therapeutic environment that this woman was exposed to was directly responsible for her death. The misguided and misinformed had destroyed another life. Therapists, with their trappings of authority, and with their assumed superior knowledge, provided the patients with information that the patients had no way of rejecting. If the therapist said it, and believed it, then it must be real. The therapist is educated and must know the truth. The therapist becomes the authority figure who validates the premise.

Are these therapists knowingly evil? With several centuries of research available, how is it possible that researchers and therapists keep making the same mistakes? Why is it that researchers continue to use hypnosis, when there is so much evidence against the value of it? Critics of the recovered memory movement, which includes a segment of the satanic ritual abuse movement, claim that therapists mislead their patients deliberately for the sole purpose of making money. While this no doubt occurs in isolated cases, it is not the prevalent motive. Most therapists are devoted to helping their patients. No one devotes eight to twelve years of their life to training for the sole purpose of cheating people. Instead, the movement hinges on two major factors: lack of adequate training and the need to reduce anxiety.

Most therapists, psychologists, and psychiatrists, have received no training about multiple personality disorder, satanic ritual abuse, or UFO abduction. In the past there were no classes on dissociation, none on confabulation, and nothing on false memory. There were no classes that invited the student to question his assumptions about the therapeutic process, or that invited alternative explanations for certain behaviors.

Most therapists receive their education on MPD, SRA, and repressed memory in the hospitals where they work. Cone, for example, said, "There, in the doctor's dining room, we would discuss each day the MPD and SRA cases that came in. Clinicians would talk over cases and share new treatment techniques. These were techniques and ideas that had never been tested. No one ever questioned the assumptions being made. The information that was shared was based on personal observation and experience. Through this interchange there came an agreement that what we were seeing was real. Not once during those years did I hear any one disputing or questioning the assumptions and techniques."

Cone continued, "I, like my colleagues, was fascinated by the gyrations and bizarre behaviors that occurred on the locked unit of the hospital. The contortions and sometimes violent behavior of the patients was truly frightening. In order to work in this kind of environment, we all needed to feel that we understood the reasons for what we were seeing. We needed to feel that we knew how to handle it. Because of this need, we latched on to every new technique and idea without question."

Cone said, "Healers hate feeling helpless. I have heard countless stories about doctors getting angry at the patients that they could not help. Not knowing what to do makes one extremely anxious. When someone offers a solution, a framework within which to work, naive and insecure therapists flock to it in droves. Such was the case with those of us who worked with MPDs. We were eager to listen to the experts. Armed with the new tools and techniques, we could walk into the consulting room feeling powerful and confident. It is this feeling of being knowledgeable, of being able to take control, that draws the therapists to accept, without question, the techniques of recovered memory therapy. Once this belief system is in place, all work becomes self-verifying. Behaviors that fit the theory are seen as evidence that it is true. Behaviors that do not fit are seen as resistance, repression, and denial.

"Therapists are reluctant to give up the things they have been taught. No one wants to hear that their deeply held beliefs are flawed. Wars are fought over differences in belief. No, it is not greed, nor avarice, but the need to be right, to feel that one knows the answer, that leads therapists into recovered memory work."

As the incidence of recovered memories and MPD increased in the community where Cone worked, several of the hospitals opened special units for the MPD and SRA patients. "It was here that I learned more about the human personality in a year than in all my years in school. Up until that point, I had accepted what I heard as truth, but on these special units, my ability to believe was stretched to its limit."

But Cone wasn't the only one to feel that way. After the publication of *Michelle Remembers* in 1980, and *Satan's Underground* in 1988, courses were developed to help both mental health care professionals and law enforcement officials deal with the growing problem of satanic ritual abuse.

In 1988, a paper, "A New Clinical Syndrome: Patients Reporting Ritual Abuse in Childhood by Satanic Cults," was published by two psychiatrists, Walter C. Young and Bennett G. Braun, and one psychologist, Roberta G. Sachs. What they did was provide another syndrome for therapists. They gave another answer to questions that seemed to have no answer.

The authors had interviewed thirty-seven people in treatment for dissociative disorder (multiple personalities) who had also told of satanic ritual abuse. The stories told by these patients, in different therapy sessions, were remarkably similar. Of course, it was noted that the stories were constructed from the elements that had been contained in *Michelle Remembers.* It was almost as if a checklist of atrocities had been created for those who believed they had been ritualistically abused.

The authors noted, in their paper, "*The lack of independent verification* [emphasis added] of the reports of cult abuse presented in this paper prevents a definitive statement that the ritual cult is true. Despite the fact that some patients have discussed ritual abuse with other patients, and the fact that patients have had contact with referring therapists who may have provided information to them, *it was our opinion* [emphasis added] that the ritual abuse was real."

In other words, though they found no outside proof that the stories were real, and that they understood patients had contaminated other patients with information, and that therapists had contaminated patients, Young, Braun, and Sachs decided that the stories were real. Braun was so impressed with his theories, that he set up a special hospital to deal with the growing problem. According to him, there were an estimated two hundred thousand Americans suffering from MPD, and of those, up to one-quarter might be victims of SRA. Everywhere he looked, he found the victims of satanic abuse.

In an interview published in *Time* Braun said that when he first began to hear the stories of satanic abuse in 1985, he was "incredulous." Now, having heard similar tales from many people from different states and countries and having treated more than two hundred of them, "he declares, 'Yes, there is satanic-ritual abuse.' "

In an article written by A. S. Ross for *Redbook,* Braun was quoted as saying that satanic ritual abuses are part of an "international organization that's got a structure somewhat similar to that of the Communist cell. It spreads from local small groups to local consuls, regional consuls, district consuls and they have meetings at different times."

Roland Summitt, a consulting psychiatrist at Harbor-UCLA Medical Center, and a child abuse expert said in 1988, "[Ritualized abuse is] the most serious threat to children and to society that we must face in our lifetime."

The myth of satanic ritual abuse now had all the elements that it needed to survive. It had been endorsed by authorities such as Braun, Young, Sachs, and Summitt. Law enforcement officials were telling local police and sheriffs that SRA was a spreading curse. Hundreds, if not thousands, of believable people were telling these horrible stories.

Finally, cementing it into the public conscious, were the books about it, especially *Michelle Remembers* and *Satan's Underground.* Television talk shows began to focus on satanic ritual abuse. Geraldo Rivera produced a number of shows dealing with satanic worship and abuse beginning in 1988. In each of these arenas, the reality of the claims was underscored, though no independent corroboration or evidence could be found. In fact, no evidence was ever offered, other than more tales told by more people who believed they had been abused.

To make it even worse, Braun found a star in the satanic world. He began to treat a woman who would eventually remember that she had been a high priestess in the cult. Others under Braun's care would remember that the woman whom we'll call Sally, was in charge at some of the meetings. One even claimed to have a photograph of Sally in robe and hood, at the head of a sacrificial table.

But like others, Sally didn't enter therapy because she believed she had been abused but because of depression. Over years of therapy, her therapist diagnosed her as suffering from multiple personality disorder. To get the help needed to treat this problem, she eventually made her way to Braun in Chicago.

Although she told the doctors on entering the hospital that she had no memories of any sort of abuse, within months she had become a high priestess in the cult, convinced of her role. In fact, she recovered memories that suggested her cult-related activities had continued up to her admission to the hospital. She had been able to conceal her secret life from her family, friends, and even her husband. There had been no clue that Sally was so important to the cult or that she had taken part, or directed, so many of the meetings.

As part of her treatment, she was forced to take a variety of medications. At first she refused, but ultimately began a regimen of medication. Eventually she was taking Xanax, Inderal, Halcion, and other medications. This in conjunction with therapy session and hypnotic regression created a state in which she was able to accept the incredible story that she was a high priestess in a satanic cult.

But because of the deterioration of her mental state, she was transferred from Braun's care to another hospital. At that point, she began to recover. Those at the new hospital were not interested in her horrific tales of satanic abuse. She was not the star of the unit. If she told these tales, or if she didn't, she was treated the same way.

Eventually she stopped taking the medication, stopped the hypnotic regression, and stopped associating with those who believed they too had been ritualistically abused. When she moved from that environment, she realized how ridiculous the stories she was telling were. There was nothing to suggest that she was part of a satanic cult, that she had ever been a member, and certainly no evidence that she was the high priestess. Apparently no independent corroboration had ever been found. That seemed to not have bothered Braun and his associates because they believed the devil was in the details.

But the lack of any kind of corroboration disturbed some of the researchers. How could the conspiracy exist if there was no trace of it? The FBI's Lanning, as he began to express his doubts, was accused of being a member of the conspiracy. And that explains it, at least, to the satisfaction of many. The cult is so powerful and so widespread that it can suppress all knowledge of itself by calling on high-ranking governmental officials when needed.

The whole phenomenon was reduced to the testimony of those who were its victims. The logic was that since no one would make up these stories, they must be true. But a problem developed there as well. Many of those who had once believed they had been abused, after they had left therapy or found a new therapist, began to doubt the reality of the memories.

Margaret Singer, a retired professor from the University of California at Berkley, interviewed fifty people who once claimed to have been members of the cults and abused by them. They had recovered "repressed" memories of ritualistic abuse. Now they believed they were mistaken. Singer told Leon Jaroff in *Time,* "These people are reporting to me that their therapists were far more sure than they were that their parents had molested them."

Singer was suggesting that the memories recovered by the patients came not from repressed memories, as the therapists believed, but from the therapists themselves. The therapists believed in the satanic ritual abuse, or the sexual abuse, and through "standard" techniques of hypnosis, visualization, and even "massage" therapy convinced the patients that the memories were real.

Braun, who brought the idea of satanic abuse into the mainstream, and who made a career of studying it and treating the victims of it, now found himself on the wrong end of lawsuits. A female patient whose medical bills totaled nearly a million dollars after treatment for SRA was suing him and the Chicago medical center for negligence.

What was becoming apparent was that the whole phenomenon of satanic ritual abuse was therapist driven. The "memories" recovered by the patients agreed with the "checklist," not because the abuse was real but because the therapists had all read the same checklists. Therapists were implanting memories through poor technique, including hypnosis and visualization. Some were asking blatantly leading questions, suggesting to the patients what they wanted to hear. A few weren't subtle, telling their patients they had been abused and it was up to them to "recover" the memories.

Law enforcement officials, such as Gunderson, would inject themselves into the controversy, touting their expertise. Those officials, and therapists who searched for satanic ritual abuse, managed to find it, even where it had not existed.

Paul Ingram, who admitted to abusing his daughters though he had no conscious memory of it, who involved two of his friends in the horror, and who finally came to believe that he was the leader of some kind of satanic cult, proves the point. Ingram was a happy man, though somewhat straitlaced, until the allegations were made by his daughters. Then, under the close questioning of fellow deputies and Richard Peterson, the memories of these events began to surface.

Jim Rabie and Ray Risch had been implicated in the abuse of Ingram's daughters, and were described as members of the cult. With Ingram confirming the allegations of his daughters, it seemed that this would become the first case where cult members were successfully prosecuted. At least that was what Under-Sheriff Neil McClanahan told his depressed investigators just before the trials were to being.

The problem was that the charges of satanic ritual abuse had disappeared. With no evidence to corroborate the tales, those charges just faded away. Ingram, to spare his daughters the ordeal of a trial, had pled guilty to six counts of third-degree rape. There were no charges of ritualistic abuse included.

Two days after Ingram pleaded guilty, the charges were dropped against Rabie and Risch. They had been held in jail for nearly six months because even with no independent corroboration everyone believed that the two young women were telling the truth about the abuse.

Richard Ofshe, a social psychologist from the University of California at Berkeley, was brought in by the prosecution as another expert on cults and mind control. Ofshe told prosecutors that no one could claim to be an expert on the satanic cults because, as far as he was concerned, no one had proved that the cults existed.

On the way in from the airport, the investigator complained that nothing in this case could be verified, and that the stories seemed to conflict. Besides all that, Ingram's confessions were filled with language that sounded more speculative than definitive. And everyone was complaining about Ingram's inability to remember the events without prodding and probing and constant assistance.

When Ofshe met Ingram, he learned that Ingram had perfectly normal recall of ordinary events during his life. But when he asked Ingram about the abuse, the answers became vague and disjoined. Ofshe knew that human memory didn't operate that way. Either Ingram was lying about the events, or he was deluded. Trying to learn more, Ofshe asked where the memories were coming from. Ingram explained the technique that he, along with Peterson and the investigators had developed. He told Ofshe that he had been using a relaxation technique as he imagined walking into a white fog. Once the memories began to flow, Ingram would write them down.

Ofshe decided to perform an experiment. He told Ingram that he had been talking to one of Ingram's children who told him that they had been forced to perform sex in front of him. Ofshe wanted to know if Ingram remembered anything like that.

Ingram, of course, said that he didn't. Ofshe insisted and one of the investigators added that it had happened at the new house. After a couple minutes of thought, Ingram said that he could "kind of see Erika and Paul Ross." Ofshe suggested that Ingram return to his cell and think about it.

The next time Ofshe saw Ingram, only hours later, the man was grinning broadly. He had written a three-page confession of the event that Ofshe had invented. Ingram now had a complete confession, with conversation among the participants, of an event that never happened. Ofshe had doubts that any of the abuse alleged by Ingram's daughters had ever taken place.

But if it hadn't happened, where did the idea of it begin? Certainly part of it came from Erika's experiences as a camp counselor. She had listened to a number of lectures from experts about child molestation and sexual abuse. So part of the story came from listening to those lectures.

And part of it came from the same source so many others had tapped. In the summer of 1988, according to Lawrence Wright in *The New Yorker,* Erika saw a copy of *Satan's Underground* while baby-sitting. She told the neighbors that she had read it all, but later told investigators that she'd read only a few chapters. It had a haunting ring of reality to her.

It should also be noted that Julie never originated ideas. Erika came up with them first. Later, as she talked to investigators, Julie would "remember" similar incidents. And although both girls talked of marks on their bodies as a result of the abuse they suffered, no medical evidence of any of their claims was ever found.

Those with a belief structure will rarely surrender it without a fight. Erika filed a negligence suit against the county for failure to properly supervise Rabie and her father. Of course, she claimed that the case was derailed by cult members including Under-Sheriff McClanahan. She suggested that the conspiracy might go as high as the governor of the state of Washington.

In December 1991, after her father had been jailed on the rape convictions, and with her family destroyed by the allegations, Erika appeared on *The Sally Jessy Raphael* show. During the program, she described, at length, the rituals she had witnessed. She told the audience, "One time, when I was sixteen, they gave me an abortion. I was five months pregnant. And the baby was still alive when they took it out. And they put it on top of me and then they cut it up. And then . . . when it was dead, then the people in the group ate parts of it."

Later, in direct conflict with the evidence, Erika said, "I spent most of my life in the hospital. And that is true. And I have scars. And, I mean, doctors were just like, looking at my body, going . . . ugh!" She was not required by the audience to prove this.

It is only now that we are beginning to understand just how easily some people can be led. It is only now that we see that the leading questions of the therapists are convincing patients that they have been abused. It is only now that we see how a desire to belong can provide a powerful motivation for patients to "remember" episodes of abuse that never happened.

And we begin to see how adults can be influenced by figures of authority. Paul Ingram was so convinced that others were right about the abuse of his daughters that he allowed them to guide him into admitting, and inventing, episodes of abuse that conformed to those told by his daughters.

Ingram, currently in jail because of his confessions, now understands what happened to him. But his family is destroyed, his life is in ruins, and his daughters believe him to be a monster who participated in a cult that used them in ritual. But the reality is there was no satanic abuse, just the overzealous therapists and psychologists who believe these things and manipulate the situation to fit their beliefs. This is a manufactured malady, infecting far too many people and destroying far too many lives.

And it all relates to the tales of alien abduction because we can see the same thing when we study those cases. There are troubled people who find their way to a therapist. Using what he or she believes to be standard techniques, the therapist, or abduction researcher, attempts to fill in blocks of missing time. Over a number of sessions, a picture begins to emerge that suggests to the researcher that an abduction has taken place.

What we find by studying cases of satanic ritual abuse and alien abduction is that the person seeking help begins to manifest the specific symptoms accepted by the researcher. In other words, if the person finds a therapist who believes in satanic ritual abuse, then those are the symptoms and stories that will be recovered. If, however, the person finds a therapist who believes in alien abduction, then those are the stories that will be recovered.

All this leads to a simple conclusion based on our research. The manifestation of the phenomenon is conditional on the researcher. If someone enters therapy believing he or she has been abused by satanists, and if the therapist believes in SRA, then the proof will be recovered. Richard Boylan demonstrated this when he took women who believed they had been abused based on their work with other therapists and convinced them they had been abducted by aliens creatures instead. It showed just how therapist driven these two phenomena are.

But more important, it showed that these two phenomena are the same thing. It all depends on the researcher or therapist and the belief structure of those around the victim.

19.

RECOVERED MEMORIES

To believe that alien abductions are real, a belief in a new psychological aberration is necessary. We must believe that memories can be pushed so far down into the subconscious that they are visible only after hypnotic regression or other memory enhancement techniques are used to recover them. We must accept the concept that the human mind can refuse to acknowledge events that it finds too painful to remember. And for it to work, we must believe that the mind can retreat into itself time and again to hide these painful memories from our consciousness. We must believe memory can be deeply repressed many times over a period of months and years, only to be recovered through therapy, hypnotic regression, or spontaneously by unprovoked visual or auditory cues.

But is there such a thing as a repressed memory? Can a memory be wiped from the conscious mind, only to return many years later? Is there, for example, a worldwide satanic organization that kills thousands of children each year as those with recovered memories repeatedly claim? Do alien abductions exist as outlined by those who believe they are innocent victims of such abductions? Is it possible to induce a false memory in someone? A false memory that is so convincing that the person will not be able to distinguish it from a real memory? Can severe abuse and mind control techniques create alternate personalities?

These are questions that are currently rocking the mental health community and that have taken on importance for those interested not only in alien abduction but in a range of other, similar mental health aberrations. Allegations of abuse and cult mind control are reported in the media quite frequently. The number of people coming forward with memories of alien abduction, satanic abuse, and multiple personalities is skyrocketing. Even worse, ritual abuse patients are now claiming that their previous therapists have ritually abused them as well. They are claiming that many mental health professionals are part of the gigantic, worldwide conspiracy to keep the secret of satanic worship hidden from the general public.[1]

---

[1]Arlys Norcross McDonald, *Repressed Memories: Can You Trust Them?* (Grand Rapids, MI: Fleming H. Revell, 1995); Elizabeth Loftus and Katherine Ketcham, *The Myth of*

Mike Lew, a self-professed expert on recovered memory therapy and the related phenomenon, has said, "A great many incest survivors have little or no memory of their childhood. In fact, this method of dealing with childhood trauma is so common that when someone tells me that they have no recollection of whole pieces of their childhood, I assume the likelihood of some sort of abuse."

Lew is typical of therapists today who make outrageous assumptions about the power and accuracy of memory. He is among those who have come to believe that these recovered memories are accurate portrayals of past situations and are valuable in understanding past patterns of abuse, whether by human agencies or alien creatures.

The question that should be asked but which is often dodged is: How good is memory? Everyone complains about memory. We all recognize that our memory does not work as well as we would like it to or as we once thought it did. And yet we have the strong conviction that what we remember is not only accurate but true. Even in the face of evidence about the poor quality of memory, we will argue the point, believing that a trick of some sort has been played.

We have been told countless times by various "authorities" that memory is the storehouse of everything we have ever seen, felt, experienced, or done. We are told that memory is like a giant videotape on which everything we have ever seen, done, or participated in is recorded with the accuracy of videotape. And while that point has been strongly contested by psychologists and scientific evidence, we learn the real purpose of memory is not to store data but to allow us to function on a day-to-day basis. Memory is not meant to be an accurate record of our past or even a storehouse of our experience but rather an inner representation of who we are and how we feel. More important, it does not have to be very accurate to carry out these everyday functions.

Most people do not realize just how bad their memory really is. They just don't think about it. But take a moment and ask yourself these questions. What did you do on your tenth birthday? What did you have for lunch three days ago? Two weeks ago? What were the names of your grade school teachers? Any of your teachers? Give a moment-by-moment summary of everything you did five weeks ago last Wednesday including what you had for lunch.

Now, think of something that happened at least ten years ago that you do remember. Perhaps it was a birthday party, or a graduation ceremony. Take a few moments and recount everything you remember about it from beginning to end, including the conversations you had at the time. Think about the events that led up to that memory, and what you did right after the event took place.

If you kept a diary, you can perform a simple, illustrative experiment. Take

---

Repressed Memory (New York: St. Martin's Griffin, 1994); Lenore Terr, *Unchained Memories: True Stories of Tramatic Memories Lost and Found* (New York: HarperCollins, 1994).

an event that you remember quite well and write down exactly how it all transpired. Search your memory for every detail that you can. In fact, ask friends and family for their memory of those events if you want to carry the experiment to its ultimate conclusion. Once you have a complete written record, check those memories with the facts as written into the diary. It will surprise you.

You will discover that memory is patchy and sporadic. We remember very little about our past. The reality is that we forget the bulk of what we do every day. It is not repressed or wiped from our mind, it is merely never recorded, or never used again, because it serves no useful purpose. Even important events are not recorded in sequence but are stored in categories or related items.

Add to this the fact the average person cannot remember anything before the age of three and a half. The brain hasn't completely formed and the language skills do not exist before that age for meaningful memory to exist. Until the age of five our memories are few and fragmentary. Memory researchers call this infantile amnesia.

Researchers Mark L. Howe and Mary L. Courage in their 1993 work, "On Resolving the Enigma of Infantile Amnesia," suggest that the onset of autobiographical memory comes with the consolidation of the sense of self. This usually occurs between the ages of two and three. Autobiographical memory is information about events pertaining to the self, so it makes sense that without a sense of self these memories could not be processed. This would also suggest that it is impossible to record memories before the age of two and extremely rare before age three.

There are those, however, who believe that they do have memories before the age of three. Some claim to remember, in some fashion, the trauma of birth. Others claim memories of events when they were one or two years old. These are events that were out of the ordinary and witnessed by older family members who talked about them with the family on various occasions.

What we see here, rather than a memory of the event witnessed by the person at a young age, is an induced memory. Having heard the story told by other family members for years, the person believes that he or she witnessed the event. Sometimes it's an event that took place before birth, but the person believes strongly that it is a memory. And there is no significant difference between that sort of induced memory and the real thing. There is no way for the person to tell whether the memory is real or induced.

Think about that for a moment. According to the experts and the scientific evidence, there is no difference between a real memory and an induced memory. The scientific data and findings bear this out in study after study. In fact, even the psychologists who do the research have been surprised by the results because facts we thought to be true about memory turn out to be myths. These myths provide what most of us believe about memory.

Let's examine that carefully from our own personal experiences. What science suggests is that the large gaps that exist in autobiographical memory, that is, our memory of our life, are readily filled by confabulation. In other words, if we believe an event to have happened to us, or we begin to attempt to answer questions about our past, our brain creates a scenario that seems likely if no substantial memory exists for that event. It might not be based in reality, but the mere process of concentrating on that event "creates" in our minds the circumstances of that event, which we believe to be the reality of the situation.

Prompting from a therapist or abduction researcher, whether using hypnosis, memory enhancement, or merely relaxation techniques, aids in the creation of the memory. When the therapist suggests, "You can remember. It's all there, you just need to access the data," the concept is accepted as true. And because an authority figure has made the claim that the memory exists, a memory is often recovered.

During the investigation into the crash of a spacecraft near Roswell, New Mexico, Kevin Randle had the opportunity to interview Jason Kellahin, an Associated Press reporter assigned to cover the story in 1947. Prior to the 1994 interview Kellahin reread the article he wrote for the AP in 1947 to refresh his memory of those long-ago events. He also read one of the books written about the case, and when interviewed, gave Randle a detailed description of his activities on that day nearly fifty years earlier.

Randle, however, didn't find those memories consistent with the established facts. For example, Kellahin said that he had received a telephone call from AP headquarters in New York early on the morning on July 8. AP headquarters did call, a fact readily available in newspaper articles written and printed in 1947. However, the call couldn't have come in during the morning because the story had not yet hit the news wires. According to all the available information, including a chronology of the news wire activity on that day published in one newspaper, the story didn't "break" until the early afternoon. But Kellahin, trying to reconstruct the entire sequence of events, including the long drive from Albuquerque to Roswell, believed the call was made in the morning, not in the afternoon. It is a minor point but does demonstrate confabulation in action.

Kellahin also told a story of searching out Mac Brazel, the rancher responsible for finding and reporting strange metallic debris, interviewing him at the crash site on his ranch, in view of military officers, and having ordered pictures to be taken on that site of Brazel and the wreckage. Kellahin, along with the photographer he identified as Robin Adair, then drove on to Roswell. Kellahin clearly believed this to be an accurate memory of what happened that day.

Again the facts that can be verified from printed sources and other eyewitnesses simply do not corroborate what Kellahin reported. Brazel wasn't at the ranch that afternoon but was already in Roswell as shown by documentation and testimony of others. Adair, the photographer, said that he met Kellahin in Roswell and not that he traveled with him to Roswell or had been with him when Kellahin claimed to have stopped at the Brazel ranch. But most important, the photographs that Kellahin claimed were taken were never printed. Had he had photographs of Mac Brazel standing in the field with the metallic debris, the AP would have made sure those pictures were printed somewhere, at some time, because they would have been more important than the pictures of Brazel taken later that day in a newspaper office in Roswell. Yet the pictures of Brazel at the newspaper office are the only ones to have been published. It means that the others simply don't exist and that provides us with a clue about confabulation by Kellahin.

The memories of the event, related to Randle by Kellahin, were confabulation. He thought about what might have happened, he read the newspaper articles, including one he had written, and reconstructed the event in his mind by concentrating on it. When this event happened in July 1947, it had been just a one-day story that collapsed like so many others had. For Kellahin, there was nothing in it to make it extraordinary, and he forgot all about it. Kellahin's lack of memory is not surprising simply because of the circumstances surrounding this event that was nearly fifty years in the past. It was too unimportant, too routine, to be maintained in his memory. When it became important nearly fifty years later, he had to "dig" for any memory of it.

But even the memories of important events, captured in what is known as "flashbulb" memories, something that we all believe in, are open to mutation, modification, and distortion. Psychologist Ulric Neisser, who taught at Emory University in January 1986 when the space shuttle *Challenger* exploded, realized he had an opportunity to study flashbulb memory. The day after the disaster, Neisser gave the students in his freshman psychology class a questionnaire about the explosion. He then filed away the questionnaires for three years, until the students were seniors. At that point he gave them the same questionnaire but added a single additional question. He wanted to know how reliable they believed their memory of the destruction of the *Challenger* to be.

Here was a chance to learn if those flashbulb memories were accurate. We could learn if those sorts of memories were more reliable than those of our everyday activities.

According to the results provided by Neisser, and his graduate assistant Nicole Harsh, a quarter of the students didn't have one memory that proved to be accurate. In one case, a student reported that he had been home with his parents when he heard the news. His earlier questionnaire, completed the day after the event, however, revealed that he had been in college at the time.

More important, however, was the reaction of the students to the proof that their memories of the events were inaccurate. None disputed the accuracy of the original statements made in the days after the destruction of the *Challenger*. One student, when confronted with the discrepancy between what she remembered three years later and what she had written only hours after the event, did say, "I still remember everything happening the way I told you. I can't help it." She was defending the memories that were clearly an invention of her own mind.

Because of studies like that conducted by Neisser, all retrieved memory, especially that retrieved using hypnotic regression is coming under more scrutiny. Some states, such as California, now refuse to let a witness who has been hypnotically regressed testify in a trial. There are just too many opportunities for that witness to be led, no matter how careful the questioner is. Any attempt to gain additional information from a witness, no matter how innocent the question is, can render the testimony suspect.

It should be noted here that those attempting to recover memories, whether of childhood sexual abuse, satanic ritual abuse, or alien abduction, believe that such memories exist in their minds. They are told by therapists or researchers that such memories are deeply buried but can be recovered. When told this repeatedly by the "authority" figure, they work very hard to find some fragment of a memory that will please that researcher. It is all that is needed to begin to build a huge case around suppressed memories of these imagined traumas.

There are other psychological factors that come into play here. It is interesting to note that people with a poor sense of self repeatedly report that they cannot remember their childhoods. Poor sense of self would seem to result in poor memory. It is also interesting that people claiming long histories of alien abduction and sexual abuse often have a poor sense of self, according to some of the research being conducted.

Most psychologists believe that a poor sense of self is the result of early deficiencies in nurturance and violations of a person's boundaries. People with a poor sense of self have difficulty getting their needs met. They are often unsure of their sexual orientation, cannot make a career choice, and cannot establish a healthy relationship. They find themselves being violated by friends and family again and again. They may resort to fantasy as a means of combating these violations.

Almost every abductee and abuse victim has made statements such as, "I have always felt like an outsider; I have always felt like I didn't belong." Many of them have had trouble forming relationships, or holding down a regular job. John Mack reinforced this when he noted of his abductees, "A lot of people give up their jobs which are often in mainstream industry."

The perception that one is different causes anxiety about the possibility that one is somehow defective. Being an abductee or an abuse victim serves at least three important purposes. First, it answers the question as to why the person has always felt different. Second, it attributes the problem to an external cause, absolving the victim for any responsibility for their problems, and third, it gives the person that long sought-after place in which to feel as if he or she belongs.

The desire to be an abductee or the victim of abuse is not a conscious choice. It's actually the combination of a weak sense of self and the desperate need to fit in somewhere. Many abductees wear their experience like a badge of honor. It is the strongest sense of identity they have ever had. Contrary to the opinions expressed by John Mack, Budd Hopkins, and others, this is, in fact, a club that people fight to belong to. This is clearly true of the satanic abuse, multiple personality, and abductee populations. Many abuse victims and abductees center their entire lives around the experience and the circle of friends developed because of it.

Colin Ross has stated all studies show that people reporting paranormal and unusual experiences have a significantly higher incidence of child abuse. Kenneth Ring and Christopher Rosing, in 1980, found the same thing. Ross believes that an abusive childhood leads to a tendency to dissociate. This was a phenomenon studied by Janet, Charcot, Freud, and Jung but later abandoned by psychiatry to pursue the repression theory of pathology.

Ross believes that it is dissociation, not repression of trauma, that predisposes people to paranormal experience. This is, of course, an example of poor sense of self. People who dissociate have a very poor autobiographical memory because different experiences are associated with different ego states. This again makes them prime subjects for confabulation. And it makes them prime candidates for the alleged recovery of long repressed memories of the horrific experiences of the past.

Says Ross, "It is clear from our previous work that paranormal experiences are not necessarily pathological in nature. They are often deliberately cultivated by psychologically healthy, high functioning individuals. We propose that paranormal experience can be an expression of a normal dissociative capacity. They can occur in highly dissociative individuals who have never been traumatized."

Maybe. But according to Hedges, the "abuse" that leads to damage to the self does not have to be physical or sexual but can be psychological. Early intrusions or withdrawals of nurturance in an individual can have devastating consequences for further development in the young. The "trauma" can be something as benign as a parent not loving the child. If this happens in the first few years of life, the person is not able to form a solid sense of self. These are people who grow up to be what we now call self-conscious and insecure.

David Hufford, author of *The Terror that Comes in the Night,* believes that we, as a culture, tend to pathologize all paranormal and unusual experience. This is a cultural bias that defines as a sign of mental malfunction any experience that cannot be verified. It is this mind-set that leads most scientists to reject the paranormal.

For example, Wallace, in 1956, published a series of papers suggesting that all prophets and shamans have visions as a result of persistent overwhelming stress. The dissociation was caused by a combination of stress, social context, and psychodynamics.

Stanislav Grof believes we cannot separate reality from fantasy, and that cure comes from exploring spiritual crises. To do this type of cure, you must enter the belief system of the patient and explore the fantasies. John Mack is doing exactly that, and says so in the beginning of his book on alien abductions. He is a proponent of Grof's technique and points out that whether the experiences are real or not is irrelevant. This is a legitimate way to do therapy, and one that many of Mack's critics do not seem to understand. However, it is in direct opposition to what most abduction researchers believe.

Biochemical factors that contribute to sexual identity could also be a factor in identity confusion and confabulation. This is supported by Shaeffer's observation that gay women are overrepresented in the MPD population. It has also been noted by a number of researchers that gay men and women are overwhelming represented in the abduction population. One estimate suggests that as many as 60 percent of those claiming abduction are also gay.

Richard Ofshe, social psychologist and expert in false memories, has pointed out that when something is a natural part of the human psyche, it gets noticed. For example, all societies recognize homosexuality as something real. By contrast, no human society has ever noticed repression. This would suggest that the concept is cultural rather that biological.

In the hundred years that we've studied the human mind and memory, there has never been suggested or discovered any mechanism by which the repression of entire sections of a person's life could take place. And there is no mechanism by which these "repressed" memories can be easily recovered. If there is no repression of memory, then the explanation for the sudden recover of memories must come from somewhere else. If they are not from an internal source, then they must come from an external source. The memories must be implanted in some fashion.

As noted earlier, Stephen Ceci has put forth the following model of how to implant a memory without hypnosis:

1. Shatter the persons' confidence.

2. Elicit one confabulation from them.

3. Develop it into a story.

Let's look at that again, carefully. When most people arrive at a therapist's office, or consult an abduction researcher, they already have had the shattering experience. Steven Kilburn was afraid of a stretch of road. Pat Roach believed that something had invaded her home and two years after the event was still terrified by it. Others tell of equally distressing, to them, experiences. While the purpose of the researcher or therapist is not to implant memories, the conditions to do so are ripe.

It should also be noted that the type of recovered memory is dependent on the belief structure of the therapist or the researcher. Those who believe they were victims of satanic ritual abuse find a researcher who shares the belief in the vast conspiracy. Under hypnosis, or other memory techniques, memories of the ritualistic abuse are recovered.

If, however, the patient seeks the services of a researcher who believes in alien abduction, then the traumatic experiences are those of a victim of abduction. John Mack noted that those who find a researcher who believes in the benign aliens have a benign experience. Those who find a researcher who believes the aliens are cold, calculating scientists report an experience that fits that description.

In other words, the type of experience is dependent on the researcher conducting the research. In psychology it is well known that researcher bias can influence results, even when the researcher is aware of the problem. Here is a practical example of that bias that has gone unrecognized by the abduction researchers. In fact, they deny that it exists.

So once the basic premise is accepted, the next phase begins. It is not a conscious plan on the part of the researcher but one that is necessary if the abduction experience is to be recovered. Some sort of a confabulation must be recovered from the patient. And from that single confabulation, the story is invented.

Remember that a confabulation, by its very nature, is not a lie. The person relating the story is strongly convinced that it is real. Paul Ingram told horrific tales of rape and abuse of his daughters, yet there is no real evidence that such abuse took place. To "please" his interrogators, fellow police officers, he created a story that "might" have been true. Once he took that step, toward confabulation, he began to invent long, involved tales of rape, abuse, and satanic worship that were all accepted as true by Ingram, his family, and the police officials.

Finally, once the initial step into confabulation has been taken, it is developed into a complete story. During therapy with Richard Boylan a tale of a pleasant dream evolved into a terrifying alien abduction. He added the details necessary, repeating them to the patient until she believed them and added them to her report. The memory of alien abduction had been implanted in her mind easily by the authority figure whose own belief structures overwhelmed those of his patient.

Once all this has been accomplished the person will accept the new memory without question. The evidence unequivocally suggests that it is not only possible to induce memories, it is much easier than originally thought. Experiments by Ceci and Dr. Elizabeth Loftus have clearly demonstrated that it is relatively easy to make people remember things that never actually happened. Both Ceci and Loftus have been able, with little effort, to induce false memories in children.

But we don't even need to look at the experimental evidence cited simply because we can examine practical demonstrations of it. Read carefully, once again, how the memories were implanted in the mind of Paul Ingram. Here is a man who was a police officer, who was a pillar of the community, and in about eight hours of routine police interrogation, was telling tales of incest and rape. He was told, time and again, that his daughters were accusing him of these crimes, and Ingram believed that his daughters must be telling the truth. If they were telling the truth, he must be guilty, and if he thought about it, he would remember the events. After just eight hours, he had the memories necessary to "please" his interrogators.

And remember, abductees are not unwilling to find tales of abduction. They come to the investigators, the therapists, and the researchers believing they have been abducted. They have found the "symptoms" in their lives, so when the tale of abduction emerges, they are not surprised by it. They have joined the club.

Experimental evidence demonstrates how easily the implantation can be accomplished. As noted earlier, Richard Ofshe, when asked to interview Paul Ingram about the tales of rape told by his daughters, implanted an idea in the officer's mind with a simple statement to him. Within hours, Ingram had created, what to him was a memory of the event, complete with dialogue and "stage" directions. All this because Ofshe mentioned that his daughter had suggested the event to be real.

This is assuming, of course, that there isn't some other motive, or that the tale or repressed memory coming to the surface isn't a lie. One California man was convicted of murder after his daughter "remembered" that she had witnessed him killing one of her childhood companions. Although she claimed under oath during the trial that she had not been hypnotized, and that the memory came back in a sudden flash of insight, that doesn't seem to be the truth. According to her, while looking at her own eight-year-old daughter, she had remembered that her father had killed her best friend twenty years earlier.

Based on her testimony, her father was convicted and sent to jail. But the California Supreme Court released the man and ordered a new trial. There had been no corroborative evidence to support the woman's claim. In fact, during her initial contact with the police, she claimed that she not only remembered the incident but that there were others who could corroborate her statements. It was only after the investigation began that police learned the memory was "recovered," though there were several conflicting stories about how it had happened.

Although the county district attorney had said they would retry the man for the crime, he eventually was forced to drop the charges. Without any corroborative testimony, they had no case against the man.

But what was most surprising is that the daughter had accused her father of other murders. She told police details that she supposedly could only have known had she witnessed the crimes. Police were able to find a case that matched the details, but they also found that her father had a solid alibi for the night of that second murder.

She told of a third murder in horrific detail. But police searches of their records, and those in many other jurisdictions, could find nothing to match the facts as reported. They had to conclude that this crime was a confabulation by a daughter filled with hate for an abusive parent.

What is most interesting is that the police, when confronted with this evidence, saw that the daughter had confabulated her father's involvement in those two murders. She had even confabulated new crimes to prove him a monster. When the evidence broke against the daughter and proved the man innocent of those new crimes, it wasn't sufficient to convince police and prosecutors that he wasn't guilty of the first murder. They were convinced that he was a murderer and it was only a technicality that had put him on the street.

And when the daughter was confronted with this information, showing that her father couldn't have committed some of the crimes of which he had been accused, she asked, "How could I have known the details if I wasn't there?" She wanted to know how she could have come up with so many small details if she hadn't witnessed the crimes in the first place.

The answers were evident to those who looked. For the first murder, every detail the daughter had mentioned was reported in newspaper and television stories published and aired at the time of the original crime. She provided no detail that hadn't been in the public arena. So the answer to the question about how she knew so much about the murder of her childhood friend was that she had read, seen, or heard about it in the media or had heard others talking about it.

For the other two crimes, she stared into the television camera and asked the audience, if she hadn't been there, how could she have known what had happened. But it was clear from the police investigation that she hadn't been there and her father hadn't been involved. There either was no crime or a crime for which her father couldn't have been the perpetrator. So how could she have known? The answer: this was confabulation elicited under duress by her therapist.

A television reporter, finishing the story, asked, "Who do you believe? One of them must be lying."

But that isn't true. Each could have been telling the truth as he or she believed it. Memory, confabulation, pleasing the operator, and a belief that everything you have ever seen or done is recorded in your mind comes into play. The daughter sincerely believed that her father murdered an eight-year-old girl and that she witnessed it. She believed that her father molested her as a child and that he committed other murders. But there is no corroborative evidence for it. Nothing.

So, the daughter is telling the truth as she knows it. Her father is telling the truth as he knows it. And both are telling the truth. The question that should have been asked was which was telling a truth that was grounded in reality and which was telling a truth that came from confabulation and repressed memory.

In another, similar case, Janice Knowlton accused her dead father of having killed the Black Dahlia, a terrible and unsolved case from Los Angeles in 1947. Although Knowlton had been exposed to movies about the Black Dahlia, she, according to her own book, repressed her memories of seeing them immediately.

There were other movies that she also repressed as she saw them. According to her book, "Ironically, the film that Jan worked hardest to repress was *It's a Wonderful Life.*" That seems to answer one of the questions we have raised. She worked to repress certain memories, which means that it is a conscious effort. How can you repress something of which you are aware? Do you just refuse to think about it, forcing your mind to move from that item to something else? And if that is what you are doing, is it repression? Does any of this make any psychological sense?

Because of these cases, a hotbed of controversy exists around recovered memory therapy. In fact, many of those who have accused family members of abuse have since recanted. Remember, Margaret Singer, a retired professor from the University of California, has interviewed fifty people who claim to have recovered memories of satanic ritual abuse. But all fifty recovered those memories while in therapy. All fifty now believe that they were mistaken in their beliefs.

Singer doesn't believe that trauma can cause memories to be repressed. The opposite happens, she says. Bits and pieces of memory can be lost to amnesia, but trauma impresses itself on the brain.

Richard Ofshe is even more vocal about the concept labeled "robust repression." Ofshe points out that therapists who believe in robust repression put no limit on it. There is no limit to the number of events repressed, the length of time over which they are repressed, or the mechanism by which the memories are recovered. These theories, according to Ofshe, can only be found in the "lunatic fringes of science and the mental health professions."

The numbers of abuse and incest victims who have begun to recant is growing daily. They realize that the memories recovered in therapy are not honest memories but are the creations of the therapists who have their own agendas at play.

Walter C. Young, Bennett G. Braun, and Roberta G. Sachs authored a paper about satanic ritual abuse, suggesting that the events described by their patients and clients were real. However, they wrote, "The lack of independent verification of the reports of cult abuse in this paper prevents a definitive statement that the ritual abuse is true." In other words, they realized the flaw in their theories but went ahead, pressing forward with their ideas. Clearly finding more victims of ritualistic abuse would help their theory. The more people they interviewed who added to the wealth of anecdotal testimony, the more likely the events were real, or so they believed. If one was shown to have imagined the events, there were always another one, or ten, or twenty who remembered similar events. And that many people couldn't be suffering from the same delusion.

Braun, a leading researcher in satanic ritual abuse, has established a hospital to deal with the victims of abuse. He teaches law enforcement officers how to recognize the abuse. But now those whom he convinced were victims of abuse are turning the tables on him. He and his hospital are being sued for millions of dollars. And it is clear that the memories were implanted by the therapists who were actively searching for victims of abuse. They all believed, with great conviction, that the victim had repressed all these horrible memories from a terrifying childhood.

Ask yourself, how exactly could this process of repression take place? Suppose a six-year-old child is playing quietly in her room. Suddenly she is attacked and brutally abused sexually by an adult. Just exactly at what point is this incident repressed? During the incident? After the incidents? A week after the incidents Nobody has adequately explained exactly when or how this remarkable process could take place.

In addition, no one has clearly defined just what is repressed. Is it only the incident itself that is repressed? Is the entire incident from beginning to end repressed, or could it be that the entire day is repressed? Is it possible that an entire portion of a child's life becomes repressed? In cases where multiple episodes of abuse are alleged, does the child learn to repress them, or is this an automatic process built into each one of us?

Let's assume, for the moment, that the repression is learned. It would seem that if a child has to learn to repress abuse memories, it would suggest that at least the first few incidents of abuse would not be repressed. The child at that point would not have learned the process.

If repression is an automatic process, then that would suggest that every time a child is about to be abused, at the moment the perpetrator begins approaching, the child automatically begins the process of repression.

Some therapists claim that children who are abused repress their entire childhoods. That is to say they have repressed their entire episodic memory. If this is true, then how do they remember things like where they lived, what they learned in school all those years, or who their friends were? The amnesia seems to be restricted to the episodes of abuse, and this goes against everything we know about amnesia. It is a logical inconsistency that we are not supposed to notice.

It has been hypothesized that what is occurring is retrograde amnesia, the type of forgetting that occurs when someone receives a blow to the head. But all of the evidence that we have about retrograde amnesia indicates that it's a result of physical trauma to the brain, not psychological trauma. In other words, it is impossible for repression to be retrograde amnesia.

Recovered memory therapy is a technique that will come and go like many oher pop theories have. During Freud's time, the therapies of choice for neuroses were bedrest, massage, hydrotherapy, and electrical stimulation. All these treatments have proven ineffective, just as recovered memory therapy is being proved.

20.

SLEEP PARALYSIS AND ALIEN ABDUCTIONS

Our review of the scientific and psychological literature, an examination of hundreds of abduction cases, and interviews with abductees suggest that 50 percent of UFO abduction reports are the result of sleep paralysis. The symptoms described, the circumstances under which the conditions occur, and the specific experiences reported mirror those of sleep paralysis exactly. Sleep paralysis, regardless of the arguments made by the abduction researchers, UFO investigators, and the abductees themselves, is clearly responsible for many of the tales of alien abduction.

It was only after the publication of David Hufford's monumental work, *The Terror That Comes in the Night,* that researchers began to understand that sleep paralysis and the related symptom of hypnagogic hallucination could affect the general population in very specific ways. Prior to his work, the idea of sleep paralysis and the sometimes accompanying hypnagogic or hypnopompic hallucinations were considered to be part of a larger syndrome that included narcolepsy. Sleep paralysis was considered to be a rare condition and linked to other maladies.

Sleep paralysis, and the attendant hallucinations, are now more fully understood. Sleep paralysis affects as many as one in five or six of the general population, but because it often is manifested only as sleep paralysis, it is rarely reported by those who experience it. Originally, and only when accompanied by other symptoms, such as narcolepsy, did a victim of sleep paralysis report it outside of the immediate family.

It turns out that there is no single, simple definition of what sleep paralysis is. There are various forms of it, such as cataplexy, and it is often associated with narcolepsy. In fact, sleep researcher K. Fukuda wrote, "Sleep paralysis has usually been described in terms of its association with narcolepsy." And Jorge Conesa wrote that when sleep paralysis occurs without the additional symptoms, it is known as isolated (idiopathic) sleep paralysis.

Sleep paralysis, as defined today, by Hufford "a period of inability to perform voluntary movements, either when falling asleep or when awakening, accompanied by conscious awareness. This condition has been ascribed both to the hypnagogic and hypnopompic states. . . . Sleep paralysis is more frequently reported in connection with falling asleep than with awakening."

Another definition of sleep paralysis was written by R. J. Campbell for his dictionary of psychiatric terms. "A benign neurologic phenomenon, more probably due to some temporary dysfunction of the reticular activating system consisting of brief episodes of inability to move and/or speak when awakening or less commonly, when falling asleep. There is no accompanying disturbance of consciousness, and the subject has complete recall for the episode. The incidence of the phenomenon is highest in younger age groups (children and young adults) and much higher in males (80 %) than females. It occurs in narcolepsy. . . . The terms by which the phenomenon has been known are nocturnal hemiplegia, nocturnal paralysis, sleep numbness, delayed psychomotor awakening, cataplexy of awakening and post dormital chalastic fits."

According Campbell, "Cataplexy differs from sleep paralysis in that it occurs during daytime activities, often following the eruption of expressions of strong emotions such as laughter."

Campbell defined cataplexy as "temporary paralysis or immobilization; loss of antigravity muscle tone without loss of consciousness, often precipitated by emotional excitement. Cataplexy is usually associated with narcolepsy."

Originally it was believed that sleep paralysis was only associated with narcolepsy, which has been described as a syndrome consisting of excessive daytime sleepiness and abnormal manifestations of REM sleep. Both sleep paralysis and hypnagogic hallucinations are believed to occur in about one-half of the narcoleptic population, or about 0.02 to 0.07 percent of the general population. It was later determined that sleep paralysis could manifest itself without the influence of narcolepsy or other related maladies.

It has also been noted that sleep paralysis has been accompanied by hypnagogic hallucinations. "A commonly reported hallucination is the feeling of a presence or entity in the room in which the individual sleeps," according to one sleep researcher.

Hypnagogic hallucinations and hypnopompic hallucinations are defined by Hufford as those hallucinations that occur just before or upon awakening from sleep. Hufford noted that dreams are technically hallucinations, so it is important to distinguish between the two. According to Hufford, "Hypnagogic hallucinations (HH) occurring before sleep, during the actual hypnagogic state, are defined by their location before stage 1; those occurring during sleep-onset REM are defined by their associations with sleep paralysis, which is experienced as wakefulness and, therefore, different from normal sleep.

"Several medical writers have suggested that sleep paralysis is a neurotic symptom. In 1965, Stephen B. Payn presented the case of a woman whose SP [sleep paralysis] he considered to be 'an extreme form of inhibition, representing a compromise between fulfillment of her sexual and hostile impulses and a defense against them.' "

Another author, Jerome Schneck, suggested a connection between the paralysis and sexual conflict. Schneck published a case in which he described the symptoms exhibited by a female patient. Because of those symptoms, he drew the conclusion that neurotic symptoms were present as underscored by the sleep paralysis.

Hufford, however, disagreed. He wrote, "[T]he entire experience seems too common, as well as too consistently patterned to be pathognomic. Any condition afflicting 15 percent of the population, but remaining largely undiagnosed, must hopefully not be too serious."

What Hufford was suggesting was that his own research had suggested that 15 percent of the American population had experienced some form of sleep paralysis and/or a hypnagogic hallucination. If it was that widespread, then it did not reflect any type of neurotic symptom. To suggest, in the absence of other symptoms, that a neurosis was present, was to miss the point. The fear of such a diagnosis, or the absence of other symptoms, could be the reason that others hadn't reported sleep paralysis or a hypnagogic hallucination. The distribution of the phenomena might be spread even further through the general population.

There is a long history of a folkloric tradition of sleep paralysis when it is defined broadly enough. Clearly, the folklore tradition is filled with accounts that when interpreted in the modern world are examples of sleep paralysis and hypnagogic hallucination. The stories of the incubus and the succubus, as described in the texts of the medieval world, match the tales told by those who experience sleep paralysis today.

Ernest M. Jones, whose major work on nightmares explored the sexual repression that led to the medieval concept of the incubus and the succubus, also suggested the idea of "helpless paralysis." The theory seemed to be that priests, as well as nuns, became celibate and this unnatural sexual repression led them to be plagued by erotic dreams. These powerful sexual feelings and fantasies were, of course, forbidden, and could not be admitted to each other or to the public. In an attempt to rationalize the sexual impulses they were feeling, they claimed that the devil slipped succubi into their beds at night to seduce them. Although they awakened only after the demons were gone, evidence of the encounter often remained. Nocturnal emissions were considered solid physical evidence that the demons were collecting sperm.

The tales of the incubi and succubi show that some aspects of sleep paralysis and the hypnagogic, as well as the hypnopompic, hallucination have been reported for centuries in the same sort of detail that is evident today. A close reading shows that all elements are present in some aspect of these tales.

In his study of sleep paralysis, Hufford identified many of the same elements that were reported in ancient tales of attacks by an incubus and succubus. He cautioned, however, "If an experience is missing some of the primary features, it cannot be identified as an Old Hag attack, although in written accounts such omissions cannot definitely rule out the identification."

Hufford's study began when he learned of the phenomenon of the "Old Hag" attack. He wrote, "Many Newfoundlanders are familiar with the Old Hag tradition and defined it as did a university student about twenty years of age: 'You are dreaming and you feel as if someone is holding you down. You can do nothing only cry out. People believe that you will die if you are not awakened.' "

The Old Hag was described as an actual event and not part of a dream. Witnesses reported that they were awakened and attacked in their beds. The two primary symptoms were a suffocating weight on the chest and a paralysis. These symptoms are often accompanied with the feeling or belief that some kind of being is in the room, though often it is only a vague impression rather than a direct observation.

Hufford reported on a number of cases in which an entity of some kind was believed to have invaded the the subject's bedroom while he or she was in a state of paralysis. He wrote of a female who explained, "In my experience I felt very numb and still. I do not know if I was able to move or not because I did not feel as if I wish to move. I was also lying on my back just staring in an odd fashion. Everything around me appeared to be at a distance and I felt as if I was enclosed by something. It was as if I could not touch reality, in my mind I was drifting in some peculiar realm. I was not dreaming but wide awake but as if in another realm but in fact was in my bedroom. I could see where I was, but as if I could not reach out or speak. My mother had the same experience. She could not come out of it, she was not dreaming but could not speak. She felt as if she was choking as she was lying on her back."

Ronald K. Siegel, who is familiar with Hufford's work, had his own encounter with an Old Hag. He reported, "I was awakened by the sound of my bedroom door opening. I was on my side and able to see the luminescent dial of my alarm clock. It was 4:20 A.M.. I heard footsteps approaching my bed, then heavy breathing. There seemed to be a murky presence in the room. I tried to throw off the covers and get up, but I was pinned to the bed. There was a weight on my chest. The more I struggled, the more I was unable to move. My heart was pounding. I strained to breathe.

"The presence got closer, and I caught a whiff of a dusty odor. The smell seemed old, like something that had been kept in an attic too long. The air itself was dry and cool, reminding me of the inside of a cave.

"Suddenly a shadow fell on the clock. Omigod! This is no joke! Something touched my neck and arm. A voice whispered in my ear. Each word was expelled from a mouth foul with tobacco. The language sounded strange, almost like English spoken backward. It didn't make any sense. Somehow the words gave rise to images in my mind: I saw rotting swamps full of toadstools, hideous reptiles, and other mephitic horrors. In my bedroom I could only see a shadow looming over my bed. I was terrified. . . . I signaled my muscles to move, but the presence immediately exerted all its weight on my chest. The weight spread through my body, gluing me to the bed. . . . A hand grasped my arm and held it tightly. . . . The hand felt cold and dead. . . .

"Then part of the mattress next to me caved in. Someone climbed onto the bed! The presence shifted its weight and straddled my body, folding itself along the curve of my back. I heard the bed start to creak. There was a texture of sexual intoxication and terror in the room.

"Throughout it all, I was forced to listen to the intruder's interminable whispering. The voice sounded female. I knew it was evil. . . .

"The intruder's heavy gelatinous body was crushing the life out of me. . . . I started to lose consciousness. Suddenly the voice stopped. I sensed the intruder moving slowly out of the room. Gradually the pressure on my chest ceased."

Siegel explained the strange vision by suggesting that he was in a state of sleep paralysis and having a hypnopompic hallucination. He wrote, "You don't have to have a medieval mind to see a succubus emerge from all these data points."

What this demonstrates is the link between modern cases of sleep paralysis and the incubi and succubi of old. It suggests that the phenomenon is not specific to a single culture or to the modern world. Instead it is a phenomenon that transcends culture and time.

Hufford, in fact, made exactly that point. He had originally believed that the Old Hag was a myth that had grown up in Nova Scotia. However, on his return to Pennsylvania, he learned that a number of his students reported experiences that were surprisingly similar to the Old Hag tradition from Nova Scotia. "There was no 'Old Hag' folkloric tradition in these students' backgrounds; they did not even know what to call the phenomenon, and had been embarrassed to tell anyone of their experiences out of fear of ridicule."

The cross-cultural aspect of the Old Hag, suggested to Hufford there had to be another cause for it. A cultural tradition that was unknown to students in Pennsylvania could not account for their tales of an Old Hag. That lead Hufford to sleep paralysis and hypnagogic hallucination to account for the similarities.

In a case reported by sleep researcher Tomoka Takeuchi, a twenty-year-old male student who was one of sixteen individuals who had experienced isolated sleep paralysis but who had not reported episodes of narcolepsy or cataplexy was asked to sleep in the lab. The student, upon being awakened by an experimenter, reported, "Before I woke, someone came into this bedroom and was here beside me."

Takeuchi reported the student could clearly see his green sweater hanging on the bedroom wall despite the darkness. He also reported, "He felt an urge to get dressed and escape from the bedroom. He heard someone come into the room and walk around the bed without speaking. He believed that he rose from the bed. He felt fearful and unpleasant during his experience. In reality, nobody came into the bedroom and he remained lying in bed without moving. During the course of making his report to the experimenter, he gradually became aware that his experience was only an hallucination."

In Hufford's survey, he noted that 17 percent of the students in Nova Scotia had a positive response to the "Old Hag" scenario. These were students who had no history of related sleep disorders and who were neither narcoleptic nor cataplexic. They were, as the abduction researchers are fond of saying, free of psychopathology.

Hufford also reported in his survey of the students that 23 percent had experienced having awakened during the night paralyzed. Although this survey could be generalized only to the population in Nova Scotia, similar results were found when other populations were tested. This is far higher than the population of narcoleptics in the general population and suggests that anyone can be affected by sleep paralysis.

Another sleep researcher, Y. Hishikawa, wrote, "Sleep paralysis may be described as a brief episode characterized by the inability to perform voluntary movements . . . on falling asleep (hypnagogic or predomital sleep paralysis). . . . Occasionally sleep paralysis is accompanied or just preceded by vivid and terrifying hallucinations." According to his study, Hishikawa suggested that the episodes lasted only a short period, could be ended by an outside stimulus such as touching, or would end spontaneously, usually within ten minutes.

Hypnagogic hallucinations, often accompanied by the belief that something, or someone, is in the room, occur as the patient is falling asleep. Hypnopompic hallucinations occur upon waking and can contain the same horrifying imagery found in the hypnagogic hallucination. Hishikawa noted that the subject "is fully aware of his condition and can recall it completely afterwards."

What we have seen are the various symptoms and manifestations of sleep paralysis as it has been reported in the psychological and popular culture literature. But how does all of that relate to alien abduction? The researchers claim that there is no connection between sleep paralysis and alien abduction.

David Jacobs wrote in *Secret Life,* "An unsuspecting woman is in her room preparing to go to bed. She gets into bed, reads a while, turns off the light, and drifts off into a peaceful night's sleep. In the middle of the night she turns over and lies on her back. She is awakened by a light that seems to be glowing in her rooms. The light moves toward her bed and takes the shape of a small 'man' with a bald head and huge black eyes. She is terrified. She wants to run but she cannot move. She wants to scream but she cannot speak. The 'man' moves toward her and looks deeply into her eyes. Suddenly she is calmer, and she 'knows' that the 'man' is not going to hurt her."

Eddie Bullard, in his 1982 massive study of alien abduction, described a number of instances of what he termed "the capture" phase of the abduction experience, or how the victim was trapped by the aliens for abduction. Of one of a series of contacts he calls the "Tujunga Canyon contacts," he wrote, "Jan was driving alone the mountain road at night with Emily beside her and Emily's 6-year old son asleep in the back. Truck traffic was heavy and both women were tired, so they pulled off at a rest stop to sleep awhile. In conscious memory both witnesses remembered seeing a light, seemingly a truck headlight at first, but along with it they heard a high-pitched whining sound and felt paralyzed by it."

John Mack also reported on a case that resembled these others. "Scott recalls reading a magazine, and before he could fall asleep he felt that the beings were 'there in my mind.' . . . Then he saw a 'round-tipped rod' pushing toward him, which Scott related to how he was anesthetized . . . Scott said, and 'they put me under' so that 'I couldn't move.' "

In another session with still another abductee, Mack wrote, "Catherine believes she was in a 'more than half asleep' state at this point."

And remember that Budd Hopkins wrote about Philip Osborne. Osborne remembered that he, Osborne, awoke in the middle the night, paralyzed. He could not move, turn his head, or call for help. It reminded him of another, earlier event in which he felt a presence in the room.

It is important to remember that Hopkins, along with others, met Osborne a few days later to explore these events under hypnosis. And the point of that hypnosis was to find out if an abduction experience had happened. Osborne, Hopkins, and the others expected to find an abduction event and, using hypnosis, they did.

Each of these researchers has overlooked the mundane in their search for the unusual. Jacobs, in writing about his "typical" abduction, has described classic sleep paralysis. He does, however, reject the idea of sleep paralysis and hypnogogic and hypnopompic hallucinations. He wrote that ". . . this explanation fails to account for those abductions that take place when the victim is awake, not tired, not in bed, and not even inside a room."

The argument being made is not that all abduction accounts can be explained by sleep paralysis but that a portion of them can be explained by the phenomenon. To reject the explanation because it doesn't fit all the cases is ridiculous. But that is exactly what Jacobs is attempting to do here. Even though his classic abduction matches, perfectly, sleep paralysis down to the fact the victim is lying on his or her back, he claims that sleep paralysis cannot explain any abductions.

In fact, all abduction researchers reject the idea, even John Mack. As a psychiatrist, he should understand that a single explanation cannot solve every aspect of the phenomenon. As a psychiatrist, he should be familiar with sleep paralysis and realize that some cases are perfect matches for sleep paralysis. Yet he seems to be saying, as are Hopkins and Jacobs, that sleep paralysis can't be the answer.

It could be argued that sleep paralysis is induced by the aliens, so rather than explaining abductions, abductions explain sleep paralysis. For this explanation to work, another step is always necessary. In other words, the victims of sleep paralysis remember, consciously and easily, the experience of sleep paralysis. They awaken to find themselves paralyzed, they have difficulty breathing, and then frequently believe there is some kind of entity in the room with them. To move this classic manifestation of sleep paralysis into the realm of alien abduction, hypnosis is necessary. Without the hypnosis, there would be no tale of alien abduction. That should provide these researchers with a clue about the cause of these tales of abduction, but they seem to have missed it. Again, remember Philip Osborne, who knew nothing of the abduction but clearly remembered multiple episodes of sleep paralysis.

What we see here, then, is an explanation for as many as half of the cases of alien abduction that do not require alien beings, flying saucers, or interstellar flight. Instead we find a common psychological phenomenon that fits, perfectly, with the descriptions offered by victims of alien abduction. Rather than embrace this explanation, abduction researchers have rejected it because it doesn't fit with their personal beliefs about abduction or their personal agendas.

Is there a valid scientific reason to reject sleep paralysis as an explanation for some cases of alien abduction? No. Clearly, it suggests a solution for some of them.

With what we now know about hypnosis, with the poorly designed research using hypnosis, and with an overwhelming belief that some sort of alien visitation is responsible for sleep paralysis, or the subject's belief in abduction, the stage is set for the recovery of abduction memories. But once again, it is the belief structures of the researchers that change a normal and somewhat common psychological phenomenon into something strange and mysterious.

Without the UFO researcher, without the use of hypnosis, and with a basic understanding of sleep paralysis, hundreds, if not thousands, of so-called abductees could sleep easily at night. They would not have to worry about alien creatures invading their bedrooms. Instead, they could learn that sleep paralysis is the cause of their "visitation." Without the UFO researcher to misinterpret sleep paralysis, a large number of people would be happier and healthier.

21.

SUPPORT GROUPS

It's a warm summer afternoon in Santa Monica, and Bill Cone is sitting quietly in a support group. One of the members, Matthew,[1] is talking about his last experience being forced to have sex with an alien. He describes, in vivid and chilling detail, the helplessness he felt as he was being raped by a creature from another world. After the meeting broke up Cone had a chance to speak to him and asked him how long he had been attending the group.

"Six years" was the reply.

"Do you feel like you're making progress?" Cone asked.

"Sure. I feel much better."

"So, when do you think you will be able to stop going?" Cone asked.

Matthew got a puzzled look on his face. "I never thought about that," he said. "I can't really imagine not going. The people in the group are the only people I can trust. They're like family."

Cone then asked him, "How about the problems that brought you to the group? Have they gotten better?"

He smiled and said, "Well, yes and no. When I first came to the group, I had only one memory of being abducted. But now I realize it has been happening all of my life. And it's still happening. When I first came to the group, I was getting abducted about once a year, but now I get abducted about once a month."

"It sounds like things have gotten worse," Cone said.

"Well, I sure don't enjoy the abductions. But the group has helped me get through it all."

Three weeks later Cone was sitting in his office in Newport Beach with a new patient. Cone asked her how long she had been attending the group.

"Fourteen years" was the reply.

"Why so long?" Cone asked, thinking he had all heard this before.

"Well, you see, I'm a survivor of sexual abuse," she told him.

Support groups can be healing. For the last several decades they have helped

---

[1]His named has been changed.

countless people deal with loss and pain. Groups were originally meant to help people recover from emotional trauma and solve personal problems. People would enter a group, talk about their problems and their pain, share the experiences of others, work it through, and move on. But something has happened. Today's so-called recovery groups no longer serve this function. The introduction of the idea of "survivor" may have changed the purpose of support groups forever. Being a group member used to be part of a healing process. But it has now become a new form of religion.

Group members are no longer people with problems; instead they are "abductees," "multiples," "survivors of incest," and "codependents." There is now a group established for every conceivable personal problem. Instead of people moving toward resolving their problems, they now learn to incorporate them into the very core of their identity. The recovery groups of today are based on the following premises: (1) All psychopathology is caused by abuse. (2) Most abuse memories are repressed. (3) The path to healing is to remember the abuse, and often to attack the abusers. (4) The person is forever to be labeled as "recovering." (5) Recovering people need group support in order to function in society.

Even though none of these premises has hard scientific evidence to back it up, each premise has become an accepted part of the recovery group movement. The most tragic element of today's recovery groups is that recovery is no longer expected to occur. Instead of being a short span of time in which a person recovers from a negative experience, recovery has now been redefined as a never-ending, lifelong process.

Today the recovery work of a group member is never finished. If a member should think of leaving the group, he or she is not congratulated for achieving recovery but is seen as defecting and returning to denial. In this way group members are conditioned to stay wounded and encouraged to remain in therapy.

Irvin Yalom is one of the leading experts on group therapy. After years of studying the healing qualities of groups, he found that successful support groups contain certain elements that lead to healing. Each of these factors plays an important part of a healthy group. However, in today's recovery groups, the function and meaning of each of them has become horribly distorted and misused.

There is no doubt that a support group can instill hope. It is in the group that people realize for the first time that they are not alone. Many others have shared their sorrow and understand their pain. For example, those who have lost a loved one realize that their pain is normal and that they can, and will, recover from the devastating loss.

However, in today's "recovery" groups the instillation of hope is gone. People do not get better; they get worse. They are told that they will never fully recover from their problem, whatever it might be, and that they will be expected to attend the group for the rest of their natural lives. The group discussions no longer center on effective methods for healing but on how well a member can display his or her pain, and share the increasingly horrific stories.

In abduction groups, hope is abandoned. As group members become more educated in the folklore of alien abductions, it becomes clear to them that the aliens are in complete control of their lives. They are, and will always be, abducted against their will as often as the aliens wish. As abductees become more and more enmeshed in the group culture, their abductions increase in frequency. After spending some time in a group, many of these people deteriorate to the degree that they can no longer hold a job or have a relationship outside of the group.

In the traditional group setting, the initial act of sharing pain leads to feelings of universality. As the members get to know each other, they share stories and ideas about how they feel, what they have tried to do to deal with their problems, and what has worked for them in the past. It is by this process that group members come to see that all people suffer in the same way and take the first steps toward healing.

But here again, things have changed. Groups no longer speak of universal suffering and the members are no longer encouraged to see themselves as normal humans in pain. Instead, each group attaches a label to its specific type of pain, using terms like "adult child" or "incest survivor." Members of the groups of today are told that they are, in fact, different from others, that they will always be different, and that no one but a fellow survivor could possibly understand them.

In abduction groups this feeling of being different is amplified. Abductees are told that they have always been different. In a vein similar to that of satanic cult group members who are told that their families have been cult members for centuries, abduction group members come to learn that the aliens have been abducting their family members for generations, and that their children will soon become victims as well. They are told that they have been chosen for some special purpose that will someday be revealed to the world. Members are encouraged to ask their families and children about any unusual experiences they have had, and to interpret all of them as possible abductions.

The members of a successful group soon become bonded by goodwill. Up until now they have felt lost and alone, but here in the group they find a new family, a place to belong. The new family provides support, knowledge, and guidance, and when the healing work is done, as in all good families, the family bids farewell to its children and sends them into the world. A successful family, and a successful group, brings about its own end.

Unlike healthy people, however, who grow to be adults, today's group members never leave their family of origin. It is in the group that they are reborn as a helpless victim, and in the group they stay.

People who enter support groups are by definition wounded people. The reason they need a support group is that they have not been able to get sufficient support in their family or social network. Nine times out of ten this is because their family has not been able to give the much needed support, or the person's wounds have made it difficult for them to form healthy social bonds. Because of this they perpetuate their own problems.

Abduction groups involve experiences that have no objective verification. A drug addict or an alcoholic attending a group has had ample confirmation from others that his problems exist, but abductees and ritual abuse victims have only each other to look to. Often even their own families don't believe them. The group becomes their only sanctuary. Many of the groups are leaderless, and consist only of people who share the same belief system, making any objectivity impossible.

As in most supportive families, members are validated and provided with an atmosphere of uncritical acceptance. There is no outside objective source to question their assumptions or methods. To these people the group becomes the much needed new family. But the recovery groups today are not based on altruism, they are based on need. The goal of today's group is to sustain itself, not to end itself. This is why most abductees never leave their groups. And it is why most abductees attack the outsiders who offer other, more mundane explanations for the abductees' stories.

Although many groups tout themselves as vehicles of recovery, they are not really interested in researching ways to recover from their problems. To recover would mean to lose the best family the group members has ever had. Instead, to maintain their place they must produce more and more abduction material. Groups like this only serve to reinforce each member's pathology. Freud said of this quality of groups, "A group is extraordinarily credulous and open to influence, it has no critical faculty, and the impossible does not exist for it. It thinks in images, which call one another up by associations . . . and whose agreement with reality is never checked by any reasonable agency."

The feelings of the group are always very simple and very exaggerated, so that the group knows neither doubt nor uncertainty. Unfortunately, even when the groups are guided by a therapist, it is always someone who wholeheartedly shares the beliefs of the group. In the majority of the abduction groups we studied, the therapist is also an abductee. In many of the groups we have observed, if the therapist has not identified himself or herself as an abductee he or she is encouraged to do so by the group members. One abduction group member told us, "I know my therapist is an abductee. Why else would she be driven to do this work? She just isn't ready to admit it, but we are encouraging her to recover her own memories."

This transformation from therapist to abductee has already happened to Richard Boylan, Leo Sprinkle, and several other high-profile abduction therapists. What is not generally known is that most of the less publicized abduction groups are run by therapists who believe that they too have been abducted.

There is a second problem operating here when the therapists believes that they, too, have been abducted. Their own agendas take over. Boylan, as we have seen, took in new patients who came to him with one set of problems, and he produced another. To Boylan, satanic ritual abuse was not real but alien abduction was. Because of his beliefs he pushed his patients into believing they had been abducted when their problems arose from other factors. Boylan appeared primarily interested in corroborating his personal belief structures. And by doing that was validating the problems of the members of the group.

As groups congeal, evolve and solidify, anyone who is not a believer soon comes to be seen as the enemy. Abductees deal with these feelings in two ways. Some learn to shut out any information that refutes or contradicts their belief. Abductees who feel this way don't care what the skeptics and unbelievers say. Many contend they know their own experiences are real. Others take a more hostile stance, and immediately attack anyone who questions their beliefs.

Both Cone and Randle had the privilege of appearing on *The Maury Povich Show* with John Mack and several of his abductees. One of the group members gave classic testimony to the problem of a self-validating support group. She said, "Even after the December incident [her last abduction experience] last year, I broke down crying about it at a support group. . . . They continue coming to my house to the point where I'm so saturated with visitation that I have no choice anymore. It's not an isolated case."

This same person, during the commercial break, began to attack Cone, telling him that he was an evil person, and that he had a closed mind because he didn't believe her. At this point he had not spoken a single word. She had never met him and knew nothing about him, but he had already been labeled as the enemy. When Cone told her that she hadn't bothered to ask if he believed her story, she refused to listen and became increasingly hostile.

The support groups for today's victims have one thing in common. They are all composed of people who feel molested and powerless. Their sexuality is controlled by a force outside of themselves. They all speak of being trapped in an unwanted, dependent relationship that is non-nurturing and sometimes hostile. They react to this by doing one of two things. They identify with the aggressor and become like their tormenters, or they withdraw into fantasy. Recovery groups supply the fantasy.

The original meaning and purpose of groups was to teach social skills that help return the group member to society. But in today's "recovery" groups, the development of socializing techniques never takes place. In fact, the groups often discourage the person from making friends outside of the group.

People who are not in the group recovering are said not to understand, and therefore they cannot be good friends. As the stories of the abduction experiences become the focus of the victim's life, socialization stops. Group members no longer strive to return to society as healthy individuals but work to identify themselves as members of a special group. The abduction group members live with a new identity, "abductee," which will most likely keep them isolated for years.

It is not uncommon for group members to become increasingly dysfunctional the longer they stay in the group. As the years go by, the members reinforce each other's ideas and pathologies. Since there is no objective outsider to question the validity of the belief system, there is no point at which recovery can take place.

Part of the recovery in a normal group is through catharsis, which is the act of discharging hurt and anger, and which can help people work through pain. In abduction groups, however, emotional discharge has become part of the ritual of membership. Rather than being healed, with each cathartic episode the victim becomes more firmly entrenched in his or her identity as victim.

True healing requires that wounds get smaller as time goes by. When treated properly, what was once a gaping wound becomes a barely noticeable scar. But in recovery groups the wound is everything. Each week, it is reopened and reexamined. No healing is allowed. It is not healing but pain and suffering that are coveted. It is not catharsis but recapitulation that is encouraged. This tactic is similar to the idea of trying to heal a broken bone by spending several hours each week talking about how you broke it. It may fill time, but it doesn't contribute to healing.

What is developing, then, rather than an environment in which healing can take place, is an arena for continued belief. Each element, as it is added to the mix, is taken on faith. No proof is offered and none is asked. Rather, the group begins to function as a religion in which it is heretical to ask for any sort of outside evidence. Testimony is everything and it is believed no matter how outrageous the claims become.

When one group member suggests that she was taken out of the house through the walls, no one questions how the laws of physics have been overstepped. Rather, other members nod and say that they, too, have been taken out through the walls.

When asked by outsiders why no photographic evidence of abduction exists, the answer is that the aliens are able to circumvent the attempts. Abductees who are taken frequently have been told to use a video camera to record the events, place it out of the way so that it "sees" what is happening in the bedroom. But rather than produce a tape that shows the aliens entering and spiriting the abductee from the room, the tape shows nothing, or the camera malfunctioned that one night, or the aliens somehow knew of its presence and turned it off before they moved into the room.

In group, these things are not questioned. Rather, everyone nods and agrees. Each believes what the other says because he or she believes that similar things are happening. And at the top is the authority figure, the therapist, who believes each of the tales. Everyone in the group is reinforcing the ideas of the individuals in the group, all without a shred of outside or physical evidence.

And in many cases, those who attend group, as observers, begin to recover memories of alien abduction as well. The slightest aberration in life is seen as proof that abduction is taking place. The victim is unable to remember because of repression or post-hypnotic suggestions to forget made by the aliens. In the group environment, these things can be carefully examined and the right spin can be put on them.

Here is the point. We have seen several people who experienced, or have seen, something strange. There is a vague memory of it or an unease felt because of it. Those people talk to friends who suggest that they see a therapist who "specializes" in abduction. The symptoms, though very general, suggest something.

At the group, as the person listens to those others express their feelings and beliefs, and relate their own tales of abduction, the person begins to search his or her own mind. It doesn't take much, but each step is rewarded with approval by the group. Just as religion offers rewards for those who have seen the light, abduction groups offer rewards for those who recover memories of their abduction.

So group sessions, once seen as a stepping-stone back into society, have changed into a society of their own. The object is no longer "to get well" but to become a functioning, productive member of the group. Once the abduction story has been told, the person is accepted into the circle.

Groups have been perverted into something they should never have become. Rather than treat those who need help, they now weaken the strong. They provide a substitute family for the victims, and they accept, on faith, the tales told by the victim. They provide a social context for a life that has moved to the edge of society, but rather than directing the person back toward the mainstream, they force them farther to the edge. Those who try to move toward the middle are seen as the enemy.

But the real point is that no matter how many people tell their stories of alien abduction, or how strongly those in the groups believe, there is still no objective evidence the events have taken place. Groups, rather than dealing with reality are now dealing with fantasy. These groups are not helping researchers learn the truth, and that is a shame. Even worse, they are not helping the individuals who need it, and that is the tragedy.

PART 5

THE EVIDENCE OF ABDUCTION

22.

SCARS, IMPLANTS, AND MISSING FETUSES

One of the major problems with the stories of alien abduction is that there is almost no way to corroborate them. The majority are single witness, and while various investigators and therapists have used hypnosis to retrieve and enhance memories, we are still dealing with individual, uncorroborated testimony, which may be poorly retrieved, suspect, or induced.

However, there are reports of physical evidence associated with some alien abductions. Many abductees allege to have scars on their bodies that are the result of the medical and reproductive procedures conducted on board the alien spacecraft. They explain that they awaken after an abduction event and find a new scar, or a wound that is beginning to scar. Photographs of these wounds and scars are offered by investigators as physical evidence of the report and proof that abductions are based in reality.

There are, however, no records available to substantiate these claims. The subject, or abductee, can provide no photographs of his or her body prior to the abduction to prove that the scar did not exist previously. We are left with the witness assertion that the wound is from the abduction, but we have no way to prove it.

Budd Hopkins has suggested these scars are a form of physical evidence. During an interview with Hopkins, conducted by Jerry Clark and published in the *International UFO Reporter,* Hopkins said, "There are, for example, certain types of physical evidence associated with these reports. One type of physical evidence consists of incisions which apparently are made when the person is abducted as a child; these are either a straight-line cut or scoop mark."

Hopkins went further, suggesting that these marks or scars were consistent in appearance from person to person. "Entirely consistent," he said. "There are basically two types. . . . One, the straight-line cut, does not seem to be jagged or torn. It looks like a scalpel cut. The other is a kind of depression which is like a little scoop mark. These two patterns turn up over and over."

John Mack, in *Abduction,* reported, "One man told me of a gash several inches deep that appeared on his leg following an abduction. Yet this cut virtually disappeared in twenty-four hours." Once again, there is no evidence for this. These blemishes, scars and scoop marks, are well known in the abduction literature. Most researchers have reported cases in which the abductee tells of the acquisition of the marks. But, as always, we have no documentation for it.

It would seem that one way to establish this is to "map" the body of a multiple abductee. A photographic record of the person, showing the clear skin, would be a valuable tool in tracking this aspect of the abduction phenomenon. It would have to be properly recorded and filed for it to be an effective form of proof. This would not require nude photographs. When dealing with the abductees, the wounds described are often on the legs, back, arms, or belly. A photograph of the subject in a swimming suit would provide the detailed map suggested.

We must also be aware that people have many "blemishes" on the skin. Most of us aren't aware of them because they are something we see on a daily basis, or they are on the back of our body, where we can't see them at all. After a reported abduction, as the subject examines his or her body, one or more of these blemishes is noticed for the first time. With this plan in effect, we could then establish if the mark was there prior to the abduction and if it was or was not physical evidence of it.

If such a study was undertaken, and all researchers could do it simply, then one aspect of the phenomenon could be documented. We would have to remember that the discovery of a new wound on the body of the abductee would not prove that case, but it would certainly provide an interesting bit of corroboration.

These marks on the body aren't the only bits of physical evidence discussed by researchers. Many abductees talk of implants. These are tiny, metallic objects, smaller than a BB, that are put into the bodies of the abductees. No one knows the purpose, though some researchers have speculated that they are "tracking" devices. The abductees, according to these researchers, are being tagged in the same way that biologists on Earth tag wild animals for study. These implants, according to the researchers, explain how the aliens can abductee a specific individual when that person is not in his or her normal surroundings.

Keith Besterfield, in the *International UFO Reporter,* wrote, "Given that alien implants are said to have been introduced into us, for what purpose are they placed there? A fairly consistent, small range of options exists." He expands on the idea of a tracking device, writing that others have speculated that the devices might not only be for tracking but for monitoring, communications, or even control.

Mack and Hopkins have produced various implant theories. Mack, for example, wrote, "Jerry [one of his abductees] recalled that she was told that some sort of tiny object was left inside of her at that time 'to monitor me' with no other explanation other than, 'We just have to do what we have to do.' "

Mack also reported that Joe (another of his abductees) told him, "[T]hey're taking a little bit of something out and they're putting something in that will make it easier to follow me."

Hopkins said, during the Clark interview, "We assume they may be locator devices of some sort—or maybe controlling or monitoring devices. Or they may have a function we can't even guess at."

The problem here is that we have no real clue about the purpose of the implants. It is speculation by both researchers and the abductees. No research has been done to validate any of these suggestions.

Dozens of these implants have been noticed by researchers. Many times they are hard nodules just under the skin. Once again, these slight bumps are something that the abductee had not noticed before. Other times they are devices or tiny objects found by medical staff and technicians using a variety of detection techniques.

For example, Hopkins reported in 1988 he had spoken to four people who had undergone magnetic resonance imagery (MRI) and who claimed to have ball-like objects near the optic nerve. Streiber, in his *Communion Newsletter,* reported that four others had undergone MRIs in which something was discovered near the site suggested by the abductee.

Still others, such as Miranda Park, a hypnotherapist, claimed to have viewed anomalies on abductee MRIs. Martin Cannon said that he had personally seen a strange, opalescent implant. And he noted the Swede, Robert Naeslund, whose doctors had found an implant on an X ray had it surgically removed. Naeslund, as noted earlier, believed it was a device for mind control. Though he had not reported an alien abduction, he did report both missing time and a nosebleed.

The problem with this case is that Naeslund seemed to believe that the implant had nothing to do with alien abduction. He believed it was some sort of mind control device placed there by evil forces in various world governments. Although it has been suggested that this case be reviewed further, there is no evidence in the current abduction literature that it has been.

Again, we are forced, in many cases, to accept the word of the abductee. No proof that the nodule had not been there for years can be provided. Like scars and scoop marks, the claim is that it is something that just appeared. But without some form of documentation, there is no reason to accept this as fact.

This does not mean that the individual is lying or misrepresenting the situation. It means that we are being asked to accept testimony with no establishing facts surrounding it. The solution is simple. As with scars and scoop marks, documentation demonstrating the lack of these sorts of bumps and nodules should be established so that when one is discovered, it can be shown to be something new.

Not all of the implants, however, can been seen or felt just under the skin.

In many cases, the implants have apparently required a surgical procedure to place them. Abductees report experiencing a nighttime event and, upon awakening, finding a pillow bloody. Many complain of nocturnal nosebleeds, believing that it is a result of the invasive surgery required for the implants.

With these, however, medical documentation could be recovered. Anything inserted into the body, especially into the nasal cavity as claimed, would be detected by X ray. And once the implant has been proven to exist, then it could be removed for study. Scientific analysis of the implant, along with the medical evidence, would be quite persuasive.

There have been numerous attempts to gather just this kind of evidence. Budd Hopkins, for example, treated Linda Cortile, who claimed alien abduction, who claimed an implant, and whom Hopkins believed to have X rays of the implant, but he failed to provide the final proof. Here was the perfect case for establishing a few facts. One of the most respected of the abduction researchers had a witness who claimed to have an implant. X rays were arranged to confirm the implant's existence, which, according to Hopkins, they did. In fact, Dr. Paul Cooper, a professor of neurosurgery at New York University, confirmed that he had seen the X rays and they did show something near Cortile's nose.

But before surgery to remove the implant could be arranged it had vanished. Cortile told Hopkins that she had awaked with a bloody nose. Under hypnotic prodding, Cortile said that she had been abducted again. The aliens, it seemed, aware of the discovery of their implant and then the planned recovery of their artifact, prevented it.

The X rays, allegedly pinpointing the location of the implant, do exist. But neither Hopkins nor Cortile were prepared, at the time the X rays were made, to recover the implant.

Others have recovered implants, however. Joel, who spoke of abductions from a very early age, found a nodule on his body. Convinced that it was an alien implant, he did have it surgically removed. Analysis, however, was less than impressive. The material was organic and of earthly origin. It seemed to be a calcium deposit that had formed naturally in his body.

At a 1991 conference in Philadelphia, information on three implant cases was presented. According to the information, two of the implants were organic, but the third seemed to have been manufactured. It was a tiny piece of wire with loops on the end and was made of carbon, aluminum, and oxygen.

John Mack has also reported the discovery of implants. "These so-called implants may be felt as small nodules below the skin and in several cases tiny objects have been recovered and analyzed biochemically and electromicrosopically. . . . I have myself studied a 1/2 to 3/4 inch thin, wiry object that was given to me by one of my clients, a twenty-four-year-old woman, after it came out of her nose following an abduction experience. Elemental analysis and electronic microscope photography revealed an interestingly twisted fiber consisting of carbon, silicon, oxygen, no nitrogen, and traces of other elements. . . . A nuclear biologist colleague said the 'specimen' was not a naturally occurring biological subject but could be a manufactured fiber of some sort. It seemed difficult to know how to proceed further."

Mack also reported that Shelia, a lifelong abductee, told of her daughter Beverly's involvement in the phenomenon. According to Mack, "When Beverly was about eight, Shelia took her to the pediatrician for a possible ear infection. The doctor removed an object about the size of a pencil eraser with 'junk all over it' and discarded it. Beverly insisted, weeping, that she had not put the object in her ear, yet she told Shelia that as far back as she can remember she covers this ear with her sheet and blanket to keep it from being exposed. As is characteristic of child abductees, Beverly had frequent unexplained nosebleeds when she was little."

John Schuessler, who has an active interest in implants, wrote in an issue of his newsletter, "A very interesting laboratory analysis of an implant was presented at the Treat II conference. No bizarre claims were made, but very detailed laboratory analysis results were presented."

On the other side of the coin, Schuessler also reported on another implant that was discovered during X-ray examination. Further review, however, showed that the implant was nothing more spectacular than a marker used during X rays.

This is a story that is replayed time and again. Dr. David Pritchard, a physicist, had the opportunity to examine an alleged implant with an interdisciplinary team, that is, a group that included biologists, chemists, and material scientists. The two small portions of the implant were designated as "Price I" and "Price II" because they had been recovered from abductee Richard Price.

Here is what everyone wanted. A group of highly credentialed scientists who were interested in studying an implant that could finally provide the evidence of alien intervention. But the analysis resulted in the revelation that the objects were not of extraterrestrial manufacture but were, in fact, of biological origin. Pritchard reported that the findings suggested that the artifact was most probably the result of successive layers of human tissue formed around some initial abnormality. The small "antenna" that were alleged to bristle from the artifact were nothing more spectacular than ordinary cotton fibers.

Pritchard wrote, "From this perspective, the result of our investigation is clear: whatever probability you initially assigned to the hypothesis that Price's artifact was of alien manufacture must be substantially decreased."

Even with these negative results, additional tests are being conducted on other implants removed from a variety of abductees. As mentioned, at the 1997 Omega Communications conference Dr. Roger Leir claimed to have removed implants from a number of abductees. Leir suggested the proper scientific tests had been performed at independent laboratories but he failed to provide any of the names of those laboratories.

Still others have come forward with implants they claim to have been removed from the bodies of abductees. Often the chain of evidence is such that it can't be proven that the metallic object shown as the implant was actually something removed from the body.

Second, there is a question about the composition of the artifact. While it is scientifically true that elements found on Earth would be found on other planets in the universe, it is also true that the implants analyzed to this point have not been of a composition or structure that is unique. In other words, there is nothing about the implant to suggest it was made on another planet. All the elements, in the combinations reported, are consistent with something manufactured on Earth.

To this point, nothing has been offered that is not explained by our current science. A more rigorous investigation, with peer review, is required, one that encompasses the proper medical documentation, the proper handling of any implant recovered, and the proper analysis of it. After all, most of the abduction phenomenon is predicated solely on the hypnotically recovered testimonies of the abductees. When the opportunity to establish the existence of an implant presents itself, we must use proper methods to investigate, which to this point has not been done.

We must use those same methods to investigate the final area where physical evidence could be recovered. In a number of cases, women have reported they were impregnated on board the alien craft, only to lose the babies before they reached full term. This has become known as the missing embryo or missing fetus syndrome.

Dr. Richard Neal, a medical doctor, researched this aspect of the abduction phenomenon before his untimely death in 1995. In *UFO* magazine he wrote, "Many researchers have claimed that they have several cases of Missing Embryo/Fetus Syndrome (ME/FS) in their files. Yet during my research into this phenomenon over the past three years, all have failed to produce one verified case. Why?"

Neal speculated that the syndrome might be the result of hysteria that is associated with the abduction phenomenon. He also suggested that it might be related to the post-traumatic experience.

Neal wrote that the missing fetus syndrome involved female abductees who alleged a pregnancy between six weeks and twelve weeks. The women claimed that the child, usually the result of some primitive form of genetic or biological experimentation, had disappeared due to alien intervention. In other words, alien abductors had taken the child.

In the article, Neal wrote, "These individuals are taken into what appears to be some type of examination room in which they are placed on a table. Abductees report they are usually administered some form of anesthesia, whether through the touching of an instrument to the head, alien hands applied to the temporal area, then telepathically told to relax and that it would not hurt, or are given some form of liquid for oral intake, ostensibly to alleviate apprehension."

In many of the cases, according to Neal, they also reported some kind of "quasi-gynecological" exam. And, as has often been claimed, these women may report that a long needle was pushed into the belly. Neal reported that much of this sounds like common medical procedures, including "a mini-laparoscopy in which there may be extraction of ova from the ovary or a procedure known as Gamete Intra-Fallopian Transfer (GIFT)."

Neal surveyed others, including Dr. Thomas "Eddie" Bullard, a folklorist, and Dr. Leo Sprinkle, one of the earliest of the abduction researchers. Both reported that they had nothing like the missing fetus syndrome in their files. Bullard did comment, "A preoccupation with reproduction is a common theme from the earliest abductions onward, and a few of the stories describe rape, sexual activity, and events that might be described as artificial inseminations or tampering with a pregnancy already underway, but I noticed nothing in the literature (up to 1986 or thereabouts) comparable to the missing embryo motif."

Sprinkle wrote, ". . . I have no medical information. . . . In a few instances, women wrote to me about strange 'dreams' of sexual activity with aliens, then describing pregnancies that resulted in birth of babies with 'big eyes' and great psychic abilities."

Neal pointed out that no photographs of these children have been presented. Nor any evidence that they have an increased psychic capability. Letters are not proof of the claims.

The difference, however, seems to be Dr. Jean Moody. She wrote in the volume 6, number 4 issue of *UFO,* that "Around nine percent of women with alien contact report medically confirmed pregnancies which disappear, usually in the fourth month. There is no evidence of miscarriage, they just were not pregnant any longer."

Moody said they were medically confirmed, which is in conflict with what Neal reported in the same issue. Neal wrote that each time he heard of a medically confirmed report, his insistence on the documentation failed to produce results. In Neal's experience as a medical doctor, he had found no persuasive evidence of the missing fetus syndrome.

Moody, however, did present one case. Claire, according to Moody, came to her "with the bits and pieces of such a story. She was unaware that she was pregnant, or perhaps she wasn't yet, when an alien being came into her bedroom. Several sessions of hypnotic regression were used to clarify her memories."

Moody related that she has two methods of obtaining data under hypnosis. One is to ask questions of the subject and the other is to say nothing once the "trigger" event has been mentioned. Interestingly, Moody referred to the first method as "the leading question method."

While under the influence of hypnosis, Claire related the trauma of an alien invasion of her room. In the mid-1970s, she was in her bedroom with her young son when a light came toward her and seemed to force her back. She described, with the proper emotion, an eye that she couldn't get away from. She could see a silver needle and then she could feel hands inside her.

Under hypnosis, she recalled that she had been told she would have a boy child and that he would have blue eyes and be a mighty warrior. She was told that she would be given the name for the boy and that she would be given instructions on how to raise him.

Claire also described being on a ship where she could see lots of lights and panels. She said that she was sick to her stomach and nude as she was prodded and probed. At one point she said, "Such pain in my uterus . . . I don't like them being inside of me."

Moody gathered more data from Claire and even talked to the son who had been in the room during the abduction. He was not present during the regression but did claim to remember the night that Claire had described in such horrifying detail.

Other evidence for the case was a scar on Claire's arm. She said that it happened on the night of the visitation and that the tissue was biopsied. No results of that were provided.

Moody also reported that Claire said that she had had two miscarriages and one missing pregnancy of a four-month-old fetus. According to Moody, this happened after she had a tubal ligation. Her doctors were incredulous when reminded of her operation. No medical evidence of the pregnancies or miscarriages was presented.

Moody writes in the conclusion to her article in UFO, "By selectively ignoring many pieces of the puzzle, the protesting reader can conclude there was no craft, no aliens, just a strange rapist in Claire's bedroom while her husband was away. By ignoring her vivid emotional scars, one could conclude it was a normal pregnancy, and the son is just 'different.' . . . But if you put all the pieces into the picture, we have one more case of alien forces reaching down into the most intimate aspects of human life."

The problem is that no evidence, beyond the emotional "scars" of Claire, has been presented. Hypnotic regression was used to retrieve the memories and no corroboration, beyond a son remembering lights, has been offered. The precise medical documentation that would help "the protesting reader" understand and believe has been neglected.

There might be a partial explanation for why there is no medical evidence to confirm these stories of disappearing pregnancies. David Jacobs, in *The Threat,* described what he termed "Extrauterine Gestation Units," which are some form of implanted device for carrying an alien hybrid child. He reported on a number of women who have experienced these devices, suggested that one had been seen on an ultrasound, but that it was not recovered. The aliens, through some unknown ability, are able to monitor the women, recovering the device before it can be removed for analysis.

Jacobs made it clear that these extrauterine gestation devices are not connected to the hosts' bodies. In a normal, human pregnancy, the mother is connected to the fetus through the placenta. The mother shares nutrients with the growing baby. But according to Jacobs, these extrauterine gestation devices are a separate system that is implanted in the women.

If they are not connected to the women, if there is no circulation of blood between the host and the device, then what purpose does it serve? Jacobs suggests that the fetus, when removed from the device, is then brought to term in glass tubes filled with liquid. The hybrid children are "decanted" much in the same way as the children in the novel *Brave New World.* At least this is the way some of his abductees recall it under hypnosis.

So why isn't there more physical evidence? In all the investigations that have been conducted, in all the reports that have been made, there has never been the documentation to prove that something had, in fact, happened. Neal wrote about the difficulties in investigation and research on missing fetus syndrome. During his research, he noted four considerations that inhibited investigation.

First, many of the abductees feared consulting with a private doctor. The women were afraid that if they requested the release of their records, and their doctor learned the reason, the doctor might believe the patient to be unstable.

Second, many of the women were unwilling to release such private information to abduction researchers. Some were afraid of humiliation, and rightly so. Too often the UFO community has demonstrated that is not to be trusted with sensitive or private information. This fear often results in a failure to provide documentation needed to corroborate the testimony.

Third, according to Neal, many of the abductees only have a vague memory of the abduction and have real concerns that the pregnancy might be normal in origin. In other words, it was the husband, boyfriend, or mate who was involved in the contraception and not alien visitors.

And finally, there is fabrication. Although Neal wrote that the majority of the alleged female abductees would not lie about the event, there may be some who were seeking notoriety for self-gratification. The production of the proper documentation, in these cases, would be extremely difficult, since it is all a hoax.

Neal provided a list of medical reasons why a woman who is not pregnant might believe she is. Many of these conditions produce symptoms that mimic pregnancy. In some cases, tests suggested the woman was pregnant when she was not. Neal advocated the proper medical documentation in these cases. Some would be explained by medical science as not involving alien abduction and intervention, but some might provide the very corroboration that the abduction researchers require.

It should also be noted that many of the women who have reported these false pregnancies, or who Jacobs believed had been implanted with his extrauterine gestation devices, were postmenopausal women, women who had had hysterectomies, or who were, for whatever reason, unable to bear children. The psychological literature is full of reports on why women who cannot conceive believe that they have, through some miracle, become pregnant. Such a belief fulfills a real psychological need in these women.

In the second part of an article appearing in the January 1992 issue of *UFO*, Neal asked, "Are [missing fetus] cases on file?" He answered his own question, writing, "Why do some of our top researchers say they have numerous cases in their files related to the ME/FS motif, yet fail to produce anything that can be investigated or documented when questioned thoroughly?"

He also noted, "As far as researchers go, we have no data for medical/scientific review, due to failure of compliance of female abductees as well as the investigators' failure to submit cases to those of us interested in this bizarre aspect of the abductee phenomenon."

To document ME/FS, Neal recommended proper medical records. The information he thought important included an established date of missed menses, a positive pregnancy test in the doctor's office, confirmation of the pregnancy by a doctor, the alleged date of the abduction, alleged date of the missing pregnancy, documentation by a doctor of the missing pregnancy or complication of the pregnancy, lab, X ray, ultrasound scan, surgical procedures or pathology reports, and finally, hypnotic regression tapes of the alleged sexual encounter or quasi-gynecological examination. In other words, he required the type of evidence that would be accepted by medical authorities.

It should be noted, however, that Jacobs's extrauterine gestation device would defeat most of the tests recommended by Neal. Because, according to Jacobs, the device has nothing to do with the host's body, is not connected to it, and is merely "warehoused" in the woman, many of the medical tests would be negative. The X rays, ultrasounds, and recovery through surgical procedures would, of course, provide the proper scientific evidence. To date that simply hasn't happened for a variety of less than persuasive reasons.

Neal pointed out that without this documentation and until the medical records can be produced for proper peer review, "the ME/FS will remain as obscure and vague in the future as it is today. Merely hear-say cases of alleged female abductees having missed pregnancies will forever remain buried in the annals of ufology."

23.

THE VALUE OF HYPNOSIS: WHAT YOU SEEK IS WHAT YOU GET

Of hypnosis and its importance, Budd Hopkins said, "Hypnosis is the only thing, that we know of, that really is effective here. It has been recommended by, and endorsed by, the American Medical Association, AMA, for use in cases of—and psychiatric groups will say this is the only real tool you have that is pretty efficient to unlocking—a period of genuine missing time."

Of hypnosis and its importance, John Mack said, "Hypnosis, in this situation, I find very believable."

Mack also said, ". . . but hypnosis becomes increasingly accurate, and there is research about this, if the matter for which you are being hypnotically regressed is of central importance to you personally, in other words, you've undergone, that was powerful, that affected you deeply. The more that is the case, the more accurate the hypnosis is."

For many, those ideas underscore their beliefs. Because the information was retrieved using hypnotic regression, that confirms, in the minds of many, the validity of the memories. They believe that people don't lie under hypnosis, and they believe that memories retrieved through hypnosis are somehow more accurate than those in the conscious mind. They believe that memory is like a video camera, recording everything we see, hear, or feel. Then, under the proper conditions, those memories can be recovered in perfect harmony with the events lived.

Because of those beliefs, many people have accepted the idea of alien abduction as real. Although nearly a quarter of the people claiming alien abduction remember some of the events without any hypnotic regression, when those people are hypnotized, it adds a dimension to their stories. Of the other 75 percent, some remember parts of the abduction, while others remember nothing until hypnosis is used to recover the repressed memories.

To some researchers, it is not the hypnosis that is important, but the emotion displayed under hypnosis as the memories are "relived." Time and again, researchers have said, "If he [or she] was faking it, then he deserved an Academy Award." According to the thinking of these researchers, people cannot fake the emotions that they exhibit if the event they recall wasn't real.

But the key here is hypnosis. If what the majority of what we believe about hypnosis is true, then we are seeing a very real phenomenon. Of course, the key word is "if." If what we believe to be true about hypnosis is inaccurate, then what we are seeing is somehow skewed. In other words, our conclusions about the event are wrong because our opening theory about hypnosis is wrong.

The earliest traces of what we now call hypnosis appear in ancient writings of Egyptian healers. The *Papyrus Ebers,* a document written circa 1550 B.C., mandated that chants and incantations be uttered before medical treatment was administered. Even then healers recognized that a mind properly prepared for treatment was more receptive to influence.

Healing in Egypt was done in temples prepared specifically for that reason. The early Egyptian sleep temples of the healing god Isis were located on the banks of the Nile. It was here that the ill and infirm came from far and wide to seek the healing sleep. The suffering hordes were told by the priests that during their sleep they would be visited by Isis, who would rid them of their maladies. A system of incantations, rituals, and laying on of hands prepared the patients for the expected cure, and guided them into the hypnotic state of receptiveness.

In Memphis, the ancient capital of Egypt, the temples were dedicated to the healing god Serapis. Serapis was actually an amalgam of the Egyptian god Isis and the local healing god Apis. The cult of Isis, Apis, or Serapis soon became so popular that it spread throughout the land, and became the prototype of the Greek healing temple.

At that time, healing in Greece was done by the method of incubation, a word that means literally "to lie down." Incubation often took place in sacred caverns. One such cavern was the oracle of Trophonius. Those who entered that cavern had frightening visions and visitations that filled them with terror. When they emerged from the cavern priests would place them in the "Chair of Memory," which enabled them to recount what they had seen. This experience bears a striking resemblance to today's hypnotic memory recovery techniques.

The cult of the temple in Egypt had become an established part of civilization. As word spread during the fourth century in Greece, the healing began to be done by a cult dedicated to the god Asklepios. Everyone soon knew about the power of the temple sleep, and the afflicted came to the healing temples in droves. Once they had made offerings of valuables and bathed in the cleansing waters near the temple, they would spend several nights at the temple entrance, praying, chanting, and listening to tales of the healing powers of Asklepios told by the priests. Properly prepared, and rife with expectation, they would be allowed to enter the magical Abaton, the sacred room where the healing sleep occurred. As was done in the Egyptian temples, the initiates were told that their cure would come in a dream. Tablets from this era tell of the visions, dreams, and miraculous cures that patients experienced in the temple. Priests would make suggestions to the patients both before and during the temple sleep. Frequently the priests would perform medical procedures, sometimes even minor surgery, while the patient was lying on the temple table in a hypnotic sleep. Afterward patients would report that Asklepios had visited them during the night and performed medical treatments on them.

Tales of the healing power of the temple traveled quickly, and soon the Romans adopted Asklepois as the latinized Aesculapius. The first Roman healing temple was built on Tibur Island, and for centuries was the center of the Roman healing cult. These healing temples lasted until the fall of the Roman empire.

Stories of mystical healing powers come to us from other ancient civilizations. In the fifth century B.C. the yogis in India described healing by making magnetic passes over the bodies of the afflicted. They also told of healing by sending thoughts through space, directing healing forces to patients miles away.

Although the healers of old would not have recognized it, the power of suggestion had become their most effective tool. It is a fascinating and relevant fact that the ingredients of abduction memories existed centuries ago: the cultural belief system, the wisdom of the healer, people in pain and distress, the induction of altered states of consciousness, the visions of the dream state, and memories of visits from the gods.

The knowledge disappeared for a period of time. The cultural elements that spawned it were lost as those ancient civilizations collapsed and disappeared. But knowledge has a way of being rediscovered, so hypnosis, or the uses of these altered states, has been learned again.

This artificial sleep, this trancelike state that provided the ancients with a healing ability, was rediscovered in 1774 by Franz Anton Mesmer when he met a Viennese Jesuit named Maximillion Hell. Hell showed him how people could be healed by exposing them to magnetized plates of steel. From that point Mesmer began to use magnets.

Soon after this Mesmer married a wealthy widow and set up a medical practice in Vienna. He gained a reputation as a great healer. He ran into trouble, however, when he began to treat a blind pianist. Although Mesmer claimed to have cured her blindness, she soon relapsed. Mesmer became the subject of ridicule and was forced to leave Vienna in disgrace.

He fled to Paris, where he established another medical practice. Here he began to use even more controversial and theatrical techniques. Mesmer was a charismatic and flamboyant man. He arranged his healing sessions as a skilled director would orchestrate a play. He devised a remarkable contraption called a baquet, which was a large tub filled with water, iron filings, and powdered glass. Patients would circle around the baquet, grasping iron rods that pierced its interior.

As music played, Mesmer would make a grand entrance. Striding through the room in flowing silk robes, he would make magnetic passes over each patient. Sometimes he would touch the patients with an iron wand.

The touch was electrifying. The patients would convulse in ecstasy. Soon the whole room would writhe in hypnotic rhythm. Some would swoon and faint. Miraculous cures occurred.

Mesmer knew the awesome power of prepared expectation.

For five years Mesmer's healing baquet was the rage of Paris. But his success aroused the ire of skeptics. In 1784 the French government established a commission to examine the powers of animal magnetism. The commission was headed by Benjamin Franklin. Franklin's report to the king stated that animal magnetism was nothing more than the power of suggestion.

Furthermore, he saw Mesmer's techniques as a menace to society because ladies under the influence of magnetism were easily seduced. Members of the commission reported that some magnetizers induced the curative state by placing their hands on the women's abdomen, and sometimes by massaging their breasts. This technique seemed to cause sexual excitement among its participants. Magnetic sessions were described as lurid sexual experiences. Mesmer left France in disgrace.

Mesmer had become one of the best known healers of his time. His methods, beliefs, and techniques were not new or unique. It's important to realize that his success stemmed not from his discoveries or abilities but from his presentation. He was able to take the discoveries and methods of others and package them in a manner that was attractive to the time.

Science has always played second fiddle to showmanship. The right man at the right time can change the tides of belief. Few people today know the story of Mesmer, but everyone is familiar with his name, which has become a part of our language. Writer Arnold Ludwig said of Mesmer, "Had his followers not caught the fire of his enthusiasm and spread his views widely, the history of hypnosis might have been different and Mesmer might have been lost in obscurity, simply relegated to the position of charlatan or quack."

Although history shows that hypnotic trance had been used since the dawn of civilization, up until the time of Mesmer no one really knew that trance was part of the method. It was the Marquis de Puysegur in 1784 who first realized that one could deliberately induce trance to effect a cure. He called this trance induction "artificial somnambulism." Puysegur found that many of his subjects when in trance could diagnose their own diseases and prescribe their own treatments. From 1784 to 1800 artificial somnambulism was considered to be the best method to access the unconscious mind.

Pierre Janet stated that people who were subject to spontaneous somnambulism, that is those who frequently entered trance states automatically, were the best subjects for artificial somnambulism. Furthermore, he claimed that people who were spontaneous somnambulists had no conscious memory of their hypnotic episodes, but the memories could be recovered under artificial somnambulism. Here is the first clear statement that hypnosis was the tool of choice for recovering memory.

The term hypnosis was coined by physician James Braid in 1843. The word originates from the Greek root "hypnos," meaning sleep. Although we now know that hypnosis in no way resembles sleep, the term has stuck.

The true scientific study of hypnosis began in the 1920s in America with the work of psychologist Clark Hull. Hull and a dedicated group of university students worked in a laboratory at the University of Wisconsin. Later he relocated to Yale to continue his investigations. In 1933 Hull published the results of his research in his book *Hypnosis and Suggestibility*. Hull was forced to abandon his study of hypnosis by the Yale School of Medicine, whose authorities felt that hypnosis was damaging and dangerous.

That view doesn't seem to have changed with additional study. The American Medical Association Report of the Council on Scientific Affairs, in conflict with what Hopkins claimed, released a statement in 1985 and again in 1994, saying, "The Council concludes that hypnosis induced recollections actually appeared to be less reliable than non-hypnotic recall."

In the UK in October 1997, the Royal College of Psychiatrists announced a ban on using recovered memories in cases of child abuse. They noted that the practice of using hypnosis for recovering memories can give rise to false memories, which are often strongly held beliefs. They also noted that there is no evidence that recovered memory techniques can reveal memory of real events or accurately elaborate factual information about past experiences.

This also seems to refute Mack's claim that the research on hypnosis has suggested that these memories, if of central importance to the subject, are accurate and the more important those beliefs the more accurate the hypnosis.

But that hasn't stopped abduction researchers. In her book *Encounters,* Dr. Edith Fiore writes, "By now you realize that your subconscious mind has a perfect memory. *Everything* you have ever experienced is recorded in your subconscious memory exactly as you perceived it. The inner mind can been seen as your own personal computer. With a computer you can access files. With your subconscious mind you can retrieve memories."

Fiore isn't alone in this belief. Many mental health care professionals make the statement that the unconscious mind remembers everything. Many laymen accept the statement without question. After all, it is what the recognized authorities have been telling people for decades. The problem is there is no evidence for the assumption. Of course, if what the unconscious remembers is truly unconscious, then we are not aware of the information anyway.

Semantics aside, there are other basic questions to be asked. If we could find a way to bring the entire contents of the unconscious into awareness, how would we know that every experience is recorded there? We would have no way of knowing if every experience has been recorded unless there were a complete and total record of everything ever done. There would have to be some outside record to corroborate everything and no such record exists. The notion is absurd and is completely untestable.

In fact, the whole idea that memory is like a tape recorder or video camera is, in and of itself, erroneous. Hundreds of studies have shown that this idea is not true. Memory is not recorded but seems, according to the research, to be stored in a highly complex manner consisting of impressions, ideas, and feelings filtered through our own belief system. Each time someone reaches for a memory, it is not "played back" but reconstructed. For this reason, no memory is ever repeated exactly the same way twice. Furthermore, according to research, as time goes by, memories are modified to fit the beliefs of the society and world around us. For this reason, stories change from one repetition to another. It is also why couples disagree on who said what during an argument.

In 1933, a neurosurgeon, Dr. Wilder Penfield, thought he had found evidence that memory is a tape recorder. During a brain operation in which the patient remained fully conscious, Penfield stimulated a portion of the patient's exposed brain. The patient said that he was reliving a memory. Penfield wrote, "The astonishing aspect of the phenomena is that suddenly he is aware of all that was in his mind during an entire strip of time. It is the stream of a form of consciousness flowing again."

While this sounds like clear evidence that memories are meticulously recorded and stored in the brain, there is a major problem. Penfield was wrong. Out of 520 patients that he studied, only 40 of them received memories. When these memories were studied closely, it became apparent that they were not memories at all, but more closely related to dreams. Sometimes the patients told of events in which they had not participated.

After reviewing Penfield's work, Dr. Elizabeth Loftus wrote, "There is good reason to believe that these reports may result from a reconstruction of past experience or from constructions created at the time of the report and bear little or no resemblance to past experience. . . . *These so called memories then appear to consist merely of the thoughts and ideas that happened to exist just prior to and during the time of stimulation* [emphasis added]."

What this means, simply, is that the thoughts the patient was having at the moment of stimulation were interpreted as memories. It doesn't mean they were, but that it was interpreted that way.

The idea of the tape recorder memory has invaded other aspects of society. In 1967, Harry Arons began to teach hypnosis to police officers in New Jersey. Arons said, "Scientific research has demonstrated that the mind—or the brain—

seems to have the capacity for retaining all impressions that enter it, like a giant tape recorder. Martin Reiser, of the LAPD, said much the same thing in the late 1970s.

In 1976, Rieser wrote, "Because the perceptual apparatus works in a cybernetic fashion, much like a giant computerized videotape recorder, the plethora of information perceived by the sensory system is recorded and stored in the brain at a subconscious level. Much of this data momentarily nonrelevant, or repressed because of emotional trauma, is difficult to recall; the problem—one of amnesia. However, hypnosis may provide the key in a significant number of cases by encouraging hypermnesia, relaxing the censorship and permitting suppressed or repressed material to return to the conscious mind."

What Reiser was trying to say was the same thing that Penfield said. Everything is remembered. The problem is to find a way to bring it from the unconscious to the conscious. To Reiser, the answer was simple. Hypnosis.

Most people who undergo hypnotic regression believe that the unconscious has recorded everything and that hypnosis can bring those memories to the surface. This becomes a self-fulfilling prophesy. They know they are supposed to remember something, and so they do.

This then is the classic definition of hypnotic regression. The subjects, or victims, under hypnosis, are regressed to a point in the past. They believe they will recover those memories and they will relive them in vivid detail. This idea is, of course, based on the erroneous theory that the memory holds everything.

The first recorded case of hypnotic age regression was in 1887 by a Spaniard identified as Colavida. In 1911, Colonel Albert de Rochas published an account of his work in regressing people into past and future lives. De Rochas's future lives material was reviewed carefully and found to have no basis in reality. But these future lives were recounted with the same emotion and conviction as past lives were. Remember, those advocating the reliability of hypnosis often mention the emotional conviction of those under the influence of hypnosis.

Sigmund Freud and Josef Brueur used hypnotic regression in their work that culminated in *Studies in Hysteria*. In the introduction they wrote, "In the great majority of cases it is not possible to establish the point of origin [of the symptoms] by simple interrogation of the patient, however thoroughly it may be carried out. This is in part because what is in question is often some experience that the patient dislikes discussing; but principally because he is genuinely unable to recollect it and often has no suspicion of the causal connection between the precipitating event and the pathological phenomenon. As a rule it is necessary to hypnotize the patient and to arouse his memories under hypnosis of the time at which the symptom made its first appearance. . . ."

This idea, however, seems to be in error. Research done on hypnotic regression suggests that people do not regress to an earlier age or time but act in the way they believe they should act. The idea that someone under hypnosis can remember being molested at six months old, for example, is of course absurd. As mentioned earlier, humans generally have very little memory of anything before age three, and they have only a smattering of memory of events before age five.

There is a second point to be made here. Remembering something like molestation at such a young age is logically impossible, even using proper hypnotic regression. Children simply don't have the language skills to label events and things in the environment. Children that young don't know what, or even who a father is. Children have no idea what it means to be molested, which is a cultural concept that will develop years later. A child of such a young age has not yet labeled body parts and no method of identifying them. No theory of memory, regardless of the technique used to recover them, can account for a narrative memory being recorded at such young ages.

Add to that the confusion over the purpose of hypnosis, and trouble develops. For generations, in fact, for centuries, hypnosis was used as a cure. The goal was to relieve symptoms, not to enhance recall or to recover memories. Originally, memories recovered in the process of hypnosis were seen as incidental to the cure. In therapeutic hypnosis, the goal is to suggest ideas, images, or memories as a tool for symptom relief. In fact, relief of symptoms in soldiers with post-traumatic stress disorder is done by suggesting false memories to soothe and comfort.

Long ago the psychologist Janet noted that hysterics, while in a trance, would remember traumatic events unavailable to them in a waking state. He concluded that the relief of symptoms was not contingent on remembering events, but rewriting them in a less traumatic form. It was the ability of hysterics to invent confabulated traumatic events and see them as real memories that both caused and later cured their suffering. This creation of personalities by the hysteric was an unwanted by-product of treatment.

It is also possible to induce memories. Experiments by Dr. Steven Ceci and Dr. Elizabeth Loftus have clearly demonstrated that it is relatively easy to make people remember things that never actually happened. Both Ceci and Loftus have been able, with little effort, to induce false memories in children.

Recent research at Ohio State University demonstrated this effect more fully. Psychologists told a group of volunteers that hypnosis could lead to false memories and that it fails to improve memory of actual events. Even with the warning, 28 percent of them believed that the false memories implanted by the research were real.

More frightening was the fact that 44 percent of those in another group believed their implanted memories to be real. They hadn't been warned before the experiment about the possibility of the implantation.

And Dr. Richard Ofshe, brought into the incest and satanic ritual abuse case of Paul Ingram, showed that it could be done with adults outside the laboratory environment. Ofshe, by the simple trick of suggesting that someone had reported a specific act of abuse, one to which Ingram had no conscious memory, induced that memory. By telling Ingram it had happened, and by suggesting only that he think about it, Ingram became convinced it had happened, and reported as much, in great detail.

Freud, in 1897, recognized that recovering unconscious memories would not necessarily lead to a cure. His patients recovered horrible memories, but over time, those memories converged. They all constructed the memories that he had predicted they would. In other words, the memories that came to the surface eventually fit his theory. Freud, it seems, was the force behind the memories and those memories did not necessarily reflect events based in reality.

Coyne Campbell, a psychiatrist trained as a psychoanalyst, rejected the psychoanalytic technique because he believed that its methods were a form of hypnosis that induced *false memories* and delusions in his patients. In his 1957 book, *Induced Delusions,* he wrote, "Hypnosis, like psychoanalysis, only accentuates symptoms already present in the individual. Hypnosis has been propagandized and has been used to fascinate people much as psychoanalysis has been. It is a residue of magical thinking and, generally speaking, the hypnotist and the psychoanalyst belong in the same group—the emotionally unstable in distress."

This idea has been expressed by other researchers as well. Eugene Bliss, a multiple personality disorder therapist, wrote, "Hypnotic experience can be as real as real events in the world, whether they are induced by a hypnotist or they are a result of spontaneous hypnosis. In fact, many of these experiences are indistinguishable from reality."

From the work that has been done, and from the research about that work, it is clear that therapeutic hypnosis has never been meant to facilitate memory. People who do "abduction therapy" should note this. These researchers, however, claim to use hypnosis in another sense. They call it "forensic hypnosis."

Unlike therapeutic hypnosis, forensic hypnosis has never been meant to cure but to enhance recall. The goal in forensic examination is not to implant ideas but to extract them. However, hundreds of studies suggest that hypnosis is a poor tool for this, and that recall can be enhanced by other methods. In fact, the American Psychiatric Association and the Society for Clinical and Experimental Hypnosis issued a formal statement that hypnosis should not be used to enhance recall. Despite this mountain of evidence, abduction researchers, including Budd Hopkins, John Mack, David Jacobs, James Harder, and John Carpenter, among many others, continue to use the technique of regressive hypnosis to extract memories of alleged alien abduction.

Clearly, regressive hypnosis is a poor tool for recovering memory, and yet, abduction researchers swear by it. In fact, they continue to endorse its use, citing the American Medical Association as having claimed hypnosis as the only tool available. In reality, the AMA has said, "Recovered memories are unreliable and should not be considered true." The Council on Scientific Affairs continued, "The council concludes that hypnosis induced recollections actually appeared to be less reliable than non-hypnotic recall."

There is an additional, and important problem with the use of hypnosis, and that is the desire to please. The prominent abduction researchers claim that the people who come to them have no knowledge of abductions. Budd Hopkins is fond of saying that one of his abductees had no knowledge of an abduction event, just an irrational fear of one stretch of highway.

The truth of the matter is that both the researcher and the client are expecting to learn something that has to do with aliens and abduction. If they have no knowledge, then why seek out an abduction therapist? And if Hopkins's client had no knowledge of why he feared the stretch of highway, why was he using Hopkins for the hypnosis? Wouldn't it be more sensible to seek out and use the services of a psychologist or a psychiatrist and not an artist who does abduction research? The truth is, both researcher and client are expecting to find an abduction memory.

The power of hypnosis in these circumstances is the subject's desire to please the hypnotist. Clinically, it is labeled as "The Desire to Please the Operator." It is out of this desire that the power of suggestion grows. It is a short step from that to leading the client or patient into the realm of alien abduction.

Most of those using "forensic" hypnosis claim they take great pains to avoid the problem of leading the witness. The truth is that it is impossible not to lead someone under the influence of hypnosis. A question as innocent as, "What happened next?" presupposes that something else happened, but more important, primes the subject to continue the narrative.

Most abduction researchers are unaware of how simple it is to lead their clients. When the client suggests, after a long session, that nothing else happened, often a new technique is employed. The hypnotist suggests that the client is on a boom with a camera to film the event, or they are viewing the event through a video camera. The client is instructed to use that camera to see what is happening as if he or she were nothing more than an observer, rather than a participant.

But the problem is that the hypnotist is telling the client that he or she is not satisfied with the results. There must be more there and the hypnotist keeps pressing until more is found. Leading the witness, then, is much more subtle than asking a question that suggests false data. It is pressing for more information even after the witness has said that nothing more is available.

When someone is told that he or she might have experienced an abduction, all objectivity flies out the window. But the therapist, an abduction researcher, and the subject are primed to recover the memory of the event.

If this is all true, why would anyone want to undergo hypnotic regression? It is a poor tool for finding the truth, it allows the subject to confabulate amazing memories and act on those memories as if they were true, and its validity is now being questioned. In fact, in many states, a witness who has been hypnotized in an attempt to learn more of an event can no longer be called as a witness. Courts, and science, recognize how easy memories and events can be reconstructed or confabulated by a clever hypnotist. Even those whose motives are a search for the truth can, and do, lead the subject into memories that are not part of reality.

Even though science has never been able to define the hypnotic experience, every year it is sought by thousands of suffering people such as those who wish to quit smoking or end other bad habits. And even though science has failed in understanding it completely, the layperson knows of it and seeks its power.

One researcher, Bernheim, said, years ago, that hypnosis does not really exist. What does exist is someone's ability to act in certain ways within specific contexts. People enter these contextually driven situations, that is, religious rituals, laboratory experiments, or healing ceremonies, for very specific reasons. The religious person goes into a trance to experience ecstasy while a suffering person does it to find a cure.

Most people have not and will not undergo hypnotic regression in their lives. It is a ritual with a long history and has been defined by some authorities as a way to unlock memories from a hidden part of the self. Those seeking hypnotic regression are often confused. They had thoughts and feelings they cannot explain. They come to a hypnotherapist because they do not feel right. They seek a cure. They believe that hypnosis can open the door into that hidden part of the self, they can find a cure, and it will make them feel whole again.

The abductee is like all those others. He or she is seeking help because of distress and enters into a highly ritualistic structure in which the expectations are clear. When the decision is reached to use hypnosis, both the hypnotist and the subject are acutely aware of the expectations and the behaviors dictated by the situation. The subject is there to recover memories of an abduction. But more important, and often hidden from the therapist, the subject is there to take on a new identity. He or she wants to become known as an abductee. Contrary to what Mack claims, this is a club that people want to join as evidenced by these circumstances.

This motivation, on the part of the subject, should not be overlooked. The subject is there to join a specific group, that is, the abductee, and the experience of hypnotic regression is an initiation rite into this new phase of life. It is a new identity that will redefine every aspect of life. It redefines belief structure, choice of friends, and mental contents. It provides a sense of self that has been absent in the past. Once the gaps are filled, through hypnosis, which is used as a tool to prove the validity of the experience, and the new identity is in place as an abductee, it is extremely difficult to shake. It is important to remember that no one relinquishes a sense of self.

The importance of hypnosis, and the credentials of the therapist, can't be ignored. It is the authority of the therapist as a psychologist or social worker, or as a recognized expert in the field of alien abduction, that underscores the reality of the session. These people would not be using hypnosis if it wasn't a pathway to the truth. They would not be conducting the research if there wasn't something to it. Abduction is a reality because the authorities say it is a reality. And authorities say it is a reality because they have "interviewed" so many victims of it. Hypnosis is the glue that holds the whole structure together.

The patients of Richard Boylan prove this in resounding fashion. One of them came to him because of her belief that she had been abused by her parents, and the suspicion of ritualistic abuse. After therapy with Boylan, however, she became an abductee. He had provided her with books and articles on the topic, discussed his beliefs in the extraterrestrial with her, and finally turned vague and meaningless dreams into an abduction experience. She hadn't been a victim of satanic ritual abuse but was the victim of alien abduction. The satanic ritual abuse was a "screened" memory.

Later, during an investigation by the California, Medical Board Board of Psychology, and Board of Behavioral Science Examiners, the woman forced into the abduction memories by Boylan recanted her testimony. The woman went from being a disturbed woman to a victim of incest and satanic ritual abuse to an abductee, all under the influence of group pressure, therapist theory, and hypnotic regression.

And Boylan isn't the only abduction researcher to engage in this sort of activity. Edith Fiore, who wrote *Encounters,* told a client that her emtional problems originally came from "entities" that had attached themselves to her. Fiore would remove these entities, but when that failed to change the client's life, Fiore suggested past life regression. When the client, under hypnosis, failed to see anything in a past life, Fiore told her, "Make it up."

After a few sessions, Fiore told the client that the "entities" had been removed and that she now thought they should explore the possibility that the client had been abducted by alien creatures. Not only that, Fiore told the client that she believed the client had been abducted, though the client had never mentioned it.

Those who have claimed that hypnosis adds to the credibility of the abduction accounts are misrepresenting the situation. Those who believe it to be the only tool available are suggesting that there is no other corroboration for it. And without that independent corroboration, we must look at it again, with a more skeptical and more objective eye.

PART 6

CONCLUSIONS

24.

ALIEN ABDUCTIONS: THE LOGICAL PARADOX

If we are to be totally honest in our research, we must look at the abduction phenomenon in the light of the evidence presented and the logic of the situation. The alien abductors have a science that allows them to cross light-years of space, a feat that our science is unable to duplicate. They fly craft, in the atmosphere, that are aerodynamically impossible. They outperform our best fighters, and when tired of the game, disappear in bursts of speed that make it seem they simply vanish. The technology they display is far superior to ours, maybe hundreds or thousand of years ahead of us. We dream of duplicating it, but we are unable to do so except in the movies and in science fiction.

Yet when we begin to discuss the abductions and the apparent research being conducted by the alien scientists, it seems to be decades behind us. Their genetic experimentation is crude, their reproductive research is unbelievably coarse, and they are apparently unable to learn anything because they conduct the same experiments over and over. Our scientists, using skill and imagination, gather their data and then build on it, designing new experiments to learn more. Their scientists can't seem to get it right the first time and must conduct the same experiments again and again rather than building on the knowledge they have gained.

What this demonstrates is the paradox of abduction research. Facts that make no sense, no matter how twisted they are. Alien scientists who have solved the problems of interstellar travel and who have been allegedly abducting us since the beginning of time and manipulating our DNA to guide our evolution are unable to cope with simple genetic research.

Or, we can twist it around so that the logical paradox works from the Earth outward. According to the Roper Poll, more than three million Americans have been abducted. This figure is accepted by abduction researchers and UFO investigators as fact. Some have taken the number and expanded it to as many as five or six million Americans. That is, one in fifty American citizens has been abducted by aliens creatures.

These figures are only good for our population of 250 million out of a world population of about five billion. If we follow the logic that one in fifty Americans has been abducted, it would seem reasonable to believe it is one in fifty throughout the world who have been abducted. This works out to an incredible one hundred million abductions. Logistically, it is impossible. Especially when it is remembered that many of the abductees claim to have been taken more than once.

It might be argued that figures for the United States do not translate into figures for the entire world. The question then is, why not? What makes Americans so different that they have been singled out for special treatment by the aliens? Would an alien race, performing the abductions, limit their research to just Americans?

We know that isn't the case. Abduction reports have come from other parts of the world. Hundreds, for example, have been reported in South America. In fact, John Mack has reported on a Zulu shaman from southern Africa who has apparently claimed to have been abducted. Researchers are always quick to point out that it is a worldwide phenomenon.

But how many alien beings, how many ships, and how much equipment is necessary to abduct so many people? A group of American researchers, looking only at the numbers for American abductions, postulated a fleet of seventy-five ships, each with a crew of sixteen. They didn't explain where the ships were hidden, why they hadn't been reported on a regular basis by thousands of citizens, or how the operation could be supported? It defies logic, but the researchers are still crunching the numbers to prove that alien abduction is not only real but common.

We could add here that the science of the alien beings appears to be magic to us because it is so technologically advanced beyond us. Here is a race of intelligent creatures who have defeated the problems of interstellar flight, who can pass through solid objects, "switch off" the spouse of their victims, and perform other feats that defy our science to explain. They then erect mental blocks in the minds of the victims in an attempt to prevent us from learning of their activities, but do such a poor job of it that anyone who has read a book on hypnosis or attended a weekend class to become a hypnotherapist is able to break through those blocks easily to gather the data. Their science in this respect is far behind our own.

It points to another real problem and that is the blind faith of the researchers into the abduction phenomenon. Each of those who engage in abduction research believes that his or her patients, clients, or subjects are telling the truth as best they can. More than one of these researchers has suggested that the emotion exhibited by the abductee is real, much too real.

This presumes that the emotions are real, the witness is relating a real event, and that outside, independent corroboration has been found for that event. John Mack, attempting to validate this notion, wrote, ". . . this intensity of recovered emotion . . . lends inescapable authenticity to the phenomenon."

According to the *APRO Bulletin* this concept is not unique to Mack. "[Jim] Harder said it would be practically impossible for the two men [Hickson and Parker, abducted near Pascagoula, Mississippi, in 1973] to simulate their feelings of terror while they were under hypnosis."

Harder later told Randle, in the Pat Roach case, that he was concerned by her lack of emotion shown in the first interview. Once it seemed that she was sufficiently frightened, once she had exhibited what Harder believed to be real emotions of fear, he decided that the case was genuine.

There is, of course, no clinical or experimental evidence that such a claim, that is, the "realness" of the emotions, means anything relevant to this study. Yet time and again, this is suggested as a "proof" that the abduction experience is a real event rather than a hallucination, delusion, or hoax. Somehow the researchers have come to believe that only those who have had a real experience will exhibit real emotion when reliving that event under regressive hypnosis.

Abduction researchers also make other claims that defy logic and reality. Mack, for example, claims that this, meaning becoming an abductee, is not a club that anyone would want to join. He is suggesting that the trauma of the event, and the subsequent problems that result from the abduction experience, are so horrific that someone would not invent the tale just to join this group. Since no one would knowingly invent an abduction story that story must be grounded in reality.

In fact, some of the abductees have begun to parrot this belief. One of the Allagash Four told a national television audience that his tale was true because anyone would have to be crazy to subject himself or herself to the ridicule of being an abductee. But such statements prove nothing. You only have to watch a few of the daytime television talk shows to realize that many people would, in fact, subject themselves to ridicule and embarrassment just for the opportunity to be on national TV.

While it can be shown that many, if not a majority of the abductees, reluctantly tell their tales at first, they soon change their minds. They are hesitant when first reporting their experiences, asking that their real names not be used and asking that they be shielded from the publicity that surrounds the claim of abduction. Within months, however, those same people are not only using their real names, but are guests on national talk shows, featured in magazine articles, and are speaking at UFO conventions, revealing all that has happened to them. In fact, they answer every question put to them, no matter how personal that question might be. When asked why, they suggest that they are doing it to spare others the trauma of the abduction. They say that if they can help one fellow abductee understand what has happened, or to deal with the trauma of abduction, then it will be worth the embarrassment of the public spotlight.

Pat Roach, in 1976 when Randle interviewed her, asked that her name not be used, and that her location be disguised. Randle invented the name Pat Price and mentioned only that she lived in Utah. Not long after the article was published in *UFO Report* Randle saw her on a syndicated television program using her real name and even undergoing more hypnosis to prove the reality of the abduction. She has been on a number of television programs and in one documentary movie.

Others, such as Charles Hickson, have said they want no publicity, but one of Hickson's first moves after being released by the aliens, and fortifying his courage with a drink or two at a local bar, was to approach a newspaper office looking for a reporter. And Travis Walton's story was out even before he had reappeared to tell his tale of abduction.

We should point out, once again, that we believe the vast majority of those claiming abduction to be sincere in their beliefs. They did not begin to tell these tales with an eye on the national spotlight, to see their name in books, or to appear on television, but came forward in an attempt to learn the truth about their experiences. It should also be noted that their sincerity does not translate into proof that the tales are grounded in reality.

But with that said, we also find in the group sessions and the psychological treatment that this is, in fact, a group that people do want to join. Cone, in his work as a clinician, reported that Janet, who finally began to believe that she had been a victim of satanic ritual abuse, didn't feel a part of the support group until she began to recover the memories of that horrific abuse. This too, should be a club that no one would want to join, yet Janet did join willingly, as have hundreds, if not thousands of others.

And for those joining the abduction club, there are free trips, television appearances, and a certain amount of notoriety. Many have been featured in books, have written their own books, or had their experiences turned into movies and miniseries on network television. It provides those who join the club with an identity they might not have had otherwise. It provides a new circle of friends who claim to have had similar abduction experiences. It provides a complete social identity for them. It provides a sense of self.

Bill Cone and Russ Estes have reported that most members of abduction support groups they have studied have no desire to leave that group. While group sessions were once seen, by psychologists, as the last stepping-stone into society as a mended person, they have evolved into social settings that take on a party atmosphere. Group members look forward to the group meetings just as many people look forward to Friday night poker games or Saturday afternoon movies with friends. Richard Boylan even attempted to hold some of his group meetings in his hot tub and held others at his home.

In fact a careful examination of Mack's arguments for alien abduction reveals quite a bit about the abduction phenomenon and the motivation, often subconscious, for a belief in it. Mack, like his fellows, believes that he has found the truth based on his personal observations of his patients who claim abduction. In his rejection of the skeptical arguments about the abduction phenomenon, he said there were five specific points they, meaning the skeptics, must explain to show why abduction is not reality. These are: (1) The high degree of consistency of detailed abduction accounts, reported with emotion appropriate to actual experiences, told by apparently reliable observers. (2) The absence of psychiatric illness or other apparent psychological or emotional factors that could account for what is being reported. (3) The physical changes and lesions affecting the bodies of the experiencers [that is Mack's term for abductees], which follow no evident psychodynamic pattern. (4) The association with UFOs witnessed independently by others while abductions are taking place (which the abductee may not see). (5) The recalled memories of abductions when the subjects were as young as two or three years of age.

Actually, what could be said is that all of Mack's points are irrelevant, and it is incumbent upon Mack to present evidence that his theories are correct, not incumbent upon us to present evidence that they are not. That, however, is engaging in the same sort of semantical arguments used by the believers. Instead, it is illustrative to examine each of the points separately.

First, we could, and have, pointed out that the media has been more than willing to provide a precise blueprint for abduction. Since the publication of Fuller's *The Interrupted Journey* in 1966, the information has been readily available in the public arena. In 1975, the NBC broadcast of *The UFO Incident* provided a national viewing audience of millions with all the information needed to invent a classic abduction tale. The publication of Whitley Strieber's *Communion* in 1987 fixed the description of an alien creature in the minds of millions who never bothered to read the book. The cover illustration did it. The refinements in the abduction phenomenon have been published and broadcast repeatedly since then.

There is, however, another point. From the very beginning, those conducting the research have known what is "real" and what is not. As we have seen, the researchers have, from the very beginning, been providing the witnesses with the information they need to sound as if they have been abducted. James Harder, during the Roach hypnotic regression sessions, provided Roach with many verbal cues as to what he expected from her. If it did not develop naturally, he was only too willing to "coach" her. How else to explain his asking about a specific type of examination when Roach mentioned she thought she had been examined? Or his suggestion the aliens had pressed a needle into her belly after she mentioned needles but had said nothing about where those needles had been inserted.

Or what about David Jacobs? According to those who have witnessed his sessions, he doesn't say much as he interrogates the victims of abduction. But for those who have been privileged to hear the tapes of those sessions, it is clear what he is doing. When the abductee strays from what Jacobs believes to be the norm, he makes no audible comment. However, when the subject touches on a point in which he believes, he nods and says, "Uh-huh." It doesn't take the abductee long to pick up on the cues and begin to massage the tale for the verbal approval of Jacobs.

Let's repeat that so that it is perfectly clear. The key to understanding how so many abductees can tell stories that are so consistent when they have never met each other is because of those who conduct the research. They are providing the clues and the links between the cases because they all know the secrets. All the abduction researchers communicate with each other as they search for validation of their own beliefs. It has nothing to do with the abductees and what they know before the sessions begin, and everything to do with the researchers and what they believe.

A way of testing that is to examine the agendas of the various researchers, those who believe one way and find victims who report the stories that match their beliefs. Hopkins finds cold, calculating aliens who are busily carrying out their genetic manipulation. Mack finds aliens that have an Eastern philosophy and who provide positive events. Harder finds aliens who conducted the same experiments as those who abducted Barney and Betty Hill. Jacobs find the Hopkins-type aliens, but they are now pursuing a specific agenda of domination. In fact, Hopkins now insists that the events are all negative, regardless of what other researchers have discovered and reported.

What this means is that the abductees don't have to be fully aware of various details of abduction throughout the country or even the world. The researchers have all read the same books, have collaborated on investigations, and have shared information with each other from the very beginning. With the sloppy techniques that we have seen, evident in so many reports, is it little wonder that abduction stories share many details? Should we point out that the details of the abduction are implanted by the researchers? Subtly and unconsciously, yes, but implanted nonetheless.

David Jacobs, among others, has suggested that the abductees are not being led because they refuse to follow the lead when specific questions are asked. Yet this attempt to lead is limited to one or two direct questions. The leading of the abductees by the researcher is certainly more subtle than that. It is in the verbal cues and responses of the researcher, as has been demonstrated time and again. It is the return to specific questions asked repeatedly until the answer desired is given.

And there is one other subtle point to be made here. Researchers seem to forget or overlook the details that don't match their specific belief structures.

They concentrate on the similarities and ignore the dissimilarities. Budd Hopkins has said repeatedly that 85 percent of all abductees report grays, but the reality is that there is a wide range of alien abductors reported. What he should have said was that 85 percent of the abductees that he interviewed have reported grays.

The coaching of the witnesses goes beyond that in the hypnotic regression sessions. In nearly all the cases, we find that the researcher supplies those who have questions with books, videos, "abduction" packages, or the names of other abductees so that the latest victim will have all the information needed to understand the phenomenon. Remember John Carpenter, after handing Leah Haley two videotapes of abductees and introducing her to another, cautioned the woman not to discuss the details of the abduction to avoid contamination. It was already much too late.

In other cases, the potential abductee attends one of the many group sessions long before he or she begins to tell his or her own story. But sitting in that group environment, listening to all the others tell of their abductions, listening to the researcher discuss the various aspects of the UFO phenomenon at length, provides a solid base of understanding. By the time the new abductee undergoes his or her first hypnotic regression session, he or she has heard it all before. It's not surprising that an abduction tale, with detail to match that of the others in the group, emerges. And, of course, once that happens, the new person is accepted as a member of the group with his or her abduction being validated by the group and the researcher who conducts the sessions.

The second part of Mack's first "law" is that the tales are reported with the appropriate emotion and he is the judge of that emotion. There is no doubt that the abductees look as if they are frightened. There is no doubt their emotions are real. However, this does not translate into a reality of the event. Emotions, under hypnosis, are often created by the hypnotist. Harder, for example, kept telling Roach she was frightened, she must have been frightened, that she was worried about her children, until she displayed the emotions he thought she should. It was then that he decided the abduction must be real. He had found the validation he wanted, but only after implanting the ideas for that validation in her mind.

The reliability and honesty of the abductees is not being questioned here. They do believe, sincerely, they were abducted. There is, however, no real evidence that they were. Just because the abductees sincerely believe in the abduction doesn't mean it happened.

Mack's second main point suggests there is an absence of psychiatric illness that could account for what is being reported. Even if this point were true, which it isn't, it would be irrelevant. An absence of any psychiatric illness does not translate into a reality of abduction. An absence of psychiatric illness doesn't not mean that a person is relating an experience that is based in our accepted reality. All it means, if it is true, is that the individual reporting abduction is free of any psychopathology that could account for a hallucination, delusion, or a belief that an abduction has taken place.

As we have seen, sleep paralysis is not a psychiatric illness. It is a real phenomenon that has been experienced, according to the best evidence, by nearly a quarter of the population. Those who report sleep paralysis also report the belief an entity, or some kind of presence, is in the room with them. And nearly half of those who report abduction are describing symptoms that mirror, exactly, sleep paralysis. David Jacobs, writing in *Secret Life,* described the typical abduction in terms that matched those of sleep paralysis. The only ingredient missing was the hypnosis to expand the sleep paralysis event into a full-blown abduction complete with big-eyed aliens, medical examination and experimentation, and an eventual return to the bedroom.

What we have learned is that sleep paralysis, which is not a psychiatric illness, can account for a large number of the tales of alien abduction. Statistical research carried out specifically for this work suggests that as many as half of the tales of abduction have sleep paralysis as the precipitating event.

Continuing with Mack's five-part argument, there is no scientific evidence that point three, "the physical changes and lesions affecting the bodies of the experiencers . . . follow no psychodynamic pattern," is true. First, it must be pointed out that the statement makes no real sense. Psychodynamics refers to "the psychology of mental or emotional forces or processes developing especially in early childhood and their effects on behavior and mental states." It is also defined as "explanation or interpretation (as of behavior or mental states) in terms of mental or emotional forces or processes." Or "motivational forces acting especially at the unconscious level." What do physical changes and lesions have to do with psychodynamics?

Psychodynamics aside, we do have stories, told by abductees, that they have scars on their bodies that they cannot remember seeing prior to abduction. They seem to happen overnight. They seem to be the result of examination on board the alien ships. There is, however, no evidence of this. No records exist to document the claim. Mack, Hopkins, Jacobs, and other researchers might believe this to be true, but they have presented no persuasive evidence that it is true. And, once again, we are given a ridiculous statement that has no scientific foundation.

His fourth point, that there are UFO sightings occurring at the time of the abduction but not always observed by the abductees is also ridiculous. Are we to believe that jet aircraft have something to do with alien abduction because jets were in the sky when some abductions took place? A better analysis of the data is required to make this sort of claim. There is no solid evidence that this is true, or if true, is even relevant.

Finally Mack suggests that children as young as two or three report abduction. If we are talking of adults who "remember" events at that age, then the statement must be rejected because other, experimental data suggest that the human brain, and the capacity for memory, isn't fully developed until about the age of three. Most memories that come from an adult claiming to be ages two to three when the event took place are most probably planted by pop culture, friends, family, or the hypnotist.

And if Mack is referring to children that he has interviewed, then the statement should be rejected. As Mack must know, children of that age are easily lead into any arena the adult wants. Experimental data suggest that children of that age are incapable of telling reality from imagination. Everything is new to them.

When these arguments break down, there is another fallback position. Hopkins has said it repeatedly, and others have used it as well. There is a minority of abductees who remember some, or all, of the abduction without the aid of hypnosis. They are suggesting that abductions must be real because of this core of individuals who remember without hypnosis.

Again, we find that such statements are without merit. We have already established that there are those who "remember" events that never took place. Paul Ingram provided detailed stories of his ritualistic abuse and rape of his daughters, though there is no evidence that he ever engaged in any sort abuse. All that was required was a session with police officers and about eight hours of intense interrogation. Because his daughters had made the claims, Ingram believed there must be something to these claims. With the guidance of the police interrogators, Ingram began to tell detailed stories of his abuse of his daughters, consciously "remembered" without the aid of hypnosis.

Richard Ofshe, when consulted by the local police, demonstrated how easily these "memories" had been implanted in Ingram. All Ofshe told Ingram was that he knew Ingram had forced his daughter and son to have sexual intercourse in front of him. Within hours Ingram produced a detailed account of the event, even though Ofshe had invented the original idea and told Ingram it was true.

Many of those who later describe abduction events remember those events with no hypnosis. They are, however, exposed to researchers who believe that abductions are real. They want to remember what happened to them, and the investigative environment is conducive to the creation of "memory" rather than the retrieval of memory.

One of the abductees said, "The memories started coming to me, not through hypnosis but through a therapeutic technique where it is very meditative and you breathe into your experience. One day as I was going through a process, the visions started to come to me."

Of course, this was after she had undergone hypnosis to learn about the events. No, she didn't remember the events under hypnosis, but she didn't remember them until she had begun to search for these sorts of memories with a researcher who believed in abduction.

There is another aspect that has not been addressed. Many of those conscious memories of abduction came from dreams. Betty Hill, Leah Haley, and Jim Weiner remembered nothing about their abductions until they had a series of dreams. These are not conscious memories of events but are dreams that have been translated into conscious memories with a belief that they somehow reflect reality.

The argument then, that some 25 percent of the abductees remember the event without the probing of hypnosis is irrelevant. And this discussion doesn't even account for the hoaxes, of which there have been several.

But there are even more elements that Mack, Hopkins, Jacobs, and all the other researchers ignore and those are from our own culture. As mentioned elsewhere, if the abduction phenomenon is real, it should be unique. We are talking of alien beings coming to Earth, abducting humans, and experimenting on them. But this is not unique. Human culture and folklore are filled with such tales. That suggests there is nothing unique to alien abduction, but that it is something that arises from the human mind and human condition.

Although the researchers reject the idea, abduction by bizarre creatures has been reported throughout human history. Trips to underground caverns where strange experiments are performed on the victims have been told for hundreds if not thousands of years. The stories of fairies, of trolls, and of underground dwellers who inflict themselves on human victims have been recorded for centuries.

The Shaver mystery, first published in a science fiction magazine in 1946, told of an underground world filled with robots that inflicted misery on the human race. They abducted innocents for their diabolical experiments in their underground world. All this ten years before Antonio Villas-Boas reported his encounter on a flying saucer, and nearly fifteen before the Hills were abducted.

For those who accept abduction as real, they must be able to explain away these tales that mimic UFO abduction. Dr. Thomas "Eddie" Bullard has attempted to do it, but his arguments seem to be little more than semantics. UFO abductions are not folklore, according to him, because the spread of the tales does not behave the way folklore behaves. That does not explain all the elements included in the folklore, nor does it explain the commonality between UFO abduction and folklore. It only suggests that folklore is now spreading in a new way, and is no longer the tales told around the campfire or by the wandering minstrel. Given the changes in mass communication, this shouldn't be very surprising.

But if the folklore traditions are ignored, for whatever reason, the cultural elements of abduction still exist. Stories of human interaction with the beings in UFOs, or in a couple of cases, the great airship of 1897, have been with us for a century. It can be argued that those early cases of abduction are the prototype for today's activities. The great airship seemed to be a craft on the cutting edge of human invention and the abduction accounts coming from that time are equally embedded in turn-of-the-century U.S. Culture.

We can, however, look beyond that and see that our society has been telling stories of alien abduction for decades. The 1908 movie *When the Man in the Moon Seeks a Wife* told a precise tale of alien abduction. The "moon man," working "feverishly in his laboratory," created a gas that allowed him to "float" to Earth. He passed "through the skylight" into a girls' dormitory but retreated as the girls screamed in terror. Later he found a woman to marry and "floated" her back to the moon, with her, and her father's, permission.

Phil Hardy, in *The Encyclopedia of Science Fiction Movies,* wrote about the short film, "Interestingly, this first alien, like so many that were to follow (especially in the fifties), was in need of female companionship." He could have also noted the similarities between that early film and the abduction phenomenon as it is reported today.

The cover of the magazine *Astounding* for June 1935 showed big-eyed aliens abducting a woman who is strapped to an examination table. For those believing alien abduction to be real, this cover that predicts the phenomenon to such a precise degree must be quite disturbing.

Or the 1954 *Killers from Space,* with its mysterious scar, the missing time, the memory blocks that are only destroyed by chemical regression, the big-eyed aliens with their mission to Earth, must also be disturbing. Here is, in a film that first appeared in theaters in 1954, but later was featured on television dozens of times, the abduction scenario as accurately portrayed as any based on abduction accounts.

Randle, as a kid, saw that movie but remembered little of it. The one aspect that stuck with him was the scene of Peter Graves driving down the road, and the huge eyes of the aliens floating in front of him. Or the scene where he looked at the window, only to see those huge eyes haunting him, providing the single, best clue about his abduction.

Equally disturbing to the abduction researchers must by the 1953 film *Invaders from Mars*. There are abductions in it, but more important, there are implants. People are taken into the hidden Martian ship and a needle is pushed into the back of their head. A small device is implanted that controls the behavior and monitors the activities of each specific person. Abduction researchers tell us that the implants reported by abductees might be for behavior control and monitoring. Derrel Sims even suggested, after the alleged mass abduction in Houston, Texas, that implants had been used to monitor the "Abduction Panel" that was going to be held.

The point is that alien abduction isn't something new, reported only after alien abductions began in the 1950s or 1960s. It is something that has been around for centuries, as folklore, which had become part of our culture in books and movies long before anyone suggested he or she had been abducted by alien beings.

So we see a tradition of these sorts of stories told over the centuries. That doesn't, of course, prove that they came from those origins. Abduction researchers would be quick to say just that. Even conceding that traditional folklore raises some interesting questions, they point out that there are the physical aspects of abduction. Folklore doesn't explain the implants seen on X rays, mysterious scars on the body, or the landing traces left by some of the alien spacecraft.

While it is true that folklore doesn't explain implants, it could be argued that popular culture does. There are a number of movies in which implants are used by alien beings. These films include *Invaders from Mars, I Married a Monster from Outer Space, Puppetmasters,* and *It Conquerered the World.*

And, of course, the answer to that argument is that even if a movie suggested an idea, it doesn't explain the physical reality of the evidence. A movie doesn't create the landing marks and the swirled grass observed by Warren Smith in the Schirmer abduction case in 1967. But the creation of testimony by a writer who needed something more to prove his case does. When only a single individual observes the evidence, when there are no pictures to prove it's existence, the only conclusion to be drawn is that it never existed. In fact, Smith has admitted that he has, in the past, created testimony to "fatten" a story or to validate it for a skeptical editor. Why should we assume that the Schmirer case is any different?

The same might be said for the implants. There are X rays that do prove that something is in the bodies of the abductees. Some of these implants have been "retrieved by the aliens" before any surgery could recover them for scientific analysis by Earth-based scientists. Other implants turned out to be nothing more than organic material probably grown by the body. In one case the recovered implant was just glass with no indication that it was anything other than terrestrially manufactured glass.

Dave Pritchard, a physicist at MIT, had the chance to examine an implant recovered from the body of abductee Richard Price, who claimed that it had been placed in his penis by aliens. Over the years it had worked its way out. Two small pieces of the implant, designated as "Price I" and "Price II," were examined by Pritchard and "the pathology group in the Wellman Laboratories of Photo Medicine at Massachusetts General Hospital."

Dr. Tom Flotte, a dermatologist who routinely examines medical samples, used both light microscopic examination and transmission electron microscopy on the samples. Pritchard, who at the MIT conference had mentioned three tiny appendages on the implant, now learned the truth about them. They were nothing more spectacular than cotton fibers.

Pritchard, in his report, noted that "All of the results obtained at MIT indicate that the Price artifact . . . is of terrestrial biological origin . . . [including the] formation of this artifact: successive layers of human tissue formed around some initial abnormality or trauma. . . ."

That is the way it is with the alien implants. Nothing has been recovered that is obviously of extraterrestrial manufacture. Many such implants have been analyzed, but there comes a point, after due diligence has been performed, when it can be said that there is no evidence that any of the abductees have been implanted by alien creatures. The evidence simply doesn't exist.

Price, according to the analysis, probably had been carrying the small bit of organic material in his body for years but hadn't noticed it until he began to "remember" alien abduction. Then, this tiny nodule was noted and a whole abduction scenario was created around it. It became an implant that would, upon analysis, prove that abduction was taking place. It was the long awaited physical evidence that Hopkins, Jacobs, Mack, and all the others have been claiming to exist. It was, obviously, none of those things.

The latest of the implant claims come from Derrel Sims and Dr. Roger Leir. They insist they are following rigorous scientific protocols and that the implants they have recovered manifest a number of highly unusual and possibly unique traits. Sims waves around lab reports, but as mentioned earlier, there have been no corroborative studies conducted. The labs that have been identified do not possess the scientific credentials to underscore the importance of their findings. In other, more mundane words, the protocols that Sims claims, and the analyses that he touts, are little more than self-promotion. They are of little, if any, scientific value.

The same can be said for the scars that abductees report after abduction. There is no evidence that the scars are the result of medical procedures on an alien craft. There is no evidence that the scars didn't exist prior to the beginning of the reports of abduction by the witnesses. There simply is not the kind of medical documentation necessary for the objective researcher to accept the testimony about the creation of the scars.

Everyone who examines his or her body can probably find scars that neither he nor she, nor parents nor relatives can remember. Anyone who examines his or her body can probably find mysterious scars, including the notorious scoop marks. Randle has a scoop-shaped scar on his hand, but unlike the abductees, he can remember how he got it. There are those in abduction research who will probably claim this is a "screened" memory.

What this means, simply, is that no evidence of any kind has been offered to prove alien abduction. All areas that have been offered have alternative explanations that fit the facts easily. Evidence has been examined, often by unbiased and objective scientists, and it has been found wanting. This fact alone should be quite disturbing, but when one implant or scar is explained, the abduction researchers always have one more to be examined. They claim that it might be the one, but it never is.

All of this leaves a single question that has, in the past, seemed to defy explanation. It is the one question that the abduction researchers, the abductees themselves, and the true believers have convinced themselves saves the day. How can witnesses in widely separated parts of the world tell, virtually, the same story of abduction, if the phenomenon is not real? Doesn't that single question suggest that what is happening is real, told by the victims of it? It is the most puzzling of the questions, and the one that seems to suggest that alien abductions are taking place.

As noted earlier, the answer to this question has been in front of everyone all along. During research for this work, all of us saw the abduction investigators dancing around that answer. They mentioned that they had been in communication with one another. Mack, for example, was once the protégé of Budd Hopkins. David Jacobs is in communication with Hopkins. Carpenter has written and met Jacobs. Boylan wrote or spoke with them all. The network of investigators is interconnected and they all talk to one another, sharing everything they know, searching for corroboration from each other, proving their points. In other words, the inside detail necessary for belief in the system was communicated among the researchers who then "found" the facts to support their theories in the minds of the abductees.

In fact Budd Hopkins established the Intruders Foundation just to facilitate the sharing of information, documentation, and other resources. Hopkins's idea, and it certainly wasn't a bad one, was to provide the information that mental health professionals would need to treat those who were suffering with the effects of alien abduction. But what he has done is share everything with other researchers, and then they wonder how the abductees know all the "secret" information.

The abduction conference held at MIT was convened with the idea of sharing information and research techniques. It was one of the few times that nearly all those who had been researching alien abduction were brought together. While this is certainly a good idea, it should be noted that the opportunity for contamination was there.

Remember that we have examined, carefully, the research techniques of all those investigators. Look at that material again. Boylan wasn't very subtle. He took innocent dreams and turned them into abduction scenarios. He provided his patients with books about alien abduction, talked to them about it, and invited them to group sessions where others who believed in abduction told their own tales of abduction. If the subject didn't know enough to please Boylan when he or she arrived as a patient, Boylan was quick to provide sources of that information. And if that didn't work, he would coach them with drawings and suggestions about what might really have happened.

Edith Fiore certainly wasn't very subtle as she attempted to implant her beliefs on her patients. In one case, with the patient under hypnosis and searching for "past life" memories, Fiore demanded that she "make it up." And it was Fiore who introduced the idea of alien abduction to the patient, telling the patient that she had "probably" been picked up by UFOs.

Jacobs was more subtle. In his sessions, as the abductee under hypnosis told him what he wanted to hear, he would vocalize in some fashion. If it was something different, he was mute. It wouldn't take the subject long to learn that Jacobs's vocalizations underscored what he wanted to hear. His sessions were seriously tainted.

C. D. B. Bryan, who sat in on a number of sessions with Hopkins, noticed that Hopkins led his witnesses as well. In his book *Close Encounters of the Fourth Kind,* he wrote, ". . . of course there were moments when I felt Budd Hopkins might have been 'leading' them. . . ."

On the *Nova* program that dealt with alien abduction, Hopkins demonstrated his leading of the witnesses when he questioned the young boy using the flash cards he had designed. These cards held the faces of figures such as Santa Claus, Batman, a skull, and a gray alien. Although the boy said he didn't recognize the alien, Hopkins returned to it and helped the boy invent a tale involving the alien face.

This then explains the last of the great questions. It doesn't matter what the abductee knows or doesn't know when he or she arrives at the home or office of the researcher. The information will be passed along quickly. The key is the researchers. They have all the answers and they provide them to the witnesses, sometimes in not very subtle fashion.

There is one other point that must be made here. Many of the researchers have convinced themselves that the government is interested in their work and is actively trying to suppress it. They talk of black helicopters hovering over their homes, or attempts to discredit them through trumped-up charges, such as those leveled against Boylan.

It could be said, and has been, that those who expressed interest in alien abduction, who are on the forefront of research to learn the truth, are being punished by the government. Mack was reprimanded for his techniques by his peers at Harvard, and Boylan and Fiore lost their licenses to practice in the state of California.

On the surface, it seems that these claims are true. But if we look beyond that, we see that the negative responses to the researcher is the result, not of the belief in alien abduction, but because of other behavior. Mack was investigated by his Harvard colleagues because they believed his research techniques were flawed, not because he undertook a study of alien abduction. They didn't criticize his belief in alien abduction but the way he conducted his research. They were concerned that the scientific method had taken a backseat to the reinforcement of his personal belief structure.

Leo Sprinkle continued at the University of Wyoming for decades after it became known that he was interested in alien abduction. He had, in cooperation with the Condon Committee, investigated the Schirmer case in 1967. He had published papers on the topic of UFOs and alien abduction for many years. He was a consultant to APRO and assisted in their investigations of UFOs. It was no secret that Sprinkle was interested in UFOs and believed them to be extraterrestrial craft. The university didn't seem to care.

In 1979, he was interviewed by Daniel Kagan and Ian Summer for their book *Mute Evidence.* Because of Sprinkle's research, there seemed to be a connection between the tales of dead and mutilated cattle and UFOs. Sprinkle, as well as others, seemed to believe that the crews of the flying saucers were responsible for many of the mutilations reported in the West.

Kagan, trying to learn more about Sprinkle's techniques, and his use of hypnosis, suggested that Sprinkle hypnotize him. What became obvious was that Sprinkle was leading Kagan under hypnosis and driving toward evidence of an alien abduction. Kagan reported that he had never seen a UFO and certainly hadn't been abducted. Sprinkle's questions, however, seemed to be designed to prove that he had been. It was all very subtle, but Kagan did find evidence that Sprinkle led his subjects. And none of this adversely affected his career at the university.

Boylan's fall had nothing to do with his belief in alien abduction but with his attempts to plant false memories in the minds of his patients. It was also his attempts to engage his patients in nude hot tubbing and other nude encounters, and his trading of his professional services for other services with his patients. It was a host of unprofessional behavior that lead to his downfall, and not his personal belief, or his research into alien abduction.

According to the "Findings of Fact" written by the state of California, one of Boylan's patients, K. G. provided massage services for Boylan in return for his treatment. This is in violation of various laws and ethics.

As with the other women who participated in Boylan's practice, K. G. was offered the opportunity of therapy in his hot tub. In the "Findings of Fact," K. G. told investigators, "He then mentioned the hot tub for water therapy to relax and let go of tension. Respondent [Boylan] directed K. G. to an area to undress and it became clear that the hot tubbing would be in the nude."

Boylan suggested that they increase the frequency of their meetings, but K. G. said that she couldn't afford it. Boylan then suggested a barter system. Eventually it was agreed that K. G. would do the massages. Boylan would exchange an hour of psychotherapy for an hour of massage. Later Boylan learned they would have to exchange money for their services. After that, they wrote checks to each other. This sort of a dodge is considered illegal by the IRS when it is done in this fashion.

Although it was determined that the evidence did not establish sexual misconduct according to the "Findings of Fact," it was during the final session with K. G. that "respondent commented to K. G. that if she happened to 'graze' his testicles, it would be okay."

So, it was these activities with K. G., and the inappropriate behavior with other patients that resulted in the revocation of Boylan's license, and not his belief in alien abductions. Although Boylan has since claimed that the licensing review was motivated by a former Air Force officer, and that the government was attempting to still his voice, such is not the case.

And Fiore stumbled not because of her belief in alien abduction but because of a complaint of unprofessional conduct. She lost her license only after she began to tell patients about her personal life during the therapy sessions, and to suggest, in at least one case, that the patient had been abducted and that patient should "make it up."

FINAL CONCLUSIONS: ALIEN ABDUCTION OR THERAPIST SEDUCTION?

Here's what it all comes down to. There is not a single shred of physical evidence that alien abductions are taking place other than the tainted testimony of the abductees. The physical evidence to support the claims is nonexistent. What has been offered as proof has been eliminated through testing by objective scientists or additional research by unbiased investigators. The scars, missing fetus, or the implants do not carry the proper medical documentation to make a strong case, and in fact, suggest something else altogether.

The abductees, as a group, show a high percentage of individuals with gender identity problems, sexual dysfunction, dysfunctional families, and broken lives. Many have been led to the conclusion that their fears, anxieties, and mental problems are the result of alien abduction by therapists who believe in alien abduction. If they had been treated instead by a therapist who believed in satanic ritual abuse, then they would have manifested those symptoms and would now believe they were victims of satanic abuse. If the therapist had believed in incest or child abuse, then the symptoms would have reflected that belief. Boylan proved it when he turned patients who believed in satanic abuse or incest into, sometimes unwilling, abductees.

The question that has yet to be answered is who is actually responsible for the alien rape? Are the beings from another planet who capture unwilling people, drag them into a ship, and then perform primitive medical procedures, and, in too many of the cases, have actual sexual relations with them (rape by clear definition) responsible? Or is it the therapists who convince those wanting and needing help that they have been abducted by aliens?

Here's exactly where the evidence, all the evidence takes us. A large number of abduction cases can be explained as sleep paralysis. There is no question that the symptoms of sleep paralysis are mirrored in reports of alien abduction. The researchers who claim otherwise are unfamiliar with the scientific literature about sleep paralysis and have failed to examine it.

Some of the cases of alien abduction are the result of poor investigative technique by the abduction researchers themselves. They have unconsciously contaminated witnesses in their drive to validate their own belief structures about the reality of alien abduction, as we have seen time and again.

Hypnosis, contrary to what we have been told by the researchers, is a poor tool for investigating abduction. Improper use of hypnosis is responsible for the implantation of memories and beliefs by researchers of alien abduction. Such implantation, because of the highly suggestible trance of the abductee, is quite simply and often inadvertent.

Using hypnosis, there is always the question of whether the subject has entered an altered state of consciousness. It is left to the interpretation of the hypnotist or researcher. There is, however, a way to determine if an altered state of consciousness exists and that is through the use of chemical agents such as sodium amytal. This is important because of an experiment conducted under the insistence of a multiple abductee.

She is a woman who, under repeated hypnotic regression, told long, involved, and detailed stories of her continuing abductions that matched those told by others. She felt that there was more to the events than hypnosis revealed. She insisted that chemicals, such as those employed in normal psychological treatments, be used. With a certified and licensed medical doctor present, the woman was given a chemical agent. Under its influence there was no evidence of any alien abduction. She remembered nothing. Absolutely nothing.

It should also be noted, in a related point, that Sherry, who had repeated abductions, who had entered a support group and was finally hypnotized so that she could remember her abductions, has not been abducted in more than a year. This was after she returned to her medication. Interviewed a final time in November 1997, she said that she no longer believed that she had ever been abducted. The abduction scenarios developed under the strain of her move, information about alien abduction provided by her friends, the influence of an abduction support group, and finally by the hypnotic regression conducted by a researcher who believed in alien abduction.

Finally, when we look at the whole picture, we can answer the last question posed by the abduction researchers. How can people in widely separated parts of the country, or the world, tell virtually the same story? The answer is that the researchers all know the so-called unpublished material and search for it in their quest for validation. The researchers contribute to the similarity of tales, not the abductees. Under the subtle—and sometimes not so subtle (Fiore's "make it up")—coaching, many of these researchers carefully direct the abductees to the tale that conforms to those told by others.

Betty Hill, for example—the first person in this country to reveal her abduction story—mentioned big-nosed aliens in her initial interviews in the early 1960s. In fact, it goes beyond a description of big noses. In her diary kept of her dreams, she wrote, ". . . their noses were larger (longer) than the average size although I have seen people with noses like theirs—like Jimmy Durante's." In today's world, according to Betty Hill, those same aliens now have small or no noses, and are, in fact, the typical grays.

This leads to a simple question. Do the abduction researchers know the truth? Yes, it seems so, though it may be subconscious. All dance around this truth, telling us things that, on the surface, make some sense. But when we look below that surface, we see the truth. Mack, for example, said, "One of the interesting aspects of the phenomenon is that the quality of the experience of the abductee will vary according to who does their regression."

In *The Threat,* Jacobs wrote, "Many hypnotists and therapists who work with abductees adhere to New Age philosophies and actively search for confirmational material. During hypnosis, the hypnotist emphasizes the material that reinforces his own world view. If both the subject and the hypnotist are involved with *New Age* beliefs, the material that results from the hypnotic sessions must be viewed skeptically, because their mindset can seriously compromise their ability to discern the facts."

Both Mack and Jacobs were attempting to tell us that if the abduction researcher believed in hostile aliens with an invasion plan, then the abductee would manifest those sorts of beliefs. If the researcher believed in cold, calculating, scientific entities, then this was the type of experience manifested. If the researcher believed in benevolent space brothers, then these were the attributes of the alien creatures. And if the researcher happened to believe in satanic ritual abuse, then those were the memories recovered rather than those of alien abduction.

But Mack and the others seemed to have missed the real point. It was not that the abductee manifested the surface beliefs of the researcher but that the abductee believed in abduction at all. The whole concept, in the vast majority of the cases, was the belief of the researcher. If we look at the history of psychology we see no case in which someone undergoing hypnosis for something else such as to lose weight or stop smoking spontaneously came up with an abduction story. In other words, as demonstrated time and again, if the researcher believed in something other than alien abduction, then the subject manifested those symptoms. The abductee sought out someone doing abduction research, so it should be no surprise that the researcher found the tale of alien abduction.

Our theories cover the facts of alien abduction. We have not had to manipulate the data, twist it, or bend it to make it fit. We understand what is being said by the abductees, we have read the reports by the researchers, and we have carefully analyzed the data. The theory of researcher manipulation, though subtle and sometimes unconscious, explains how and why some people believe they have been abducted by aliens.

We understand how the belief begins, as an episode of sleep paralysis or in vivid dreams or sometimes as nondirectional anxieties that are focused by hypnosis and directed discussion. We understand how the continued probing, under hypnosis, can influence and suggest ideas to those being interviewed and regressed. And we understand that in the vast majority of the cases, it is a real desire of the investigators to learn a truth that inadvertently creates that unreal truth.

The evidence for this is clear and overwhelming. We have seen it as each of the abduction researchers touches on it and then drifts away. To understand alien abduction we have not had to create a new psychopathology, change a worldview, and ignore the data. We have only had to put it all together in one consistent whole.

There is, of course, an exception to this. Some of the victims of alien abduction have suggested that it is not a physical act but one in the psychic or supernatural world. The abduction does not take place in our reality but in another reality, one that is viewed solely through the mind of the victim. If this is the case, then there is no way to prove it, no way to disprove it, and certainly no way to test it.

Scientists are just beginning to understand the workings of the human mind. They have seen, on CAT scans and other examinations of the brain, just how much of the brain is used and which areas are used during variation activities and thought processes. We have seen that low-level electromagnetic fields introduced into the brain, or around the brain, mimic some forms of epilepsy or temporal lobe lesions and produce a stimulation of the brain that causes the victim to perceive things that are not there. Michael Persinger, in his work, found that the belief structure of the victim was important in understanding how the "visions" would manifest themselves. It was not a psychopathology that he was investigating, but the effects of stimulation to various parts of the brain and the reaction of the subject to that stimulation.

The point here, however, is that if the abduction phenomenon is more of an extrasensory perception than a physical one, the research needs to move in a different direction. It seems that the physical reality of abduction has been eliminated by the preponderance of the evidence . . . or a lack of that physical evidence. In other words, there is no solid evidence that alien beings are physically abducting people from their cars, beds, or homes.

Science hasn't failed us here. Science told us how to act, how to research, and how to draw conclusions. When the conclusions were not what we wanted, we rejected the science and invented reasons to do so. It is a government conspiracy. It is science that has narrow vision and blinders. It is the enlightened amateurs who drag science with them by their pioneering efforts.

But those making the claims often forget the real point. Science rejected rocks from the sky (meteorites) until a proper scientific study was completed in France that proved rocks could fall from the sky. Science scoffed at the idea of the living fossil, rejecting reports that a coelacanth had been captured in a fisherman's net. But it was science, when presented with the proper evidence that reversed itself, amazed that the fish had survived its alleged extinction seventy-five million years earlier.

For too long we have been persuaded by "authorities" who tell us that alien abduction is real. They present case after case, demanding that we prove that the abduction isn't real. But that isn't the way science is supposed to work. The researchers who claim the abnormality are required to prove that it exists.

They have failed to do so.

APPENDIXES

APPENDIX A:

TRANSCRIPT OF THE PAT ROACH INTERVIEW

To fully understand how the leading of the witness takes place, at least under hypnosis, it is necessary to listen to, or to read, the complete transcripts of the sessions. The following are the transcripts from the taped interviews conducted with Pat Roach in July 1975. The "I" is for the interviewer James Harde, and the "R" is, of course, Roach.

(I) The feeling of concentration, going back in time, get that feeling you had that day, that you were going to bed. Get that feeling while you were asleep, you're going to sleep. When you get that feeling, tell me, tell me that . . . you've got the feeling of being on air. . . . And now remember that you woke up. Get that feeling that you woke up. Now you can talk. Get that feeling when you woke up. You can tell us about it now. What was the feeling you had.

(R) I'm surprised by . . .

(I) Let's just get that feeling again of what it was like. You felt surprised, did you wake up then, or was that before you woke up?

(R) It was a bright light . . .

(I) Did you go to the window?

(R) No. I was in the living room and I was on the couch.

(I) You were on the couch.

(R) I sleep there occasionally.

(I) I see.

(R) When I woke up the cat wasn't with me, my little boy. You know two figures were standing over me.

(I) It's all right, no one is going to hurt you.

(R) I wasn't really afraid. Maybe I felt they must know what they were doing.

(I) What happened then? You can get that memory back. Just feel yourself being there.

(R) I remember looking up, I was lying down, you know, and they're bright. They're skinny. Whatever they were, they're skinny and they look like they've dressed up in all white. People that would be in the service or something?

(I) What gave you that idea?

(R) Their uniforms.

(I) They seemed to have a uniform?
(R) Yes.
(I) Did it seem to have any buttons or anything?
(R) No.
(I) Did they talk to you at that time?
(R) No.
(I) Didn't hear anything?
(R) I wondered about my children.
(I) Then what happened
(R) It looked like they're organized, you know, I don't want to go with them.
(I) Get that feeling of concentration of what happened as if you were there and tell us what happened.
(R) My arms ache. I'm worried about my kids.
(I) I see.
(R) Everybody's in the room. Not everybody but . . .
(I) Let's go around the room and see who's there. You can see them, just get that feeling, just picture what was there.
(R) Bonnie's standing there, Debbie's standing there and can't see Kent, that's my little one that was lying by me. I have a feeling somebody's carrying him but I don't see how that could be. They're holding him or something. But they have us all together."
(I) Was Shawn there?
(R) I see Shawn.
(I) All right, Shawn was there now. Was Mary there?
(R) Don't see Mary.
(I) You don't see Mary.
(R) It's a group, you know, it's in a group and the one in the corner comes closer to us. He's standing by us.
(I) All right.
(R) I don't remember what happened.
(I) You can remember what happened. You can tell us what happened next after that. Were you told you wouldn't remember?
(R) They have a machine that they carry. They're very businesslike, and they hurt my arms because I don't want to go anywhere. They seemed to grasp me on each side of my arms. I see all my children are fighting now.
(I) All right, they're going to be all right, but you can remember very well. Get that feeling of remembering. It may be a little bit frightening but you can . . . 'Cause you got out of it all right. You were all right afterwards. So it's all right to remember.
(R) That's true.
(I) Debbie was much better afterwards for it.
(R) Yes.

(I) Maybe they didn't understand that they were hurting you, but you can tell us how it felt.
(R) They seemed to grasp me on my upper arms. And . . . was so scared that I guess I, I don't remember. I don't remember going out the door.
(I) What is the next thing you remember?
(R) I see bright room, big bright room. Big bright room, yes. They're standing around.
(I) What else can you see in this big bright room?
(R) That's all I see.
(I) You can just see the brightness?
(R) I can see people, the figures. They have uniforms on.
(I) How many people are there that you can see?
(R) Well, can see about four or five all that have uniforms.
(I) How many of your children are there now?
(R) I don't see them.
(I) How do you feel at the moment?
(R) I wonder if I really see all of this, you know.
(I) You are wondering at the time . . .
(R) If I really see it or if it's my imagination, you know.
(I) Go on, tell us. That's a very intelligent thing for you to recognize. Do you remember everything you told that you would remember something. Is there something that was put in your mind that you would remember?
(R) They weren't as nice as they wanted me to think.
(I) Did you get that impression?
(R) Yes.
(I) What impression did you have?
(R) That they were cold-blooded. That they didn't want me to know it.
(I) You had that intuition?
(R) Yes.
(I) Is there something especially that happened that made you think that they were cold-blooded?
(R) No, 'cause they treated me like I was a guinea pig.
(I) You felt that they treated you like a guinea pig.
(R) Yes. They really didn't care about people, you know, as people.
(I) Now let's get that feeling as you came through the door or whatever that place was, did you get a feeling of coming through a door?
(R) Yes, into the big room, yes.
(I) What shape was the room?
(R) Round.
(I) What does it look like? Besides being round, is there anything you see in this room?
(R) It's very bright, it is. Door's on my right-hand side and I look out, you

can see out at the stars, not the top but through the side toward the top . . .

(I) You can see stars. I see. Is it clear?

(R) No, I can see stars. It's as if you could see the stars. It looked like a lot of technology. It's all machines and buttons and on the wall.

(I) How big were the buttons?

(R) They're on machines.

(I) What kinds of machines? Did they look like typewriters, computers?

(R) They looked like computers.

(I) What made you think they looked like computers?

(R) Because they had wavy lines going through them.

(I) It moved? On a little screen?

(R) Yes.

(I) Did you see any books?

(R) No.

(I) Lighted buttons? Lights?

(R) Yes. Lights and buttons. That's all I want to remember.

(I) Is there something that you think would be frightening to remember?

(R) Yes. I know it.

(I) You know that there would be something that is frightening?

(R) Yes.

(I) Now you got out of this perfectly safe, nothing happened to you that hurt you, but it might have been a very frightening experience at that time. Is there something you think that would really . . .

(R) No, I don't remember being examined but I know I was and that's what bothers me.

(I) You think you have been physically examined?

(R) Yes.

(I) Probed? Somebody touched you?

(R) Yes.

(I) Did you feel there there was a machine involved in this?

(R) And people.

(I) Did you get the impression that you were up on a table?

(R) Yes.

(I) Were your clothes on? Did they take your clothes off?

(R) I don't remember.

(I) It might be hard for you to remember, it probably wasn't necessary or essential. Did they tell you that you wouldn't remember this? Did you get that impression that you wouldn't remember?

(R) They really didn't talk to me.

(I) They really didn't?

(R) No. They seemed to have a special procedure and everybody did what they were supposed to, and they didn't really communicate with me. If they did, I don't remember it.

(I) So you really don't remember that they said anything in English to you?

(R) I wanted to like them but I ended up not liking them.

(I) At this time there was no one else that you saw in the ward? Human beings?

(R) No. 'Cause I'm sure I went into another room. It looked like the rooms were on the right-hand side and this big room wasn't really all that round because the doors were along the side.

(I) Are you remembering more than you did before? Is it all coming back to you clearly?

(R) I don't like it.

(I) I understand that you wouldn't. This is what you want to remember.

(R) Yes.

(I) Because it's really very curious and interesting to know what actually happened to you. Is it possible that you were persuaded that you might not remember but you really can if you want to?

(R) I know I can, but I don't want to remember it now.

(I) Is it because it was too frightening?

(R) I was upset.

(I) You didn't like their attitude.

(R) No. I had to do what they wanted me to do regardless and then I didn't know what was happening with my children and they didn't communicate with me. You know they just took us all like that. But you know, they seemed to have helmets on, something on their head. I don't see them like I thought they were.

(I) You see them differently now than you thought you saw them?

(R) I didn't seem them before, I didn't see them, but my children said that the face is the same but they have something on their head.

(I) Can you remember what the face looks like?

(R) I remember the big eyes.

(I) And do you remember a pupil in the eyes, a round pupil, or was it a slitted pupil like a cat?

(R) It doesn't matter . . . let me think. 'Cause they looked at me closer in my face.

(I) Did they? How big would you say their eyes were? The size of a quarter?

(R) They were big.

(I) A fifty-cent piece?

(R) No. Quarter.

(I) Was it round?

(R) No.

(I) How did it move?
(R) Oval.
(I) Sideways?
(R) It had a big pupil. It was a round pupil.
(I) Was it black?
(R) Yes. They looked like us but the eyes went around the side.
(I) They did? Have you read any books about this sort of thing?
(R) No. Not anybody that looked like that.
(I) What about the nose? Do you remember anything about the nose?
(R) Don't remember a nose.
(I) Was there anything that looked like holes there?
(R) No.
(I) What about a mouth?
(R) A fish.
(I) It looked like a fish? Does that mean it didn't have any lips?
(R) Yes.
(I) Did the mouth look wide?
(R) It looked long from side to side.
(I) Did it go around the side of the face too?
(R) No. It was just long, you know.
(I) Did you see any teeth?
(R) No.
(I) Did you see inside the mouth?
(R) No
(I) Did they move their eyes?
(R) Oh, yes. They moved their eyes a lot.
(I) Sideways?
(R) They moved their eyes all over. It was really weird.
(I) What about . . . with their heads . . . with the rest of their body?
(R) They were pretty much what you'd expect them to be.
(I) How would you expect them to be?
(R) Like normal people.
(I) Like normal humans? Were they tall, short?
(R) They were little.
(I) Like how high were they?
(R) When they stood above me when I was lying down . . . Yes and also when I was lying down being examined, they weren't five foot. They were little.
(I) Were they taller than three feet?
(R) Yes.
(I) Were they as tall as four feet?
(R) Yes.

(I) They were a little taller than four feet?
(R) I think so.
(I) How did their arms seem in proportion to their bodies. Were they long, short?
(R) They had a pan or something they put me on.
(I) A pan?
(R) With something in it. Their arms were more angular. Their bottom of their arms looked long.
(I) You mean between what would be the elbow and their hands?
(R) Yes.
(I) Remember their hands. What they looked like.
(R) I'm trying to think. They have those funny hands like Bonnie said but they're orange.
(I) Orange color? Did they seem to have fingers?
(R) Didn't look like fingers.
(I) Did they move their hands ever?
(R) Yeah.
(I) Did they open their hands ever?
(R) Yes. They opened . . . it was almost like a clasp.
(I) Like there were two fingers, or three?
(R) I wouldn't call them fingers, they were big . . .
(I) That's all right. You can remember it.
(R) I didn't like those . . .
(I) I can understand that you didn't like them. Did they seem to have feet that looked like ours? You really didn't have a chance to see them?
(R) No. Club feet.
(I) You mean they were wide feet, relative to our long?
(R) No, they were round. But they were in a boot.
(I) Did their feet seem to be as wide as they were long, or were they longer than they were wide?
(R) I don't remember because the pants went down almost to the, their trousers, or pants, whatever they were, they wore clothes like us.
(I) Did they seem like clothes like us? Were they colored?
(R) They were fluorescent.
(I) What color fluorescent?
(R) Silvery and blue.
(I) Did they have a belt?
(R) Yes.
(I) Was the upper part the same color as the lower part?
(R) I don't remember.
(I) When they put this pan on you, did you get any, was it a little pan?

(R) They didn't put the pan on me, they took something out of it and wiped me with it.
(I) Wiped you, just like a piece of cloth? Did it hurt?
(R) I don't remember.
(I) It probably didn't hurt you. They probably were just taking a little skin sample or something superficial. Cells or something.
(R) I don't know.
(I) You can really remember, you just don't want to remember.
(R) I don't want to.
(I) That's okay. I understand.
(R) It makes me angry.
(I) I can imagine, you were worried about your children. Your children may remember what happened and then afterwards you may want to. You will want to remember what happened to the children so that you can reassure them, probably. So it would be a good idea if you remembered what happened to you, if you can possibly do that without its bothering you too much.
(R) After I left that room, I wasn't with the children.
(I) I see. But they may be worried a little bit about what happened to them and you'll want to make sure it isn't too frightening. You don't want to upset them unnecessarily.
(R) No.
(I) I want to ask you one question, and you don't have to answer it. Did they put a needle in your stomach or anything like that? You can just answer with your fingers, you don't have to say.
(R) I'd rather say, I don't remember anything like that.
(I) You don't remember any blood samples that they took?
(R) Nothing. They hooked me up to a machine. Checked everything, examined me from top to bottom. They put needles in me in places.
(I) Do you remember what places?
(R) No.
(I) Perhaps in your arms or legs?
(R) They put needles everywhere it seemed like.
(I) Was it Chinese acupuncture do you suppose?
(R) I don't know.
(I) Did they hurt?
(R) No, they didn't hurt me, I was numb all over.
(I) You were feeling numb everywhere?
(R) Yes, but I knew what was going on.
(I) Did you have your eyes open?
(R) Yes.

(I) Did you watch them?
(R) Yes, I saw them moving around.
(I) How many of them were around the table?
(R) There were only two that were doing the examining. And there was one that was helping.
(I) And he never talked to you? Did you talk to him?
(R) No, I was angry with him.
(I) You didn't say anything?
(R) No, they seemed to be so clinical. I didn't feel they were very friendly. In the beginning I thought they might be but . . .
(I) What happened when you got up off the table? Did they let you put your clothes back on?
(R) Yes. Then they just went about their business and I was upset.
(I) Did you go out the door then?
(R) Yes, they took me there.
(I) How many took you there?
(R) Two.
(I) How did they take you back? Did they carry you?
(R) Yes, more or less. I don't know how it was. I wasn't really walking.
(I) Did you feel yourself floating? Did they take you back to the living room?
(R) I was so angry I just stood there anyway. They couldn't have made me move once I got outside. If they hadn't moved me, I would have stood right there.
(I) Did they take you back into the house?
(R) Yes.
(I) Now, you still feel there are some things that [you] could remember that you are not really wanting to remember.
(R) I don't want to remember it all.
(I) Is it because they said you wouldn't be able to remember?
(R) I didn't want to communicate with them.
(I) Did they tell you that?
(R) I don't know.
(I) Did they tell you you were going to feel like that? Is it possible they put that into your mind? That you would be angry with them?
(R) They knew I didn't like what they did.
(I) How do you know that they didn't? How do you think they knew that?
(R) Probably because I wouldn't talk to them. They wanted to talk with me and I didn't talk to them.
(I) Did they ask you some questions?
(R) I don't remember. I'm tired.
(I) Are you feeling tired now?
(R) Yes. It was horrid.

(I) Then you went back to bed, you felt very sleepy. Did you see your children after that that night?
(R) Yes, when I woke up.
(I) Did you wake up right away or . . .
(R) I didn't know they went out the back.
(I) And you were worried about them?
(R) Yes.
(I) So did you wake up in the night and look for them?
(R) We all woke up at the same time.
(I) Do you remember what time it was?
(R) I didn't see any clock. It was in the other room.
(I) Is that when you called the police?
(R) Yes. My children came to me. We all woke up and there was a lot of noise.
(I) What kind of information do you think they wanted from you?
(R) . . . At least I think they did.
(I) I see. So they did say some things.
(R) I can't quite remember. I just have that feeling, that they didn't really want a thing. I was angry with them . . .
(I) It would be very helpful for me to know, as a scientist, what kinds of things that they are looking for. That would be very helpful if you could remember that . . . if it wouldn't be too much trouble.
(R) They wanted to know how our minds work.
(I) That's very interesting.
(R) They want . . . to give them certain information that they don't understand yet.
(I) What kind of information?
(R) How we think, how we feel, our emotions. They don't know about us.
(I) That's very interesting. Do you think that makes them interesting?
(R) No . . . I don't like what they want.
(I) You thought you were being intruded upon.
(R) Yes. They didn't care, because they don't have an understanding of emotions like ours. Maybe they're trying to understand our emotions. I may be wrong but . . .
(I) You know, Pat, you're one of the more intelligent people that have been in touch with this thing.

That ended the first session. Roach had awakened at that point. On the morning of July 9 the second hypnotic regression session was held.

(I) Now, we're strengthening that part of your will that's resisting their suggestion that you are going to forget. Now you feel that part of your mind strengthening.

(R) Yes.

(I) And as time goes on, that part of your mind is going to be stronger and stronger and you're going to resist your original suggestion of what you should forget. Do you understand that? Because we have more control and power now than they had then. Understand that? And we are going to be able to be back again and you're going to feel what they're trying to do to you that you are going to resist, understand that. So we are going to start from the beginning, the first impression that you got. Now you can start to talk and tell me what was the first impression that you got from them? You can talk now and tell us.

(R) They wanted me to go with them.

(I) They wanted you to go? Did you go willingly?

(R) I went at first.

(I) What was the first impression you got from them besides that they wanted you to go with them?

(R) They wouldn't explain anything.

(I) What impression did you get from they that you were never to remember? What impression did they put into your mind? Do you remember specifically?

(R) They had lots of machines.

(I) Were some of these machines attached to you by wires? It's hard for you to remember. It's a very disturbing thing for you, but you can still remember. You out out of it all right. You are safe. Nothing happened that hurt you, so it's all right to remember. Let's just get that feeling and remember.

(R) They put me on a table and they hooked me up on one leg and arm, and I didn't like their examination. They . . .

(I) Was it like a G-Y-N examination?

(R) That's part of it. They. I don't like what they do with my head.

(I) What are they doing with your head.

(R) Taking my thoughts.

(I) They were taking your thoughts?

(R) Yes.

(I) How did you know they were taking your thoughts?

(R) I could feel it. They don't have the right to take them.

(I) Of course they don't but you can tell them. You see, you can help us know what they're up to if you can tell us what kind of things they did to take your thoughts.

(R) I see a pan and a rubber glove, a regular hand and they have a like a blood pressure test. They put a needle in and they take my mind, my thoughts.

(I) Where was the needle?

(R) I see it coming toward me but I don't know where it goes.

(I) Where is it coming towards you?

(R) To the front of me.

(I) In you stomach, your bottom, in your face?

(R) No. In the front of me. Don't know.

(I) Yes, you do know. Was it your chest? You do remember. You can remember. You're strong enough. Your willpower can make you remember. It's over. It's all over now. It doesn't hurt you anymore. You are safe and sound. It would be very good to know what they did to you.

(R) I'd better not.

(I) You see that was two years ago, Pat. You know we're just trying to find out. There's nothing we can do about it except know what happened. Nothing serious happened. Debbie got better. She wasn't as sick as she used to be. You know they just didn't understand that they were hurting you and I think that must have been it. You see, they just don't understand certain things. They didn't understand what willpower you had.

(R) I hope they all crash.

(I) What was that?

(R) I hope they all crash.

(I) Tell me more about how they seemed to be taking your thoughts, what went through your mind? Were you able to think that you just couldn't think what you wanted to think, is that the problem?

(R) . . .

(I) You're not sure. Did you try to think about something and then were stopped from thinking about it. It's over. Now what's happening?

(R) I'm getting dressed.

(I) Did they help you out of the spacecraft now?

(R) They sent me.

(I) They sent you?

(R) With somebody else.

(I) Back into your home?

(R) Yes. They don't know.

(I) Don't know what?

(R) They don't know how we humans are.

(I) What is it that they don't know. Did you try to tell them what we are like?

(R) I called them stupid.

(I) You called them stupid? Did they understand that?

(R) Yes.

(I) What did they say to that?

(R) They weren't angry. They just do what they want to do anyway.

(I) Did you see anybody with anything that looked like a smile on his face?

(R) No. The man was a regular man.

(I) There was a regular man?
(R) There were a couple of those.
(I) How many fingers did he have on his rubber glove?
(R) Just like ours
(I) Did he look like a regular human being?
(R) Yes.
(I) Was he taller? Bigger?
(R) Yes. He was bald on top.
(I) What kind of clothes was he wearing? Was he the one who did the examining?
(R) Yes. He helped.
(I) Did you clearly distinguish him from . . .
(R) Something black. Some black clothing.
(I) How did you know he was a human being? Did he have a nose?
(R) He looked like a person.
(I) Did he have regular eyes?
(R) He kept bending over me. Bending over me all the time.
(I) How old was he?
(R) Fifty-five.
(I) How about the other people who were around? Did they have noses?
(R) They just stand around.
(I) Tell us more about the needle that bothered you so much that comes toward you.
(R) No. I don't like it.
(I) Of course you don't like it. Can you tell us about it?
(R) I don't know.
(I) Tell us about how you thought they were taking your thoughts. Were you trying to think about something and found that you couldn't think about it? What I need to know . . . how you knew. Did you find yourself thinking some things that there wasn't any reason for you to think about?
(R) They were probing me in different places. My head. It was something I couldn't see. Because they put the needle in and after that happened . . .
(I) Where did they put the needle in? Perhaps in your stomach? Perhaps in your navel?
(R) I don't know.
(I) Did it hurt?
(R) No.
(I) You were numb. After they put the needle in, then you had feelings that they were taking your thoughts.
(R) Yes.
(I) What did they keep doing?

(R) Bringing up things in my mind.
(I) Like what?
(R) Like birthdays.
(I) They wanted to know about your birthday? What did they want to know?
(R) What I love, what I don't love, what animals I liked.
(I) That's very interesting. They were making some kind of psychological profile. Did they ask you about your husband?
(R) They asked me about my people, my family, and where I was and how I felt about it, what I liked to eat. Even manipulated me.
(I) What made you think they were manipulating you?
(R) I had no choice.
(I) You felt that you had to answer their questions.
(R) No, I had to.
(I) Did you answer their questions truthfully?
(R) Yes.
(I) Did they ask you about things that you wouldn't tell them?
(R) No. But I don't care to answer. Just want to go away.
(I) Did you particularly dislike any one of them?
(R) Don't like people like that.
(I) But after all they were interested in what happened to you. They did seem to have some interest in what your life had been like.
(R) They didn't care about me.
(I) What do you think they care about?
(R) Their information.
(I) What kind of information do you think they were trying to find out?
(R) They need us.
(I) What did they say about it?
(R) I don't know why they need us.
(I) Is that something they said or just an impression you have.
(R) They're very intent. They need information quickly. I don't know if it's my imagination but they limit in time and we don't.
(I) Would you say that again.
(R) They limit time and we don't.
(I) They have a limited time? What does that mean? Is that what they gave you the impression, that they had a limited amount of time.
(R) Not amount of time but limited time.
(I) That we don't have?
(R) Yes . . . I'm tired
(I) You feel that there are still some things that you should be able to tell us that you haven't been able to?
(R) . . . I don't want to know.

(I) Let's not ask you what you want to know, let's ask what your subconscious wants to know.
(R) It doesn't want to know.
(I) Your subconscious doesn't want to know what happened. What is the reason you don't want to know? Because you were told that you didn't want to know?
(R) I don't know.
(I) Is it possible that you were just told that you wouldn't want to know? Were you given pills?
(R) I was given shots.
(I) With a needle?
(R) Yes.
(I) Did it look like a hypodermic needle?
(R) Yes.
(I) What part of your body was it given?
(R) Here somewhere.
(I) Did you see any marks on your body afterwards?
(R) I didn't. Except on my arm.
(I) So what we can know now is that they gave you some powerful hypnotic drug that made you forget, but that doesn't mean that every part, or some part that still controls and you see that they persuaded you that you wanted to forget. Now your will is strong enough that you can overcome that. You're not sure, but it could be that you could really overcome this spell. Tell us more about the person that looked like a human being. He bent over you.
(R) He had gray hair. He wore glasses. He acted nice.
(I) Did he talk to you?
(R) I see the sky and stars. He let me see them.
(I) Did he tell you anything about the stars in the sky?
(R) He let me think I was riding.
(I) He let you think you were riding.
(R) In the sky.
(I) Did you like that?
(R) It's okay.
(I) Was this just an imagination?
(R) Yes.
(I) He wanted you to imagine this?
(R) And standing out on a cliff by the sea, winds blowing. He made me feel things like that.
(I) Was it pleasant?
(R) Yes.

(I) Was it relaxing?
(R) No, because I knew he was . . .
(I) Do you think he was putting you in hypnosis?
(R) I don't know.
(I) Was it similar to this feeling that you have now?
(R) That's why I don't like this.
(I) Because it's something that you had before that you didn't like.
(R) My head hurts.
(I) Your head doesn't hurt now.
(R) You know it does. It hurts.
(I) I'm going to put my finger down over your finger and as soon as I have stroked your finger three times, it will stop. One, two . . . It's going away. Now three. It's stopped hurting. Is that better now?
(R) Yes. It doesn't hurt now.
(I) I'm glad for that. Now you feel better?
(R) Yes.
(I) Would you like to talk to me a little later today when you are less tired?
(R) Yes.
(I) Would you like to come back and help remember what you were talking about now . . . Now, Pat, it's been a bad experience that you've been through but we're very thankful that you have been able to do this and I think that you will find that it's very valuable in the long run. I want you to feel positive about something that we're going to learn from. That you are going to feel refreshed and going to feel glad that you could remember some things. If you have any trouble with headaches or something like that, we will be able to take care of that. Do you feel good now?
(R) I'm crying.
(I) You're crying? It's all right. It's natural for you to feel this. It was a very emotional experience for you. But it's over.
(R) My kids.
(I) Your kids are all right. If you can get that feeling it is going to be all right. . . .

The session ended at that point. The final session was held was held on the evening of July 9. The session was short.

(R) Medicine. I'm tired.
(I) The medicine made you feel weak or tired?
(R) Tired.
(I) We talked about that before. I want you to go back and think about the

fourth person who was there. Do you still remember what you said before about describing him?

(R) His hair. I remember his hands . . . A rubber tie like . . .

(I) A rubber tie?

(R) . . . to tie on me, my arm except he didn't tie it on my arm. They won't leave me alone.

(I) Let's go back to the time when you said you could see the stars . . . and what he said. Do you remember his voice? What was his voice like? Did it sound mechanical? Did others talk . . . ?

(R) Voices seemed garbly. He talked regularly. He kept talking and talking.

(I) Asking you questions? He didn't command you, did he?

(R) Yes.

(I) What did he tell you to do?

(R) Take them off.

(I) He told you to take something off?

(R) Take a walk. Take a walk. When I was sad.

(I) He made you go back through times you were sad?

(R) Yes . . . do all the things like a woman does.

(I) He made you feel things you didn't want to feel. Did he make you feel happy, too?

(R) Yes.

(I) Did you feel guilty? What were some of the things you wanted to feel?

(R) Nothing subtle.

(I) Is this before they put the thing into you stomach? Before? You don't remember?

(R) I do remember. I think.

(I) It doesn't matter.

(R) They took a long time.

(I) How long was it? Do you think it was half an hour?

(R) Yes. A little more.

(I) Now what's happening?

(R) I see a clock, like a dial. It's going around.

(I) Does it have numbers on it?

(R) I can't see too well. I can see it, but I can't see it too well.

(I) What else do you see?

(R) Glass cases.

(I) You see glass cases? What are in them?

(R) Glass tubes. They're on the wall and they look like they have liquid running through them.

(I) Colored liquid?

(R) I don't know. It's bright. Everything is bright. Let me get down.

(I) You'd like to get down but they won't let you? Did they let you get down then?
(R) Not yet.
(I) What's happening now?
(R) They're wiping me with a cloth on my body. I can feel them. See them. I think it's over.
(I) Are they helping you sit up?
(R) No, they're not doing that. They just got a lift.
(I) They lifted you?
(R) Yes. Just a lift.
(I) Did you lift off the table?
(R) Yes.
(I) Did you feel yourself floating?
(R) Yes.
(I) Then what happened?
(R) I'm standing.
(I) You felt you were in a standing position?
(R) Yes.
(I) You suddenly felt yourself pressing against the floor or are you still floating?
(R) I guess I'm putting my clothes on.
(I) Did you put your own clothes on?
(R) Yes. Nobody helps me. They're gone. There is somebody outside. Standing outside.
(I) Outside the room or outside the ship?
(R) Outside where I am, waiting for me.
(I) You feel he's waiting for you?
(R) Yes. . . . No, I fell asleep with my clothes on. On the couch.
(I) I see. Did you have slacks on?
(R) Yes.
(I) Are you going back to bed?
(R) I'm going home.
(I) You're very tired.
(R) Going past the fence. Where's my children?
(I) Don't worry about your children. They'll be all right. Your children will be all right. You just relax.

The tape and the interview end at that point.

APPENDIX B:

THE ZETA RETICULI CONNECTION

There is one piece of circumstantial evidence that has impressed many people. It is a piece of evidence that was borne of the Betty Hill abduction and which points to a home world of at least some of the alien creatures who many believe are abducting people. Let's look at this evidence and see if it is as persuasive as it seems.

Betty Hill, during one of the hypnotic regression sessions with Dr. Simon, claimed to have seen a star map while on board the alien craft. According to John Fuller, author of *The Interrupted Journey,* Betty kept precise notes of her dreams, writing them down while the details were fresh in her mind. These notes, according to Fuller are very similar to the hypnotically regressed testimony recovered by Dr. Simon.

According to the notes, as published by Fuller and later by Jerry Clark in volume three of his UFO encyclopedia, Betty ". . . asked where he [the leader of the alien crew] was from, and he asked if I knew anything about the universe. I said no, but I would like to learn. He went over to the wall and pulled down a map, strange to me. Now I would believe this to be a sky map. It was a map of the heavens, with numerous sized stars and planets, some large, some only pinpoints. Between many of these, lines were drawn, some broken lines, some light solid lines, some heavy black lines. They were not straight, but curved. Some went from one planet to another, to another, in a series of lines. Others had no lines, and he said the lines were expeditions. He asked me where the earth was on this map, and I admitted that I had no idea. He became slightly sarcastic and said that if I did not know where the earth was, it was impossible to show me where he was from; he snapped the map back into place."

Simon had suggested that Betty draw the star map but she was reluctant to do so, afraid that her poor artistic skills would not allow a proper duplicate. Simon then suggested that she should draw the map when she felt ready to do so. Not long after the session, she produced a map with twelve points on it showing the connections among the stars. The solid lines were for trade routes and the broken, or dotted, lines were expeditionary routes. Fuller published the map in *The Interrupted Journey*.

In April 1965, the *New York Times* printed a map of the constellation Pegasus because Russian astronomers had found what they believed to be an artificial radio source near it. Betty Hill, seeing the map, was surprised by how closely it resembled the star map she had seen. She even applied the star names from the *Times* map to her sketch suggesting that the alien creatures home star was either Homan or Baham. This map, of course, did not show our sun on it.

Marjorie Fish, a third grade teacher and amateur astronomer from Ohio and later a research assistant at the Oak Ridge National Laboratory, became intrigued with the Hill star map. Fish believed that she could figure it out and learn which star was the home to the aliens.

There were few clues for her other than what Betty Hill told her during the investigation by Fish. She assumed one of the stars that was connected to the others with lines belonged to our sun. She assumed that the map represented our section of the galaxy, that the aliens would be interested in stars of the same type as our sun, the travel patterns would make some sense, and the travel patterns would avoid the largest stars and those that are not on the main sequence (that is, stars that are basically stable for long periods of time and like our sun).

Fish built a number of three-dimensional representations of our section of the galaxy and then viewed them from different angles, searching for the Hill pattern. Eventually she found one with the stars Zeti 1 and Zeta 2 Reticuli as the base. They are separated by "light weeks" but are far enough apart that planetary orbits would be stable and life could evolve on those planets.

But others were also searching for the pattern. Charles W. Atterberg found a pattern that had the stars Epsilon Eridani and Epsilon Indi as the base stars. It too fits with the Hill map, and two of the stars on it, Tau Ceti and Epsilon Eridani, were targeted by Project Ozma, one of the first of the SETI searches. In other words, astronomers involved in the search for extraterrestrial intelligences believed that two of the stars in the Atterberg interpretation were likely candidates for planetary systems and intelligent life. Tau Ceti was also one of the candidates on the Fish map.

Suddenly we have three published interpretations of Betty Hill's star map, all of which make sense. But Marjorie Fish disagreed. Of the Atterberg interpretation, she noted that he had included some red drawfs as stars visited by the aliens. She said that she had ruled out red drawfs because there are so many of them and if she used red drawfs in a logical construction, then all the lines would be used before the travel patterns reached Earth. She had assumed that the sun would be one of the stars connected to the others on the map, although the "leader" of the alien crew had provided no indication that this was true.

She also assumed that if they, the aliens, were interested in red drawfs, that is, that they visited some, then there should have been lines connecting other red drawfs, but there were not. Her assumption was that one red drawf would be as interesting to a spacefaring race as the next. But it could be that some red dwarfs were more interesting because of things we cannot see. Because we can detect no difference between one red drawf and another doesn't mean that there aren't differences.

She makes other, similar assumptions in her rejection of Atterberg's model. She notes that a number of relately close double stars, such as 61 Cygni, Struve 2398, Groombridge 34, and Kruger 60, are part of Atterberg's pattern but that there is no line to Alpha Centuri. Once again, she assumes that the alien race would be visiting Alpha Centuri if they were visiting the other double star systems, and once again we can point out that there might be something of great interest in the systems visited but not in the one that is bypassed. It should also be noted that according to some astronomy texts, Alpha Centuri is a triple star system and that might be the reason for exclusion by the aliens. It is highly unlikely that a triple star would have any planetary systems. All this is, of course, guesswork.

What we have, and what we find interesting, is that the UFO community has embraced the Fish interpretation of the Hill star map and rejected the others, including the one found by Betty Hill herself. The acceptance of the Zeti Reticuli interpretation is based on an earthbound logic that presupposes we can understand the workings of an alien mind. It presupposes that we can apply our logic to a star map without having the necessary information to make that logic valid.

If we look at our own history of exploration, we find what would be logical gaps. But these gaps can be explained by the circumstances of the expeditions, the reasons for the exploration, and the benefits that one trade route held over another. If we don't have that information, then the logic of the situation defies us and we creates erroneous interpretation.

According to Jerry Clark, the solid lines on the Hill star map denotated trade routes and the broken lines denotated expeditions. It seems to us that no trade exists with these alien creatures so that the sun should not be noted as a trade route. A case can be made for the sun being on an expedition route but not on a trade route. This would appear to be another logical fallacy of the Fish interpretation. If the sun is connected at all, it should be as part of an expeditionary route.

This leads to another important question. Why was the assumption made that the earth, or rather the sun, would be among the stars that are connected by the lines? There is no evidence that it should be. There was nothing said to Hill to suggest it, yet this is one of the basic assumptions used by Fish. If it is eliminated, then the other interpretations are as logical and legitimate as that made by Fish.

There is another point to be made here. This star map was "discovered" during one of the hypnotic regression sessions, at least according to some of those interested in it. It came from Betty Hill's memory during one of those sessions. Although Frank Salisbury of the University of Utah did say that the story and map came to light under hypnosis, and that is good evidence that it actually took place, we know know that such claims are in error. As we have seen repeatedly, hypnosis is a poor investigative tool that is more likely to create false memories than actually to access real ones. The hypnotically retreived memories are less reliable than those accessed in other ways.

Of course, according to Fuller, Betty Hill had "dreamed" the star map long before she had said anything to Simon under hypnosis. To some that has suggested a conscious memory of the event prior to the hypnosis, but that isn't actually the case. The memories returned to her in dreams and she carefully logged them. There is no scientific evidence to suggest that "memories" that surface in dreams, especially those not remembered in a conscious or waking state, are any more reliable than those retrieved with hypnosis.

So where are we on this? One prominent UFO researcher has suggested that we know some of the "grays" come from the Zeta Reticuli system. But we don't know that. What we have is a bit of evidence, retrieved under hypnosis, that has at least three good interpretations for it. There is no logical reason to accept one interpretation over another. All fit the pattern created by Betty Hill.

In fact, over the years, these other interpretations have slipped from the UFO literature. Clark, in his UFO encyclopedia, made no mention of either Betty Hill's interpretation or that by Atterberg. Instead, he focused solely on the Fish model.

Clark does, however, explain where the theory that Zeta 1 and Zeta 2 Reticuli is a double star originated. According to him, a French astronomer suggested precisely this. However, later observations by that same astronomer proved this not to be true and he withdrew his claim.

What we really know is that Betty Hill's star map was created during her dreams and reinforced by hypnotic regression. No evidence has been presented to suggest that the map is valid. It is, in essence, twelve dots connected by various lines. It is not a very good clue as to the home world of the aliens, it is not solid evidence that the event took place, and it has misled millions who believe that here is evidence of alien visitation and alien abduction.

We can argue the statistical significance of the Fish interpretation of the Hill model, we can argue that Fish's criterion for selection of the stars in the map is solid, and we can argue that hypnotically regressed testimony is solid and valuable, but when all is said and done we are once again left with no solid evidence. What we have is an interesting aberration in the abduction phenomenon that does nothing to advance our understanding but certainly clouds the issue with seeming corroborative data. In the end we are left where we began, with nothing in the way of hard evidence. Instead we are left with the nightmares of a kind, sincere woman who believes she was abducted by aliens.

BIBLIOGRAPHY

"Abduction in Western Kansas." *International UFO Reporter* 1, no. 2 (Nov./Dec. 1976): 12.

"Abduction in Western Kansas." *International UFO Reporter* 2, no. 10 (Sept./Oct. 1977): 4–7.

Adams, Jean. "Contactee: Firsthand." *UFO* 3, no. 1 (1988): 31–32.

Adamski, George. *Behind the Flying Saucer Mystery*. New York: Paperback Library, 1967.

———. *Inside the Flying Saucers*. New York: Paperback Library, 1967.

American Medical Association. "Scientific Status of Refreshing Recollection by the Use of Hypnosis." *International Journal of Clinical and Experimental Hypnosis* 34, no. 1 (1986): 1–12.

Anderson, Michele. "Biospex: Biological Space Experiments." NASA Technical Memorandum 58217, NASA, Washington, D.C., 1979.

Appelle, S. "The Abduction Experience: A Critical Evaluation of Theory and Evidence." *Journal of UFO Studies*, new Series, 6 (1995/1996): 29–78.

———. "Federal Policy for the Protection of Human Subjects and Its Applicability to Abduction Research," in *Alien Discussion: Proceedings of the Abduction Study Conference*, 1996.

———. "On Behavior Explanation of UFO Sightings." *Perceptual and Motor Skills* 32, no. 3, (June 1971): 994.

Arndt, William B., John C. Foehl, and F. Elaine Good. "Specific Sexual Fantasy Themes: A Multidimensional Study." *Journal of Personality and Social Psychology* 48, no. 2 (1985): 472–80.

Asimov, Issac. *Is Anyone There?* New York: Ace Books, 1967.

Atic UFO Briefing, April 1952, Project Blue Book Files.

"The Aurora, Texas, Case," *APRO Bulletin* (May/June 1973): 1, 3–4.

Baker, Robert A. "Are UFO Abductions for Real? No, No, a Thousand Times No!" *Journal for UFO Studies* 1 (1989): 104–10.

———. "Hypnosis, Memory, and Incidental Memory." *American Journal of Clinical Hypnosis* 25, no. 4 (1983): 253–300.

———. *They Call It Hypnosis*. New York: Prometheus Books, 1990.

Balch, R. W., and D. Taylor. "Seekers and Saucers: The Role of the Cultic Milieu in Joining a UFO Cult." *American Behavior Scientist* 20, no. 6 (July/Aug. 1977): 839–60.

Barber, T. X., N. P. Spanos, and J. F. Chaves. *Hypnosis, Imagination, and Human Potentialities*. New York: Pergamon Press, 1974.

Barker, Gray. *They Knew Too Much about Flying Saucers*. New York: University Books, 1967.

Bartholomew, R. E., K. Basterfield, and G. S. Howard. "UFO Abductees and Contactees: Psychopathology or Fantasy Proneness?" *Professional Psychology Research and Practice* 22, no. 3 (June 1991): 215–22.

Basterfield, Keith. "Abductions and Post-Traumatic Stress Disorder." *International UFO Reporter* (May/June 1994): 20, 23–24.

———. "Implants." *International UFO Reporter*. (Jan./Feb. 1992): 18–20.

Baxter, John, and Thomas Atkins. *The Fire Came By*. Garden City, NY: Doubleday, 1976.

Beckley, Timothy Green. "Kidnapped by Aliens! The True Story of Carl Hignon's Incredible Contact." *UFO Report* 2, no. 5 (Fall 1975): 41–43, 71–79.

Bell, C. "Panic Attacks: Relationship to Isolated Sleep Paralysis. *American Journal of Psychiatry* 143 (1986): 1484.

Bender, Albert K. *Flying Saucers and the Three Men*. New York: Paperback Library, 1968.

Benjamin, Sara. "Contactee: Firsthand." *UFO* 4, no. 4 (1989): 37–38.

Bennetts, Leslie. "Nightmares on Main Street." *Vanity Fair* 56, no. 6 (June 1993): 42–51.

Bergman, J. S., H. Grahm, and H. C. Leavitt. "Rorschach Exploration of Consecutive Hypnotic Age Level Regression." *Psychosomatic Medicine* 2 (1947): 20–28.

Binder, Otto. *Flying Saucers Are Watching Us*. New York: Tower, 1968.

———. "The Secret Warehouse of UFO Proof." *UFO Report*.

———. *What We Really Know About Flying Saucers*. Greenwich, Conn.: Fawcett Gold Medal, 1967.

Blackmore, S. "Alien Abduction: The Inside Story. *New Scientist* 44 (1994): 29–31.

Bliss, E. L. *Multiple Personality, Allied Disorders and Hypnosis*. New York: Oxford University Press, 1986.

Bloecher, Ted. *Report on the UFO Wave of 1947*. Washington, D.C.: Author, 1967.

Blum, Howard. *Out There: The Government's Secret Quest for Extraterrestials*. New York: Simon and Schuster, 1991.

Blum, Ralph, with Judy Blum. *Beyond Earth: Man's Contact with UFOs*. New York: Bantam, 1974.

Bowen, Charles (ed). *The Humanoids*. Chicago: Henry Regency, 1969.

Boylan, Richard J. *Close Extraterrestrial Encounters*. Tigard, OR: Wild Flower, 1994.

Brooksmith, Peter. "Do Aliens Dream of Jacob's Sheep?" *Fortean Times* 83 (Oct. 1995): 22–29.

———. *UFO: The Complete Sightings*. New York: Barnes & Noble, 1995.

Brunvand, Jan Harold. *The Choking Doberman and Other "New" Urban Legends*. New York: W. W. Norton & Co., 1984.

———. *The Vanishing Hitchhiker: American Urban Legends and Their Meanings*. New York: W. W. Norton & Co., 1981.

Bryan, C. D. B. *Close Encounters of the Fourth Kind: Alien Abductions, UFOS, and the Conference at M.I.T.* New York: Alfred A. Knopf, 1995.

Buckle, Eileen. "Aurora Spaceman—R.I.P.?" *Flying Saucer Review* (July/Aug. 1973): 7–9.

Bullard, Thomas E. "Epistemological Totalitarianism: The Skeptical Case Against Abductions." *International UFO Reporter* (Sept./Oct. 1994): 9–16.

———. "Folkloric Dimensions of the UFO Phenomenon." *Journal of UFO Studies*, new series, 3 (1991): 1–58.

———. "Hypnosis No 'Truth Serum.' " *UFO* 4, no. 2 (1989): 31–35.

———. *UFO Abductions: The Measure of the Mystery, Vol. 1: Comparative Study of Abduction Reports. Vol 2: Catalog of Cases*. Mt. Rainier, MD: Fund for UFO Research, 1987.

Campbell, R. J. *Psychiatric Dictionary*. New York: Oxford University Press, 1996.

Canadeo, Anne. *UFO's The Fact or Fiction Files*. New York: Walker, 1990.

Cannon, Martin. "Contactee: Firsthand." *UFO* 3, no. 5 (1988): 30–32.

Carpenter, John. "Abduction Notes: Reptilians and Other Unmentionables." *MUFON UFO Journal* 300 (Apr. 1993): 10–11.

———. "Commentary," in *Unusual Personal Experiences: An Analysis of the Data from Three National Surveys*. Las Vegas: Bigelow Holding Co., 1994.

———. "Double Abduction Case: Correlation of Hypnosis Data." *Journal of UFO Studies*, new series, 3 (1991): 91–114.

———. "The Significance of Multiple Participant Abductions." In *MUFON 1996 International UFO Symposium Proceedings*. Seguin, TX: MUFON, 1996.

Catoe, Lynn E. *UFOs and Related Subjects: An Annotated Bibliography*. Washington, D.C.: Government Printing Office, 1969.

Chariton, Wallace O. *The Great Texas Airship Mystery*. Plano, TX: Wordware, 1991.

Chavarria, Hector. "*El Caso Puebla*." OVNI: 10–14.

Clark, Jerome. "Airships: Part I." *International UFO Reporter* (Jan./Feb. 1991): 4–23.

———. "Airships: Part II." *International UFO Reporter* (Mar./Apr. 1991): 20–23.

———. "The Bizarre Sandy Larson Contact: UFO Abduction in North Dakota." *UFO Report* 3, no. 3 (Aug. 1976): 21–23, 46–53.

———. "Conversation with Budd Hopkins." *International UFO Reporter* (Nov./Dec. 1988): 4–12.

———. "Dr. Berthod Eric Schwaz—Psychiatrist–Researcher." *UFO Report* 3, no. 4 (Oct. 1976): 27–29, 52–54.

———. *The Emergence of a Phenomenon: UFOs from the Beginning through 1959*. Detroit: Omnigraphics, 1992.

———. "The Great Unidentified Airship Scare." *Official UFO* (Nov. 1976).

———. *High Strangness: UFOs from 1960 through 1979*. Detroit: Omnigraphics, 1996.

———. *UFO's in the 1980s*. Detroit: Apogee, 1990.

Cohen, Daniel. *Encyclopedia of the Strange*. New York: Avon, 1987.

———. *The Great Airship Mystery: A UFO of the 1890s*. New York: Dodd, Mead, 1981.

———. *UFOs—The Third Wave*. New York: Evans, 1988.

Conesa, Jorge. "Relationship Between Isolated Sleep Paralysis and Geomagnetic Influences: A Case Study." *Perceptual and Motor Skills* 80, (1995): 1263–73.

Conn, J. H. "Is Hypnosis Really Dangerous?" *International Journal of Clinical and Experiemental Hypnosis* 20 (1972):61–69.

Cooper, Vicki. "*The Psychology of Abduction.*" *UFO* 3, no. 1 (1988):14–18.

Crawford, H. J., and S. N. ALLEN. "Enhanced Visual Memory During Hypnosis as Mediated by Hypnotic Responsiveness and Cognitive Strategies. *Journal of Experimental Psychology: General* 112, no. 4 (1983):662–85.

Curtis, William. "Contactee: Firsthand." *UFO* 4, no. 3. (1989):36–38.

Dahlitz, M. "Sleep Paralysis." *Lancet* 341 (1993):406–7.

Davenport, M. *Visitors from Time: The Secret of the UFOs*. Tigard, OR: Wildflower Press, 1992.

Davidson, Leon, ed. *Flying Saucers: An Analysis of Air Force Project Blue Book Special Report No. 14*. Clarksburg, VA: Saucerian Press, 1971.

Davies, John K. *Cosmic Impact*. New York: St. Martin's, 1986.

Davis, L. "A Comparison of UFO and Near Death Experiences as Vehicles for the Evolution of Human Consciousness." *Journal of Near Death Studies* 6, no. 4 (Summer 1988):240–57.

Davis, Sally Ogle. "Sympathy for the Devil." *Los Angele Magazine* 38, no. 7 (July 1993):48–56.

De Felitta, F. *The Entity*. New York: Warner Books, 1978.

Dittburner, T. L., and M. A. Persinger. "Intensity of Amnesia During Hypnosis Is Positively Correlated with Estimated Prevalence of Sexual Abuse and Alien Abduction: Implications for the False Memory Syndrome. *Perceptual and Motor Skills* 3 (1993):895–98.

"DoD News Releases And Fact Sheets," 1952–1968.

Druffel, Ann. "Missing Fetus Case Solved." *MUFON UFO Journal* 283 (1991): 8–12.

———. "Resisting UFO 'Entities'." *UFO* 4, no. 4 (1989):16–19.

Druffel, Ann, and Rogo, D. Scott. *The Tujunga Canyon Contacts*. Englewood Cliffs, NJ: Prentice-Hall, Inc., 1980.

Eberle, Paul, and Shirley Eberle. *The McMartin Preschool Trial*. Amherst, NY: Prometheus Books, 1993.

Ecker, Vicki Cooper. "Alien Implants, Part II: Fate of the Evidence." *UFO* (May/June 1996):24–28.

Editors of *Look*. "Flying Saucers." *Look* (1966).

Edwards, Frank. *Flying Saucers—Here and Now!* New York: Bantam, 1968.

———. *Flying Saucers—Serious Business*. New York: Bantam, 1966.

———. *Strange World*. New York: Bantam, 1964.

Edwards, Nick. "Contactee: Firsthand." *UFO* 5, no. 2. (1990):40.

Estes, Russ (producer). "Faces of the Visitors." Crystal Sky Productions, 1997.

———. "Magic and Mystery of Giant Rock." Crystal Sky Productions, 1997.

———. "Quality of the Messenger." Crystal Sky Productions, 1993.

———. "Roswell Remembered." Crystal Sky Productions, 1996.

Evans, Hilary. "Abduction—What Do We Tell Abigail?" *UFO* 3, no. 4 (1988): 10–11.

Everett, H. C. "Sleep Paralysis in Medical Students." *Journal of Nervous and Mental Disease* 135 (1963):283–87.

Fawcett, Bill, ed. *Making Contact*. New York: Morrow, 1997.

Fawcett, Lawrence, and Barry J. Greenwood. *Clear Intent: The Government Cover-up of the UFO Experience*. Englewood Cliffs, NJ: Prentice-Hall, 1984.

Feldman, Gail Carr. *Lessons in Evil, Lessons from the Light: A True Story of Satanic Abuse and Spiritual Healing*. New York: Crown, 1992.

Fenny, Tom. "The Satan Factor." *Maclean's* 105, no. 25 (June 22, 1992):29.

Festinger, Leon, Henry Riecken, and Stanley Schachter. *When Prophecy Fails*. New York: Harper Torchbooks, 1964.

Finney, Ben R., and Eric M. Jones *Interstellar Migration and the Human Experience*. Berkeley, CA: University of California Press, 1985.

Fiore, E. *Encounters*. New York: Ballantine, 1989.

———. *The Unquiet Dead: A Psychologist Treats Spirit Possession*. New York: Ballantine, 1991.

———. *You Have Been Here Before: A Psychologist Looks at Past Lives*. New York: Ballantine, 1991.

First Status Report, Project STORK (Preliminary to Special Report No. 14), Apr. 1952.

Fischer, David Hackett. *Historians' Fallacies: Toward a Logic of Historical Thought*. New York: Harper Torchbooks, 1970.

"Flying Saucers Are Real." *Flying Saucer Review* (Jan./Feb. 1956):2–5

Ford, Brian. *German Secret Weapons: Blueprint for Mars*. New York: Ballantine, 1969.

Foster, Tad. Unpublished articles for Condon Committee Casebook, 1969.

Fowler, Raymond E. *Allagash Abductions*. Tigard, OR: Wild Flower Press, 1993.

———. *The Andreasson Affair*. Englewood Cliffs, NJ: Prentice Hall, 1979

———. *The Andreasson Affair, Phase Two*. Englewood Cliffs, NJ: Prentice-Hall, 1982.

———. *Casebook of a UFO Investigator*. Englewood Cliffs, NJ: Prentice-Hall, 1981.

———. *The Watchers*. New York: Bantam, 1990.

———. *The Watchers II: Exploring UFOs and the Near Death Experience*. Tigard, OR: Wild Flower Press, 1993.

———. "What about Crashed UFOs?" *Official UFO* (April 1976):55–57.

Fraser, G. A. "Satanic Ritual Abuse: A Cause of Multiple Personality Disorder." *Journal of Child and Youth Care*. Special Issue (1990):55–66.

Friedman, W. J. "Memory of the Time of Past Events." *Psychological Bulletin* 113, no. 1 (1993):44–64.

Fukuda, K. "One Explanatory Basis for the Descrepancy of Reported Prevalence of Sleep Paralysis Among Healthy Respondants." *Perceptual and Motor Skills* 77 (1995): 803–7.

———. "Personality of Healthy Young Adults with Sleep Paralysis." *Perceptual and Motor Skills* 73 (1991): 955–62.

Fuller, John G. *Aliens in the Sky*. New York: Berkley Books, 1969.

———. *Incident at Exeter*. New York: G. P. Putnam's Sons, 1966.

———. *The Interrupted Journey*. New York: Dial, 1966.

Garry, M., and E. Loftus. "Pseudo-memories without Hypnosis." *International Journal of Clinical and Experimental Hypnosis* 42, no. 4 (Oct. 1994):363–78.

Gillmor, Daniel S., ed. *Scientific Study of Unidentified Flying Objects*. New York: Bantam, 1969.

Gindes, B. C. "Delusional Production Under Hypnosis." *International Journal of Clinical and Experimental Hypnosis* 11, no. 1 (1963): 1–10.

Goertzel, Ted. "Measuring the Prevalence of False Memories: A New Interpretation of a 'UFO Abduction Study.' " *Skeptical Inquirer* 18 (Spring 1994): 266–72.

Goldsmith, Donald. *Nemesis*. New York: Berkley Books, 1985

———. *The Quest for Extraterrestrial Life*. Mill Valley, CA: University Science Books, 1980.

Goldstein, E. *Confabulations*. Boca Raton, FL: SIRS Books, 1992.

Goldstein, E., and K. Farmer. *True Stories of False Memories*. Boca Raton, FL: SIRS Books, 1993.

Good, Timothy. *Above Top Secret*. New York: Morrow, 1988.

———. *Alien Contact*. New York: Morrow, 1993.

———. *The UFO Report*. New York: Avon, 1989

Goode, G. B. "Sleep Paralysis." *Archives of Neurology* 6 (1962):228–34.

Gotlib, David A. "Abductions: Imagined or Imaginal?" *International UFO Reporter* (July/Aug. 1992): 16–20.

Gotlib, David, Stuart Appelle, Georgia Flamburis, and Mark Rodeghier. "Ethics Code for Abduction Experience Investigation and Treatment." *Journal of UFO Studies*, new series, 5 (1994):55–81.

Grabowski, K., N. Roese, and M. Thomas. "The Role of Expectancy in Hypnotic Hypermnesia: A Brief Communication." *International Journal of Clinical and Experimental Hypnosis* 39, no. 4 (Oct. 1991):193–7.

Gregory, Janet. "Similarities in UFO and Demon Lore." *Flying Saucer Review* 17, no. 2 (Mar./Apr. 1971):32.

Grof, Stanislav. *Adventures in Self Discovery*. Albany, NY: State University of New York Press, 1988.

———. *Beyond the Brain: Birth, Death, and Transcendence in Psychotherapy*. Albany, NY: State University of New York Press, 1985.

Hadfield, J. A. *Dreams and Nightmares*. Baltimore: Penguin Books, 1971.

Haisell, D. *The Missing Seven Hours*. Markham, Ont, Canada: Paperjacks, 1978.

Haley, Leah. *Lost Was the Key*. Tuscaloosa, AL: Greenleaf Publications, 1993.

Hall, Robert L., Mark Rodeghier, and Donald A. Johnson. "The Prevalence of Abductions: A Critical Look." *Journal of UFO Studies*, new series, 4 (1992): 131–5.

Hardy, Phil. *The Encyclopedia of Science Fiction Movies*. Minneapolis, MN: Woodbury Press, 1986.

Hishikawa, Y., in *Narcolepsy: Proceedings of the First International Symposium on Narcolepsy, Advances in Sleep Research,* vol. 3. New York: Spectrum Publications, 1976.

Hoed, J., E. A. Lucas, and W. C. Dement. "Hallucinatory Experiences During Cataplexy in Patients with Narcolepsy. *American Journal of Psychiatry* 136 (1979):1210–11.

Holzer, Hans. *The Ufonauts*. New York: Fawcett Gold Medal, 1976.

Hopkins, Budd. *Intruders: The Incredible Visitation at Copley Woods*. New York: Random House, 1987.

———. "Invisibility and the UFO Abduction Phenomenon." In *MUFON 1993 International UFO Symposium Proceedings*. Seguin, TX: MUFON, 1993.

———. "The Linda Cortile Abduction Case, Parts I, II." *MUFON UFO Journal*. New York: Intruders Foundation.

———. *Missing Time: A Documented Study of UFO Abductions*. New York: Richard Marek Publishers, 1981.

———. "One Ufologist's Methodology." *UFO* 4, no. 2 (1989): 26–30.

———. "The Sixth Witness in the Linda Cortile Abduction Case." In *MUFON 1996 International UFO Symposium Proceedings*. Seguin, TX: MUFON, 1996.

Hopkins, Budd, David, Jacobs, and R. Westrum. *Unusual Personal Experiences: An Analysis of the Data from Three National Surveys*. Las Vegas: Bigelow Holding Co., 1991.

Hudson, P. S. "Ritual Child Abuse: A Survey of Symptoms and Allegations" and "In the Shadow of Satan: The Ritual Abuse of Children. *Journal of Child and Youth Care*. Special Issue (1990):27–54.

Hufford, David. *The Terror That Comes in the Night*. Philadelphia, PA: University of Pennsylvania Press, 1982.

Houran, J. "Tolerance of Ambiguity and Perception of UFOs." *Perceptual and Motor Skills*. (1997): 973–970.

Houran, J., Large, R., and Crist-Houran, M. "An Assessment of Contextual Mediation in Trance States of Shamanic Journeys." *Perceptual and Motor Skills*. 85(197) 59–65.

Huyghe, Patrick. *The Field Guide to Extraterrestrials*. New York: Avon Books, 1996.

———. "The Great High Rise Abduction." *Omni* (April 1994): 62–7, 96–7.

Hynek, J. Allen. *The UFO Experience: A Scientific Inquiry*. Chicago: Henry Regency, 1975.

Hynek, J. Allen, and Jacques Vallee. *The Edge of Reality*. Chicago: Henry Regency, 1972.

Jacobs, David M. *Secret Life*. New York: Simon and Schuster, 1992.

———. *The Threat: The Secret Agenda: What the Aliens Really Want . . . And How They Plan to Get It*. New York: Simon & Schuster, 1998.

———. *The UFO Controversy in America*. New York: Signet, 1975.

Jacobs, David, and Budd Hopkins. "Suggested Techniques for Hypnosis and Therapy of Abductees." *Journal of UFO Studies*, new series, 4 (1992):138–50.

Jaynes, J. *The Origin of Consciousness in the Breakdown of the Bicameral Mind*. Boston: Houghton Mifflin Co., 1976.

Jaroff, Leon. "Lies of the Mind." *Time* (November 29, 1993):52–55.

Jones, E. M. *On the Nightmare*. International Psycho-Analytical Library, no. 20. London: Hogarth Press, 1931.

Jordan, Debbie, and Kathy Mitchell. *Abducted!* New York: Dell, 1995.

Jung, Carl G. *Flying Saucers: A Modern Myth of Things Seen in the Sky*. New York: Harcourt, Brace, 1959.

Kagan, Daniel, and Ian Summers. *Mute Evidence*. New York: Bantam, 1983.

Kahn, E., A. Edwards, D. M. Davis, and J. Fine. "Psycho-physiological Study of Nightmares and Night Terrors: Mental Content and Recall of Stage Four Night Terrors." *Journal of Nevous and Mental Disease* (1974):174–88.

Kasten, Kathy. "Contactee: Firsthand." *UFO* 4, no. 1 (1989):33–35.

Keel, John. *The Complete Guide to Mysterious Beings*. New York: Doubleday, 1994.

———. *Strange Creatures from Space and Time*. New York: Fawcett, 1970.

———. *UFOs: Operation Trojan Horse*. New York: G. P. Putnam's Sons, 1970.

Keyhoe, Donald E. *Aliens from Space*, New York: Signet, 1974.

———. *Flying Saucers from Outer Space*. New York: Henry Holt, 1953.

Klass, Philip J. *UFOs Explained*. New York: Random House, 1974.

———. *The Public Decieved*. Buffalo, NY: Prometheus Books, 1983.

Kinder, Gary. *Light Years: An Investigation into the Extraterrestrial Experiences of Eduard Meier*. New York: Atlantic Monthly Press, 1987.

Knowlton, Janice, and Michael Newton. *My Daddy was the Black Dahlia Killer*. New York: Pocket Books, 1995.

Kottmeyer, Martin. "Abduction and the Boundary Deficit Hypothesis." *Magonia* 32 (Mar. 1988).

———. "The Creative Fire." *Magonia*, 29 (Apr. 1988).

———. "Entirely Unpredisposed." *Magonia*, 36 (Jan. 1990).

———. "Ufology as an Evolving System of Paranoia." *UFO* 7, no. 3 (1992): 28–34.

———. "A Universe of Spies." *Magonia*, 39 (Apr. 1991).

Lange, R. and Houran, J. Context-induced paranormal experiences: Support for Houran and Lange's model of haunting phenomena. *Perceptual and Motor Skills*, 84, (1997) 1455–1458

———. Delusions of the paranormal: A haunting question of perception. *Journal of Nervous and Mental Disease*. (1998) 637–645.

———. Fortean phenomena on film: Evidence or artifact? *Journal of Scientific Exploration*, Vol. 11, No. 1. (1997) 41–46.

———. Role of contexual mediation in direct versus reconstructed angelic encounters. *Perceptual and Motor Skills*. 83. (1996) 1259–1270.

Laibow, Rima E. "Clinical Discrepancies Between Expected and Observed Data in Patients Reporting UFO Abductions: Implications for Treatment." Manuscript, 1989.

———. *Experienced Anomalous Trauma: Physical, Psychological and Cultural Dimensions*. New York: Brunner Mazel, 1992.

Lanning, K. V. "Ritual Abuse: A Law Enforcement View or Perspective. *Child Abuse and Neglect* 15 (1991): 171–3.

Lawson, A. H. "Perinatal Imagery in UFO Abduction Reports." *The Journal of Psychohistory* 12, no. 2 (Fall 1984).

Leach, Robert Terrell. "Abduction Pioneer." *UFO* 11, no. 3 (May/June 1996): 28–36.

Lee, Aldora and Ronnal L. "Evaluation of the Status Inconsistancy Theory of UFO Sightings." *Catalog of Selected Documents in Psychology* 2 (Spring 1972): 66.

Leiman, E. B., and S. Epstein. "Thematic Sexual Responses as Related to Sexual Drive and Guilt." *Journal of Abnormal and Social Psychology* 63 (1961):169–75.

Leir, Roger K. "Medical and Surgical Aspects of the UFO Abduction Phenomenon." In *MUFON 1996 International UFO Symposium Proceedings*. Seguin, TX: MUFON, 1996.

Levitt, E. "A Reversal of Hypnotically Refreshed Testimony: A Brief Communication." *International Journal of Clinical and Experimental Hypnosis* 38, no. 1 (Jan. 1990): 6–9.

Lewis, James R., ed. *The Gods Have Landed: New Religions from Other Worlds*. Albany, NY: State University of New York Press, 1995.

Library of Congress Legislative Reference Service, "Facts about UFOs," May 1966.

Loftus, R. *Eye-Witness Testimony*. Cambridge, MA: Harvard University Press, 1979.

Loftus, E., and K. Ketcham. *The Myth of Repressed Memory*. New York: St. Martin's Griffin, 1994.

Lore, Gordon, and Harold H. Deneault. *Mysteries of the Skies: UFOs in Perspective*. Englewood Cliffs, NJ: Prentice-Hall, 1968.

Lorenzen, Coral. *The Shadow of the Unknown*. New York: New American Library, 1970.

———. "The Disappearance of Rivalino Mafra da Silva: Kidnapped by a UFO?" *Fate* 16, no. 6 (June 1963): 26–33.

Lorenzen, Carol and Jim. *Abducted!* New York: Berkley 1977.

———. *Encounters with UFO Occupants*. New York: Berkley, 1976.

———. *Flying Saucer Occupants*. New York: Signet, 1967.

———. *Flying Saucers: The Startling Evidence of the Invasion from Outer Space*. New York: Signet, 1966.

Low, Dr. Robert J. "Letter to Lt. Col. Robert Hippler," Jan. 27, 1967.

Ludwig, A. L. "An Historical Survey of the Early Roots of Mesmerism. *International Journal of Clinical and Experimental Hypnosis* 12 no. 4 (1964): 205–17.

Lynn, S. J., M. Milano, and J. R. Weeks. "Hypnosis and Pseudomemories: The Effects of Prehypnotic Expectancies." *Journal of Personality and Social Psychology* 60, no. 2 (1991): 318–26.

Lynn, S. J., J. W. Rhue, B. P. Meyers, and J. R. Weeks. "Psuedomemory in Hypnotized and Stimulated Subjects." *International Journal of Clinical Hypnosis* 42 (1994).

Maack, Iris. "Truck, Rig Abducted" *The APRO Bulletin* (Dec. 1979): 1–3.
Mack, John E. *Abduction*. New York: Charles Scribner's Sons, 1994.
———. "Helping Abductees." *International UFO Reporter* 17, no. 4 (1992): 10–15, 20.
———. "Studying Intrusions from the Subtle Realm: How Can We Deepen Our Knowledge?" In *MUFON 1996 International UFO Symposium Proceedings*. Seguin, TX: MUFON, 1996.
Maclean, Harry N. *Once Upon a Time: A True Story of Memory, Murder and the Law*. New York: Dell Publishing Group, 1994.
Mayer, R. S. *Satan's Children*. New York: Avon, 1991.
"McClellan Sub-Committee Hearings," Mar.1958.
"McCormack Sub-Committee Briefing," Aug. 1958.
Mcshane, C. "Satanic Sexual Abuse: A Paradigm." *Journal of Women and Social Work* 8, no. 2 (1993): 200–12.
Menzel, Donald H., and Lyle G. Boyd. *The World of Flying Saucers*. Garden City, NY: Doubleday, 1963.
Menzel, Donald H., and Ernest H. Taves. *The UFO Enigma*. Garden City, NY: Doubleday, 1977.
Michael, Allen. "Contactee: Firsthand." *UFO* 5, no. 1 (1990):31–2.
Michel, Aime. *The Truth about Flying Saucers*. New York: Pyramid, 1967.
Montgomery, Ruth, *Aliens Among Us*. New York: G. P. Putnam's Sons, 1985.
Moreault, D., and D. R. Follingstad. "Sexual Fantasies of Females as a Function of Sex Guilt and Experimental Response Cues." *Journal of Consulting and Clinical Psychology* 46 (1978):1385–93.
Moore, Steve. "Close Encounters: A Mythic Perspective." *Fortean Times* 37 (Spring 1982):34–9, 64.
Mundy, Jean. "A Case of Unnatural Pregnancy." *UFO* 6, no. 4 (1991):23–4, 44.
Musser, Dale. "Anatomy of a Mass UFO Abduction." *UFO Universe* (Fall 1993): 24.
Mutter, C. B. "Hypnosis with Defendants: Does It Really Work?" *American Journal of Clinical Hypnsis* 32 no. 4 (1990):257–62.
Nathan, Debbie, and Michael Snendeker. *Satan's Silence*. New York: Basic Books, 1995.
Neal, Richard M. "Medical Explanations, Not 'Alien,' " *UFO* 7, no. 1 (1992): 16–20.
———. "Missing Embryo/Fetus Syndrome." *UFO* 6, no. 4 (1991):18–22.
Newton, Michael. *Raising Hell: An Encyclopedia of Devil Worship and Satanic Crime*. New York: Avon Books, 1993.
NICAP, *The UFO Evidence*. Washington, DC: NICAP, 1964.
Niemtzow, Richard C. "Paralysis and UFO Close Encounters." *The APRO Bulletin* (Mar. 1975):1, 6.

Niemtzow, Richard C., and John Schuessler. "Evaluation of Medical Injuries Resulting from UFO Close Encounters." *The APRO Bulletin*. (Nov. 1979):5.

Noll, R. "Satanism, UFO Abductions and Clinicians." *Dossociation* 2, no. 1 (1989):251–4.

Nurcombe, B., and J. Unutzer. "The Ritual Abuse of Children: Clinical Features and Diagnostic Reasoning." *Journal of the American of Child and Adolescent Psychiatry* 30, no. 2 (1991):272–6.

Oberg, James. "UFO Update: UFO Buffs May Be Unwitting Pawns in an Elaborate Government Charade," *Omni* 15, no. 11 (Sept. 1993):75.

O'Brien, Glenn. "The Saucer Men and the Unspeakable Things They Did to Stanley Ingram's Daughter." *Oui* 6, no. 8 (Aug. 1977):90–8, 106, 108.

Ofshe, R. J., and M. T. Singer. "Recovered-memory Therapy and Robust Repression: Influence and Pseudo-memories. *International Journal of Clinical and Experimental Hypnosis* 42 no. 4 (Oct. 1994):391–410.

Ofshe, Richard, and Ethan Watters. *Making Monsters*. Berkley, CA: University of California Press, 1996.

Olive, Dick. "Most UFO's Explainable, Says Scientist." *Elmira (NY) Star-Gazette* (Jan. 26, 1967):19.

Orne, M. T. "The Mechanisms of Hypnotic Age Regression: An Experimental Study." *International Journal of Clinical and Experimental Hypnosis* 46 (1951): 213–25.

Palmer, Raymond, and Kenneth Arnold. *The Coming of the Saucers.* Amherst, 1952.

Parnell, J. O. "Personality Characteristics on the MMPI, 16PF and ACL of Persons Who Claim UFO Experiences." Ph.D. diss., University of Wyoming, University Microfilms International, order no. DA 8623104, 1986.

Patterson, Randall. "I Was Abducted by Aliens." *Houston Post* (June 21, 1993): B-1.

———. "Space Case." *Houston Press* (May 30–June 5, 1996):14–20.

Payn, S. B. "A Psychoanalytic Approach to Sleep Paralysis: Review and Report of a Case." *Journal of Nervous and Mental Disease* 140 (1965):432.

Peebles, Curtis. *The Moby Dick Project*. Washington, DC: Smithsonian Institution Press, 1991.

———. *Watch the Skies*. New York: Berkley, 1995.

Peebles, M. J. "Through a Glass Darkly: The Psychoanalytical Use of Hypnosis with Post-Traumatic Stress Disorder." *International Journal of Clinical and Experimental Hypnosis* 37, no. 3 (July 1989):192–206.

Penfield, W. *The Mystery of the Mind: A Critical Study of Consciousness and the Human Brain*. Princeton: Princeton University Press, 1975.

Perrin, Robin D., and Les Parrot. "Memories of Satanic Ritual Abuse: The Truth Behind the Panic." *Christianity Today* 37, no. 7 (June 21, 1993):18–23.

Persinger, M. A. "Elicitation of Childhood Memories in Hypnosis-like Settings Is Associated With Complex Partial Epileptic-like Signs for Women but Not for Men: Implications for the False Memory Syndrome." *Perceptual & Motor Skills* 78, no. 2 (Apr. 1994):643–51.

———. "Geophysical Variables and Behavior: IX. Expected Clinical Consequences of Close Proximity to UFO Related Luminosities." *Perceptual & Motor Skills* 56 (1983):259–65.

———. "Neuropsychological Profiles of Adults Who Report 'Sudden Remembering' of Early Childhood Memories: Implications for Claims of Sex Abuse and Alien Visitation/Abduction Experiences." *Perceptual & Motor Skills* 75 (1992):259–66.

———. "Religious and Mystical Experiences as Artifacts of Temporal Lobe Function: A General Hypothesis." *Perceptual & Motor Skills* 57 (1983):1255–62.

Pettinati, H. M. *Hypnosis and Memory*. New York: Guilford Press, 1988.

Petzak, Val. "Contactee: Firsthand." *UFO* 3, no. 4 (1988):39–40. "Possible Kidnapping by UFO Averted by Youth." *Skylook* 98 (Jan. 1976):15.

Powers, Susan Marie. "Thematic Content of the Reports of UFO Abductees and Close Encounter Witnesses: Indications of Repressed Sexual Abuse." *Journal of UFO Studies*, New Series 5 (1994):35–54.

Press Conference—General Samford, Project Blue Book Files, 1952.

Pritchard, Andrea, John E. Mack, Pam Kasey, and Claudia Yapp. *Alien Discussions: Proceedings of the Abduction Study Conference Held at MIT, Cambridge, MA*. Cambridge, MA: North Cambridge Press, 1994.

Project Blue Book Files (microfilm). Washington, D.C.: National Archives.

Pulver, S. "Delusions Following Hypnosis. *International Journal of Clinical and Experimental Hypnosis* 11, no. 1 (1963):11–22.

Putnam, F. W. "The Satanic Ritual Abuse Controversy." *Child Abuse & Neglect* 15 (1991):175–79.

Randle, Kevin D. "Does Pop Culture Affect Our Views?" In *MUFON 1996 International UFO Symposium Proceedings*. Seguin, TX: MUFON, 1996.

———. "The Family That Was Kidnapped by Ufonauts." *UFO Report* (June 1976):21–3, 50–2, 54.

———. "The Flight of the Great Airship." *True Flying Saucers and UFOs Quarterly* (Spring 1977).

———. *The October Scenario*. Iowa City, Iowa: Middle Coast Publishing, 1988.

———. *Project Blue Book—Exposed*. New York: Marlowe & Co., 1997.

———. *The UFO Casebook*. New York: Warner, 1989.

———. "The UFO Kidnapping That Challenged Science. *UFO Report* (Spring 1975):15–17, 54.

Randle, Kevin D., and Robert Charles Cornett. "Project Blue Book Cover-up: Pentagon Suppressed UFO Data," *UFO Report* 2, no. 5 (Fall 1975).

Randle, Kevin D., and Russ Estes. *Faces of the Visitors*. New York: Simon and Schuster, 1997.

Randle, Kevin D., and Donald R. Schmitt. *The Truth about the UFO Crash at Roswell*. New York: M. Evans & Co., 1994.

———. *UFO Crash at Roswell*. New York: Avon, 1991.

Randles, Jenny. *Abduction: Over 200 Documented UFO Kidnappings Investigated*. London: Robert Hale, 1988.

———. "The Fencehouses Humanoid Encounter." *MUFON UFO Journal* 113 (Apr. 1977):8.

Randles, Jenny, and Paul Whethall. *Alien Contact: Window on Another World*. London: Neville Spearman, 1981.

———. "Report of Air Force Research Regarding the 'Roswell Incident,' " July 1994.

Ring, Kenneth. *The Omega Project: Near-Death Experiences, UFO Encounters and Mind at Large*. New York: Morrow, 1992.

Rodeghier, M. "Hypnosis and the Hill Abduction Case." *International UFO Reporter* (Mar./Apr. 1994):4–6, 23–24.

Rodeghier, Mark, Jeff Goodpaster, and Sandra Blatterbauer. "Psychosocial Characteristics of Abductees: Results from CUFOS Abduction Project." *Journal of UFO Studies*, new series, 3 (1991):59–90.

Rogo, D. Scott. *Beyond Reality: The Role Unseen Dimensions Play on Our Lives*. Wellingborough, Northamptonshire, England: The Aquarian Press, 1990.

Rogo, D. Scott, ed. *UFO Abductions: True Cases of Alien Kidnappings*. New York: Signet, 1980.

Ross, A. S. "Blame it on the Devil. *Redbook*. 183, no. 2 (June 1994):86–90, 114–6.

Rudford, T. "Influence and Power of Media." *Lancet* 347 (1996): 1533–5.

Ruppelt, Edward J. *The Report on Unidentified Flying Objects*. New York: Ace, 1956.

Sabin, T. R., and W. Coe. *Hypnosis: A Social Psychological Analysis of Influence Communication*. New York: Holt, Rhinehart & Winston, 1972.

Sagan, Carl. *Demon-Haunted World*. New York: Ballantine, 1996.

Sagan, Carl, and Thornton, Page, eds. *UFO's: Scientific Debate*. New York: Norton, 1974.

Saunders, D. R. "Factor Analysis of UFO-related Attitudes." *Perceptual & Motor Skills* 27, no. 3, pt. 2 (1968): 1207–18.

Saunders, David, and R. Roger Harkins. *UFOs? Yes!* New York: New American Library, 1968.

Schacter, Daniel L. *Searching for Memory*. New York: Basic, 1996.

Schilder, P. *The Nature of Hypnosis*. New York: International Universities Press, 1956.

Schneck, J. M. "Sleep Paralysis. *American Journal of Psychiatry* 108 (1952):923.
———. "Sleep Paralysis without Narcolepsy or Cataplexy: Report of a Case." *Journal of the American Medical Association* (1960):1129–30.
Scully, Frank. "Scully's Scrapbook." *Variety* (Oct. 12, 1949):61.
———. *Behind the Flying Saucers*. New York: Henry Holt, 1950.
Sheaffer, Robert. *The UFO Verdict*. Buffalo, NY: Prometheus, 1981.
Shenefield, Marianne. "Contactee: Firsthand." *UFO* 5, no. 3 (1990):37–39.
Shopper, M. "Fear of Alien Abduction." *Journal of American Academy and Adolescent Psychiatry* 35 (1996). 555–6.
Siano, Brian. "All the Babies You Can Eat." *The Humanist* 53, no. 2 (Mar./Apr. 1993):40–41.
Sims, Derrel. "TV Show Fakes Abduction & Presents as Real: Formal Answer and Apology to the UPN Disaster." *The HUFON Report* (Jan./Feb. 1998):1, 5.
Sinclair, Michael. "Possible Canadian Abduction." *The APRO Bulletin* (Mar. 1977).
Slate, B. Ann. "Contactee Supplies New Clues to the UFO Mystery." *UFO Report* 3, no. 1 (Apr. 1976):26–30, 58.
———. "Is Earth an Extraterrestrial Laboratory?" *UFO Report* 2, no. 1 (June 1974):12–14, 64–66, 68–70.
Slattery, W. J. *The Erotic Imagination*. New York: Bantam, 1975.
Smith, Michelle, and Lawrence Pazder. *Michelle Remembers*. New York: Simon and Schuster, 1981.
Smith, Yvonne. "Anatomy of an Abduction." In *MUFON Symposium Proceedings*. Seguin, TX: MUFON, 1994.
"The Space Men at Wright-Patterson." *UFO* Update.
Snyder, S. "Serotoninergic Agents in Treatment of Isolated Sleep Paralysis." *American Journal of Psychiatry* 139 (1983):1202–3.
Spanos, N. P., P. A. Cross, K. Dickson, and S. C. DuBreiul. "Close Encounters: An Examination of UFO Experiences." *Journal of Abnormal Psychology* 102, no. 4 (1993):624–32.
Special Report No. 14 (Project Blue Book) 1955.
Spencer, John. *The UFO Encyclopedia*. New York: Avon, 1993.
Spencer, John, and Hilary Evans, eds. *Phenomenon*. New York: Avon, 1988.
Sprinkle, Leo. "Hypnotic and Psychic Implications in the Investigation of UFO Reports." In Coral Lorenzen and Jim Lorenzen, eds. *Encounters with UFO Occupants*. New York, Berkley, 1976:256–329.
———. "Psychotherapeutic Services for Persons Who Claim UFO Experiences." *Psychotherapy in Private Practice* 6, no. 3 (1988):151–7.
Status Reports, "Grudge—Blue Book, Nos. 1–12."
Steiger, Brad. *Alien Meetings*. New York: Ace, 1978.

———. *Project Blue Book*. New York: Ballantine, 1976.
———. *Strangers from the Skies*. New York: Award, 1966.
———. *The UFO Abductors*. New York: Berkley, 1988.
———. "UFOs and Our Women." *UFO Report* (Aug. 1976):33–35, 68–72.
Steiger, Brad, and Hayden Hewes. *Inside Heaven's Gate*. New York: Signet, 1997.
Steiger, Brad and Francie. *The Star People*. New York: Berkley, 1981.
Steiger, Brad, and Sherry Hanson Steiger. *The Rainbow Conspiracy*. New York: Pinnacle, 1994.
Stone, Clifford E. "*UFO's: Let the Evidence Speak for Itself.*" Manuscript, 1991.
———. "The U.S. Air Force's Real, Official Investigation of UFO's." Manuscript, 1993.
Story, Ronald D. *The Encyclopedia of UFOs*. Garden City, N.Y.: Doubleday, 1980.
Strieber, W. *Communion: A True Story*. New York: Morrow, 1987.
———. *The Secret School*. New York: HarperCollins, 1996.
———. *Transformation: The Breakthrough*. New York: Morrow, 1988.
Stupple, David. W. "Historical Links Between the Occult and Flying Saucers." *Journal of UFO Studies*, new series, 5 (1994):93–108.
Sullivan, Walter. *We Are Not Alone*. New York: Signet, 1966.
Sumner, Donald A. "Skyhook Churchill 1966," *Naval Reserve Reviews* (Jan. 1967):29.
Tabu, J. M., M. Kramer, D. Arand, and G. A. Jacobs. "Nightmare Dreams and Nightmare Confabulation." *Comprehensive Psychiatry* 19 (1978):285–91.
Taff, Barry. "Abduction Reports Mount." *UFO* 4, no. 1 (1989):12–14.
———. "Anatomy of an 'EBE.' " *UFO* 4, no. 3 (1989):18–22.
Takhar, J. "Alien Abduction in PSTD." *Journal of American Academy of Child and Adolescent Psychiatry* 35, 1995:874–5.
Talarski, Chris. "Film Review: Fire in the Sky." *International UFO Reporter*. (May/June 1993):21.
Technical Report, "Unidentified Aerial Objects, Project SIGN," Feb. 1949.
Technical Report, "Unidentified Flying Objects, Project GRUDGE," Aug. 1949.
Terry, Maury. *The Ultimate Evil*. New York: Bantam, 1989.
Thalbourne, M and French, C. "The Sheep-Goat Variable and Belief in Non-Paranormal Anomalous Phenomena." *Journal of Society for Psychical Research*, Vol. 62, No. 848. 41–45.
Thompson, Keith. *Angels and Aliens: UFOs and the Mythic Imagination*. Reading, MA: Addison-Wesley, 1991.
Thompson, Tina D., ed. *TRW Space Log*. Redondo Beach, CA: TRW, 1991.
Tilton, Christa. "Contactee: Firsthand." *UFO* 4, no. 2 (1989):36–7.
Turner, Karla. "Alien Abductions in the Gingerbread House." *UFO Universe* (Spring 1993).

———. *Into the Fringe*. New York: Berkley, 1992.

U.S. Congress, House Committee on Armed Forces. *Unidentified Flying Objects*. Hearings, 89th Congress, 2nd Session, Apr. 5, 1966. Washington DC: U.S. Government Printing Office, 1968.

U.S. Congress Committee on Science and Astronautics. *Symposium on Unidentified Flying Objects*. July 29, 1968, Hearings. Washington, DC: U.S. Government Printing Office, 1968.

Vallee, Jacques. *Anatomy of a Phenomenon*. New York: Ace, 1966.

———. *Challenge to Science*. New York: Ace, 1966.

———. *Dimensions*. New York: Ballantine, 1989.

———. *Forbidden Science*. New York: Marlowe & Co. 1996.

———. *Passport to Magonia: From Folklore to Flying Saucers*. Chicago: Henry Regenery, Co. 1969.

———. *Revelations*. New York: Ballantine, 1991.

Van Tassel, George W. *The Council of Seven Lights*. Los Angeles: DeVoross and Company, 1958.

———. *I Rode a Flying Saucer! The Mystery of the Flying Saucers Revealed*. Los Angeles: New Age Publishing Company, 1952.

Victor, Jeffrey S. "Satanic Cults' Ritual Abuse of Children: Horror or Hoax?" *USA Today* 122, no. 25 (Nov. 1993):58–61.

———. *Satanic Panic: The Creation of a Comtemporary Legend*. Chicago: Open Court, 1993.

———. "The Search for Scapegoat Deviants." *The Humanist* 52, no. 5 (Sept./Oct. 1992): 10–14.

"Visitors From Venus." *Time* (Jan. 9, 1950):49.

Walton, T. *Fire in the Sky*. New York: Marlowe & Co., 1996.

———. *The Walton Experience*. New York: Berkley, 1978.

Wagstaff, G. B. *Hypnosis, Compliance, and Belief*. New York: St. Martin's, 1981.

Ward, D. "The Little Man Who Wasn't There: Encounters with the Supernormal." *Fabula: Journal of Folktale Studies* 18 (1977):213–25.

Warren, D. "Status Inconsistency Theory and Flying Saucer Sightings." *Science* 153 (1970):1213–20.

Watkins, Leslie, and David Ambrose. *Alternative Three*. London: Sphere, 1978.

Watson, Nigel. *Portraits of Alien Encounters*. London: Valis B, 1990.

Watters, E. "*The Devil* and Mr. Ingram." *Mother Jones*. (Jan./Feb. 1991).

Webb, David F. 1973: *Year of the Humanoids*. Evanston, IL: CUFOS, 1974.

Westrum, R. "Witnesses of UFOs and Other Anomalies." In R. R. Haines (ed.) *UFO Phenomena and Other Behavioral Sciences*. Metuchen, NJ: Scrowehow Press. 89–112

Wheeler, David R. *The Lubbock Lights*. New York: Award, 1977.

White, Dale. *Is Something Up There?* New York: Doubleday & Co. 1968.

White, Don. "Kentucky Women Report Close Encounter." *Skylook* 101 (Apr. 1976):3–4.

Wilkins, Harold T. *Flying Saucers on the Attack*. New York: Citadel, 1954.

———. *Flying Saucers Uncensored*. New York: Pyramid, 1967.

Wilson, John P. "Post-Traumatic Stress Disorder (PTSD) and Experienced Anomalous Trauma (EAT): Similarities in Reported UFO Abductions and Exposure to Invisible Toxic Contaminants." *Journal of UFO Studies*, new series, 2 (1990):1–17.

Wise, David, and Thomas B. Ross. *The Invisible Government*. New York: 1964.

Wolf, Fred Alan. *Parallel Universe: The Search for Other Worlds*. New York: Simon and Schuster, 1988.

"Women Say They Were Abducted by UFO." *Commonwealth Journal* (Somerset, KY) (Feb. 8, 1977).

Wright, Lawrence. "A Reporter at Large: Remembering Satan, Part I," *The New Yorker* 69, no. 13 (May 17, 1993):60–81.

———. "A Reporter at Large: Remembering Satan, Part II," *The New Yorker*. 69, no. 14 (May 24, 1993):54–74.

———. *Remembering Satan*. New York: Knopf, 1994.

"Wyoming—Another Abduction?" *The APRO Bulletin* (Mar. 1980):6.

Yapko, Michael D. *Suggestions of Abuse*. New York: Simon and Schuster, 1994.

Young, W., R. Sachs, B. G. Braun, and R. T. Watkins. "Patients Reporting Ritual Abuse in Childhood: A Clinical Syndrome, Report of 37 Cases. *Child Abuse & Neglect* 15 (1991):405–19.

Zaleski, Carol. *Otherworld Journeys: Accounts of Near-Death Experience in Medieval and Mordern Times*. New York: Oxford University Press, 1987.

Zeidman, Jennie. "I Remember Blue Book." *International UFO Reporter* (Mar./ Apr. 1991):7.

Zimmer, Troy A. "Social Psychology Correlates of Possible UFO Sightings." *Journal of Social Psychology* 123, no. 2 (Aug. 1984):199–206.

INDEX

A

B

C

D

E

F

G

H

N

O

P

V

W

X

Y

Z